Technology and Culture

Technology and Culture
An Anthology

Edited by
Melvin Kranzberg and William H. Davenport

SCHOCKEN BOOKS · NEW YORK

Contents

PART IV: INVENTION AND INNOVATION

Introduction: At the Start

I

The Society for the History of Technology (SHOT), incorporated in May 1958, held its first annual meeting the following December. A year later the Society published Volume I, Number 1 of its new international quarterly *Technology and Culture,* from the files of which, covering and celebrating its first decade in print, the present editors have compiled this anthology. The material reflects the original purpose of SHOT: to encourage the study of the development of technology and its relations with society and culture. Professor Melvin Kranzberg, a prime mover in the establishment of the Society and editor-in-chief of its quarterly since its inception, wrote an introductory article for the first number (Winter 1959/60) which he called "At the Start."[1] We have thought it fitting to reprint portions of that article "at the start" of this introduction, for most of it has held up surprisingly well. Readers of the original will note a few necessary changes in tense, the omission of private dated references, and the addition of a few editorial remarks to prepare all readers—old and new—for their journey through the pages to come.

II

"Fragmentation!" [wrote Kranzberg] was the offhand critical judgment of a distinguished historian when he first heard of the formation of the Society for the History of Technology. Like most scholars, he rightly deplored the modern tendency to "learn more and more about less and less," and he was delighted when he realized that the nature of our subject matter requires an "interlacing" of

9

disciplines rather than a further "fragmentation" of knowledge.

Indeed, the development of technology and its relations with society and culture are such broad subjects that our new Society and its publication had to be interdisciplinary in scope. We were, and are, concerned not only with the history of technological devices and processes, but also with the relations of technology to science, politics, social change, economics, and the arts and humanities. For the first time an effort was being made to bring together the engineer, the scientist, the industrialist, the social scientist, and the humanist to promote the study and interpretation of developments which are of mutual interest and concern.

The title of our publication, *Technology and Culture*, was not lightly chosen. It reveals the breadth of our definition of culture and indicates our awareness of the complex and intricate interrelationships of all aspects of technology. We use "culture" in the broad anthropological sense defined by Edward B. Tylor almost a century ago: "Culture is that complex whole which includes knowledge, belief, art, morals, laws, custom, and any other capabilities and habits acquired by man as a member of society." Technology itself is one of the most distinctive and significant of man's capabilities, and it is essential that we learn how it developed in order to analyze its relations with the other elements of culture.

The interdisciplinary nature of our subject matter inevitably dictated the form and scope of our journal. It accounted also for our audience. We intended to appeal to the engineer, to the social scientist, to the scientist, to the humanist—to the academic scholar as well as to the intelligent layman—for our subject is of importance to all these. The engineer should realize that his professional activities impinge upon all elements of our culture—that a bridge or telephone fulfills economic and social needs and possesses aesthetic and cultural values as well as technological elements. Similarly, the social scientists and the humanists should be aware of technological influences on society and the individual, but they cannot understand these influences unless they have some knowledge of how technical devices came into being and how they function.

Anyone who is at all interested in understanding the past, in learning how the present got to be the way it is, or in speculating about the future—and this would include every thinking man—must be concerned with the development of technology and its relation with society and culture.

The justification, then, for our Society and for this publication is our subject matter: technology. There is little point in belaboring the obvious importance of technology: the use of tools, together with the development of moral sensibility—and the interrelation between changing moralities and changing technologies is by no means clear—has enabled man to advance from an apelike creature through the Stone and Bronze ages eventually into an industrial society whose objects we see all around us and which conditions our daily lives. Furthermore, our hopes and fears for the future of mankind are largely bound up with technology.

III

Despite this tremendous and acknowledged significance of technology, it is curious that in twentieth-century United States, one of the most technologically minded of all nations in history, there has been no organized group nor scholarly periodical specifically devoted to the study of technology as a human activity and of its relation to other concerns.

How can we account for this neglect? It has its roots deep in the past, and we can trace it as far back as Plato's distinction between brain and hand, i.e., the notion that thinking is man's highest activity whereas manual labor lacks dignity and is confined to lower-class individuals of inferior capacity. This Platonic notion corresponded to the social system of antiquity when work was left largely to the slaves; Aristotle merely expressed a characteristic ancient attitude when he denied slaves the rational qualities of man, thereby classifying them as inferior beings.

This Platonic dualism of brain and hand persisted through the Middle Ages, even though the monks gladly performed manual labor as a means of extolling God.[2] We owe many great technological advances to these cloistered brethren, who believed that "to work is to pray" (*laborare est orare*). But the word "servile," from the same root as "serf," betrays the low esteem in which manual work was held by the medieval aristocracy.

As the bourgeois industrial society began coming into existence in modern times, there were indications that the Platonic aristocratic dualism was beginning to wane. The self-made man who had started as a menial laborer and worked his way up the ladder of success came to be a respected member of the community, while the aristocrat who disdained to soil his hands by honest toil became a target of social criticism. This revolution in social attitudes went farthest in America; the development of social democracy—caused by the influence of the frontier, the disciples of Frederick Jackson Turner would say—elevated the role of the worker. Indeed, the American myth, from log cabin to the White House, fostered the feeling that manual labor was not a thing to be despised but rather an indispensable prerequisite for the great American dream of success. What a far cry from the ancient attitude toward work!

As the public attitude toward work changed so did the prevalent attitude toward technology. The magnificent achievements of the Industrial Revolution in supplying man's material wants and creature comforts served to develop an awareness of the role of technology in civilization.

Paradoxically, the widespread use and appreciation of the products of technology did not result in greater esteem for the engineer, the man responsible for this progress. Despite the fact that our civilization has become overwhelmingly dependent on technology, despite the fact that the products of technological development are used and admired, despite the fact that engineering has become an increasingly complex field requiring specialized education, the engineer, even today, has not received adequate recognition for his training and for his contribution to society. The constant efforts of the American Society for Engineering Education and the Engineers Council for Professional Development to elevate the professional status of the engineer attest to this.

The reason for this paradox is not hard to find: the American success story glorified the man who *began* his career by working with his hands; his success lay in progressing *beyond* that stage so that he no longer need perform tasks requiring muscle or technical skills. Thus the man whose lifetime work was designing and developing tools for work or using these tools to create things—in other words, the engineer—did not find his status elevated proportionally to the high opinion held of his products by the public. In fact, the word "engineer" itself was, and still continues to be, loosely applied to both the technologist who deals with highly complex nuclear devices and the man who drives a locomotive, to the designer and builder of intricate machine tools and computing devices as well as to the plumber who fixes a leaky faucet or unplugs a sewer and labels himself a "sanitary engineer." All too often technology was associated with mere "gadgetry." Although high above the level of the unskilled manual laborer, the modern engineer still suffers from the anachronistic attitude toward the men who make and work with tools, which is part of our heritage of social attitudes from classical antiquity.

For a time the academic world lagged behind popular opinion in its appreciation and awareness of the role of technology. Until relatively recent times the world of scholarship had concentrated its attention upon the humanities, particularly the classics. Influenced by Plato's separation of "ideas" from "things" and his insistence that true virtue lay in knowledge of the "ideas," the humanities tended to denigrate technology as being "materialistic." Nor is it surprising that devotees of the intellectual life should have continued to adhere to Plato's elevation of things of the mind over the labors of the hand, especially when the world about them was increasingly devoting itself to material concerns, neglecting both the spiritual and intellectual aspects of life.

Under these circumstances, the very successes of technology probably militated against the study of its background and development, even when the "cult of work" developed during the nineteenth century. Humanistic scholars of the Victorian era sometimes looked with horror upon the rising industrial society which they believed destructive of the beauty of life and of nature. Ostrichlike, they tried to conjure away industrialism by burying their heads in the sandy desert of more conventional scholarship. The ivory-towered life of contemplation, stressed by ancient philosophers, seemed incompatible with the study of contemporary changes in society, especially those concerned with the feared, and often hated, technology.

Even when the scholars and educational institutions could no longer ignore technology, and even after the foundation of engineering colleges devoted to training in this field, the study of technological growth languished. Why bother with the past? Why investigate what has already been superseded? The study of political or intellectual history admitted no such questions; past politics, past philosophy, past literature—all were believed to teach valuable lessons, as well as having intrinsic value. No such claims were made for the history of technology; not only was technology itself viewed as an inferior subject, but the study of its past was considered irrelevant. Besides, technological advances occurred so rapidly that both scholar and student were hard-pressed to keep up with the newest developments, let alone peer into the lessons of the past.

The history of science suffered under much the same disability; both scientists and the public were too concerned with the "latest" scientific discoveries to devote any time to the past. Only the constant efforts of a small group of scholars, particularly those of the late George Sarton, succeeded in making the history of science into a reputable scholarly discipline, and only recently has the history of science achieved some measure of academic acceptance. In this respect, the history of technology has lagged several decades behind the history of science.

IV

The neglect of the study of technology was of course not absolute, and the findings of modern scholarship were eventually to make the scholarly world take a keen interest in the role of technology. Modern physiology, psychology, evolutionary biology, and anthropology—all combined to make the scholarly world aware that *Homo sapiens* could not be distinguished from *Homo faber*. Indeed, it gradually was realized that man could not have become a thinker had he not at the same time been a maker.

As the fallacious Platonic distinction of brain and hand lost support, the study of man's technological creativity became as respectable as the study of man's artistic or literary creativity. Many scholars advanced the proposition that man's creative instincts were of the same provenance, no matter in what fields these were exercised. In other words, scholarship finally began to absorb the implications of human physiology and human psychology, just as the public had already assimilated the implications of social democracy with its emphasis upon the dignity of work.

Individual scholars, organizations, and periodicals began to concern themselves with technological studies. Scientific journals and societies, including those dealing with the history and philosophy of science, produced programs and invaluable articles on certain aspects of technology; engineering periodicals published material on the history of their respective fields; and journals in the social sciences and the humanities frequently contained articles dealing with the relations between their specialized disciplines and technology.

Nevertheless, all this worthwhile activity scarcely constituted a major attempt to explore all facets of technology. The scientific journals did not have as their main purpose the assessment of the role of technology in the past, present, and future; hence there remained a wide gap in our knowledge of the relationships between man's attempts to understand nature and his efforts to control it, between his questionings of nature and his making and doing things. The readers of engineering publications tended to be more interested in blueprints than in historical insight, and the sporadic contributions on the history of technology in engineering journals did not represent any systematic attempt to make clear the relations between engineering and other aspects of culture. Similarly, the periodicals of the social scientists and humanists were more properly concerned with sociological data, with anthropological interpretations, with economic problems, with philosophical speculations, or with aesthetic considerations than with technological elements.

This failure to study thoroughly the development of technology and its relations with society and culture was not due to indifference on the part of the scholars—they were well aware of its significance—but rather to the lack of personnel qualified to deal with these studies. Serious historical scholars, with but a few notable exceptions, shied away from the field because of a feeling that they lacked the requisite technical knowledge to treat it properly. It was largely left to the engineers to write technological history, and despite the fact that many engineers wrote gracefully, others found it easier to express themselves in blueprints, or in steel and concrete, than in words. Even when an engineer was articulate, his efforts would frequently reflect the fact that just as few historians are learned in technology, few engineers are skilled in the rigors of historical research. The same considerations applied to the other social sciences and humanistic disciplines: lack of an adequate technical background often made scholars in those fields neglect very important relationships between social and cultural developments and technological change. However, the contributions of these scholars had the merits of pointing out that technology could not be studied in a vacuum and that its history must also comprise the social and cultural elements in civilization.

Hence there remained an obstacle to the study of technology and its role in society. This obstacle was no longer the indifference of scholars nurtured on the Platonic dualism of ideas and things, of brain and hand; rather, it was the lack of basic knowledge regarding the course of technological development itself.

Anthropologists, sociologists, economists, historians, humanists, scientists—all those who were concerned with the relations of technology with other elements of society and culture—were forced to the realization that their studies had no solid foundation because of the lack of a substantive knowledge of technological development. Not enough was known about the "hardware"—the actual devices and techniques which constituted technology. Valid generalizations could not be made regarding the aesthetic impact of technology, its relations with processes of social change, its effect upon industrial organization, its role in the growth and decline of civilization, or regarding any of a number of other fields of investigation, without accurate and detailed knowledge of the history of technology as such.

This consciousness as to the deficiency of substantive data accounts in large measure for the welcome accorded the five-volume *History of Technology,* edited by Charles Singer, E. J. Holmyard, A. R. Hall, and Trevor I. Williams. This encyclopedic work pioneers in a field where individual scholars, such as R. J. Forbes, Franz Maria Feldhaus, Joseph Needham, have already shown the way. *A History of Technology* represents a beginning, a stimulus to further research, rather than a culmination of scholarly efforts. It provides a necessary foundation upon which scholars can build, or, to change the metaphor, it is similar to the first efforts of the cartographers of the fifteenth century shortly after the voyages of Columbus and his immediate successors: the existence of a new world is known, but there still remains much *terra incognita;* and much further exploration is necessary to fill in the exact outlines and to explore the interiors in depth.

A systematic and continuing study of the history of technology as such is necessary, and that was one of the reasons for the founding of the Society of the History of Technology and for the publication of its journal. To be sure, *Technology and Culture* is not the first nor the only journal in this field. The British Newcomen Society has long and successfully encouraged the history of technology through its *Transactions; Technikgeschichte* of the Verein Deutscher Ingenieure, first edited by Conrad Matschoss in 1909, suspended publication in 1941 and resumed in 1965; the Centre de documentation d'histoire des techniques in Paris publishes *Documents pour l'histoire des techniques* at irregular intervals; the Technisches Museum in Vienna publishes an annual volume, as does the Tekniska Museet in Stockholm; the Polish Academy of Sciences sponsors *Kwartalnik Nauki i Techniki* (Quarterly Journal of the History of Science and Technology); and museums and institutes in Italy, Hungary, Czechoslovakia, and other countries also issue serial bulletins or journals. Virtually all these publications focus almost exclusively on national contributions. *Technology and Culture,* therefore, is the only international quarterly serving the needs of scholars in America and throughout the world.

Despite the great merits of the Singer volumes and other works, these publications do not obviate the necessity for further research in the history of technology. Instead, they encourage it. What is more, they increase the need and desirability of studying the relations of technology with other elements of society and culture. *A History of Technology,* for example, deals with technology only within the framework of its own subject matter, as if it were insulated from the rest of society. Such a treatment derives inevitably from the definition of technology as "how things are commonly done or made" and "what things are done and made."[3] But many other questions immediately come to mind: *Why* are things done and made as they are? What effects have these methods and things upon other areas of human activity? How have other elements in society and culture affected how, what, and why things are done or made? The five volumes of *A History of Technology* codify the present state of scholarship, but they do not answer these further questions.

Yet we cannot hope to understand our civilization, or even technology itself, unless we grasp the manifold relations between technology and society. In other words, an *interdisciplinary* approach is needed if we are properly to study the history of technology. This multifaceted approach accounts for the unique character of our organization and this publication.

No existing American journal or organization gives primacy to the study of the history of technology; no existing journal or organization in the Western world gives primacy to the study of the relations of technology with society and culture. It is these serious gaps in organized scholarship which the Society for the History of Technology and its journal, *Technology and Culture,* attempt to fill.

V

So, with a few alterations, wrote Professor Kranzberg in 1959. He went on to describe the threefold educational mission of *Technology and Culture,* the title we have preserved for this anthology: "to promote the scholarly study of the history of technology, to show the relations between technology and other elements of culture, and to make these elements of knowledge available and comprehensible to the educated citizen." With the present volume in mind, we would stress the application of this "mission" to the education and enjoyment of the undergraduate student and the thinking, reading layman as well. In that same history-making Winter 1959/60 number, Howard Mumford Jones had already recognized this further challenge and opportunity for service:[4]

> If . . . the importance of the history of technology in general culture is to be recognized, if its contribution to humane learning is to be just, active, and full, then it cannot leave writing for the lay reader to popular journalism. The popular book on the history of technology, at least in my experience, tends to a false dramatization of the subject, it adopts a romantic conception of "genius," it falsifies the relation between technological advance and the economic order, it is frequently inaccurate in its facts, and it ends by uncritically assuming that current industrial culture is the best possible culture in this best of all possible worlds. . . .
> The historian of technology is under some obligation to the general public—otherwise, the importance of his subject remains forever in the hands of journalists and their kind of popularizer—and he may well ponder the qualities that make popular manuals readable and appealing. He must also, it seems to me, open relations with the general historian if technology is to be properly treated in general histories (political, social, and economic histories especially). . . . But, no doubt because of the difficulties of the subject, the discussion of technology in most of these books is, I think, naïve. The opportunity for the historian of technology, either as collaborator, contributor, or consultant, to redress the balance in such volumes is an open one, and I trust that some scholars in the field will see it.

VI

We hope that this anthology will help "to redress the balance." In the first decade of publication many pieces appeared in *Technology and Culture* which were aimed at other than specialists; we have picked selections which illustrate the original principles upon which the Society was founded while at the same time keeping the general reader in mind, in the hope that he may acquire a better understanding of the technological age in which we all live.

Technology is advancing so fast today that, as Emmanuel Mesthene has written, we are now moving the fences back while the baseball game is going on. It is difficult for the spectator to keep up with the rule changes. Willy nilly, we find ourselves for one reason or another, sometimes merely emotionally, taking sides—with Jacques Ellul, who finds man the plaything of technique; or with Zbigniew Brzezinski, who hymns the imminent arrival of the Technetronic Age; or with Lewis Mumford, who, mindful of the beauty and power of the Machine, warns us nevertheless to keep man first, machine second. The reader may already have made up his mind. In that case, in studying the selections to come he may change it—or find better reasons for going on believing as he does. At any rate he should find food for thought and fodder for argument. He will encounter topics ranging from bursting boilers to harvesting tomatoes. Hopefully he will be excited into going farther afield into the cross-references and the bibliographies. Although any anthology is an eclectic synthesis of a good deal of material, it must not be merely an end, but a means.

Finally, there remain many reasons for studying such a collection as this. As Peter Drucker says elsewhere in these pages, "Without study and understanding of work, how can we hope to arrive at an understanding of technology?" The student and the layman should develop a historical perspective, learn the impact of the machine on events, appreciate the meaning of technological revolutions, perceive the various roles of technology in society, and become aware of its interplay with such topics as art, values, and international relations. Sir Philip Sidney wrote that the purpose of poetry is to teach and delight. We hope that *Technology and Culture* will perform a similar function in its own modest way.

Cleveland, Ohio　　　　　　MELVIN KRANZBERG
Claremont, California　　　WILLIAM H. DAVENPORT

August 1971

REFERENCES
[1] Much of the material of this article derives from a meeting of the Advisory Committee for Technology and Society, held at Case Institute of Technology on January 30–31, 1958, which led to the formation of the Society for the History of Technology. Participants in this conference, either by correspondence or in person, were Russell L. Ackoff, Carl W. Condit, George A. Gullette, Morrell Heald, Thomas P. Hughes, Melvin Kranzberg, Edward Lurie, Hugo A. Meier, Robert P. Multhauf, Lewis Mumford, William Fielding Ogburn, Stanley Pargellis, John B. Rae, Thomas M. Smith, and Lynn White, Jr. At the same meeting it was decided that SHOT would publish a quarterly journal, *Technology and Culture,* and arrangements were made with the Wayne State University Press for its publication. Since 1965 (Volume 6), the University of Chicago Press has published *Technology and Culture* for the Society.

[2] See Lynn White, Jr., "Dynamo and Virgin Reconsidered," *The American Scholar,* XXVII, No. 2 (Spring, 1958), 183–194.

[3] Charles Singer, E. J. Holmyard, A. R. Hall, and T. I. Williams, *A History of Technology* (5 vols., New York and London, 1954–1958), Vol. I, Chap. vii.

[4] Howard Mumford Jones, "Ideas, History, Technology," *Technology and Culture,* I, 1 (Winter 1959/60): 27.

Part I
Technology and Society

Introduction

Every society seeks for some explanations—or, rather, scapegoats—when things go wrong. In primitive, animistic societies, the evil spirits are blamed; in ancient Greece, the fates and gods had to be propitiated; in the Christian era, the blame could be placed on man's inherent sinfulness; in the eighteenth-century Enlightenment, outmoded political and social institutions, such as the absolute monarchy and the remnants of the medieval Church, were regarded as the barriers to man's progress and happiness; and in the nineteenth century, human ignorance of, and hence resistance to, the advances of science and technology were regarded by some as the bane of mankind's existence, while others felt that the capitalistic social structure stood in the way of human betterment. In our own times, especially among some members of the intellectual community, the blame is placed on technology.

Blaming technology for all the troubles of Western civilization in the mid-twentieth century is, of course, a "cop out"—it is an answer, but not an explanation. By placing the blame for all the ills of the human predicament in the last decades of the twentieth century on some impersonal force—"technology"—men can hope to avoid responsibility for their own actions. But this begs some essential questions: does technology really determine the nature, values, and institutions of society? If so, to what extent? Furthermore, how has technology molded society? Does society have any control over technology and its applications? If so, to what extent can technology be controlled and by what mechanisms?

In brief, we are faced with a question of "technological determinism." How and to what extent does technology determine the nature and structure of society and human thought and actions within that society? Is there a "technological imperative" which demands that society be structured in certain ways and that men act along certain lines? Is it possible that human values and economic institutions and social structure can help determine the course of technology rather than the other way around? Or, to put the question in terms of a familiar metaphor: are machines the masters of man, or is man the master of his own machines?

In Part One of this volume our authors deal with the question of technological determinism by investigating the interrelationships between technology and society. As might be expected, scholars concerned with the history of technology tend to play up the role of technology in its impact on human affairs, considering it one of the prime determinants of social values and institutions, but they differ in their emphases and approaches—and they arrive at different answers to the questions involving technological-societal relationships.

Robert L. Heilbroner poses the question directly: "Do machines make history?" Looking at this question in terms of technology's impact on the nature of the socioeconomic order and assuming that technology clearly affects that order, he asks if there is a rigid sequence in technological evolution which requires societies developing along technological lines to travel in a certain fixed path. His answer is that technology does display a "structured" history. And in answer to his second question concerning the ways in which the mode of production affects the superstructure of social relationships, he concludes that the "technology of a society imposes a determinant pattern of social relations on that society," and indeed he traces such patterns. Nevertheless, Heilbroner could hardly be termed a technological determinist. His entire article is framed in the context of a refutation of Karl Marx's simplistic and deterministic view of technology as "making" history; Heilbroner concludes that such technological determinism is "peculiarly a problem of a certain historic epoch—specifically that of high capitalism and low socialism. . . ."

Peter F. Drucker looks even farther back into history to trace the impact of technology on sociopolitical developments—to what he calls the "first technological revolution," which produced the irrigation societies of ancient Egypt, Mesopotamia, India, and China. The technological innovation of irrigation systems, claims Drucker, first established the concept of government, social classes, and the organization and institutionalization of knowledge, and even created the individual as the "focal point" of society. With an eye to today's problems, he concludes that technological innovation demands sociopolitical innovations and that proper social and political responses to technological innovations "are largely circumscribed by new technology." Yet, like Heilbroner, Drucker is not a strict technological determinist, for he concludes that the human and social purposes to which technological innovation is applied, and hence the new sociopolitical arrangements which arise from it, are within human control.

Lewis Mumford is both more optimistic and more pessimistic than either Heilbroner or Drucker. Looking back through history, he sees constant tensions between two opposite forms of social organization—"between small-scale association and large-scale organization, between personal autonomy and institutional regulation"—and he finds this same conflict "deeply embedded in technology itself." There is what he calls "democratic technics"—a small-scale method of production involving small-scale human relationships and allowing for a great deal of personal autonomy—contrasted with "authoritarian technics," involving centralized political control over large-scale units and including forms of compulsion and physical coercion for the performance of technical tasks. Citing examples of both forms of technical and human organization during the past, he warns against the acceptance of authoritarian technics in the present time, with its promise of material plenty, and urges that contemporary large-scale technology be "redeemed" through the democratic process.

From a different vantage point, Lord Ritchie-Calder views the problem of technological challenges and opportunities in terms of the developing, or emerging, nations of today. He asserts the capacity of science and technology to "solve all the material problems of the world today," but he questions their opportunity to do so because the nations possessing the material "know-how" are more concerned about matters of political power and prestige than they are about helping their brethren in poor, less-developed countries. He attacks the arrogance of the industrially developed states by showing the efficacy of simpler and more primitive technologies in finding answers to the problems of the emerging nations.

John B. Rae is more concerned about the structure of technology itself within the story of the growth of American society. American technology, he points out, is overwhelmingly pragmatic in character; Americans have been more interested in knowing how than in knowing why, for Americans wanted to get things done in the shortest possible time and with the least possible human work. A second element in the American experience which he finds of importance is the interrelationship between technological development and industrial application; in the American context, the effectiveness of technical invention is dependent on economic time and circumstance, as expressed in industrial applications. Because industrial development has played so great a role in the American story, a study of the ways in which technology has effected industrial growth is required for an intelligent appreciation of technology's impact on American society.

All our authors, then, accept the notion that technology has a strong influence on social developments, although they differ in their emphases and disagree on whether certain social consequences must inevitably accompany specific technical changes or whether technology creates certain options for society which society's own values must determine. This last point raises the question of the possibility of society's control over technology, and here John G. Burke tells the story of the first attempts by the United States government to regulate the deficiencies of one particular technological development, steam boilers. The application of steam engines to water transportation posed a new problem for governmental authorities: steam boilers had a distressing tendency to explode, with great danger to both life and property. Did the federal government have the right to "invade the rights of private property" by regulating the structure and operations of steam boilers? Burke relates the political struggle over the rights of the federal government to intervene in such matters, then the scientific and technical problems of determining the proper method of making and operating steam boilers, and finally the long fight to bring about effective enforcement of the regulations. The early nineteenth-century arguments about federal regulation of technical advances are still echoed today in the public disputes over governmental regulation of environmental quality, for the bursting-boiler legislation described by Burke is the prototype of much of today's legislative endeavors. The long struggle to achieve such regulation may be disheartening to today's activists who demand instant results, but Burke's story illustrates how the misuse of certain technical devices can be controlled by society if sufficient technical knowledge and expertise are made available and if the public has the will to make its feelings felt in the halls of Congress.

These selections by no means provide a final answer to

the question of whether or not technology is the sole and prime determinant of the state of society, but they do show the complexities of the question and the many subtle and different ways in which technology and society react. Above all, they indicate the possibilities for social development opened up by technological advances, and they make clear to us the human responsibility for the use, misuse, or abuse of technology in fulfilling human needs and social wants. It is clear from these selections that the answer to the question of whether machines are the masters or servants of mankind is to be found out not by questioning the machines but by posing the question to man himself.

Do Machines Make History?

Robert L. Heilbroner

"The hand-mill gives you society with the feudal lord; the steam-mill, society with the industrial capitalist."

Marx, *The Poverty of Philosophy*

That machines make history in some sense—that the level of technology has a direct bearing on the human drama—is of course obvious. That they do not make all of history, however that word be defined, is equally clear. The challenge, then, is to see if one can say something systematic about the matter, to see whether one can order the problem so that it becomes intellectually manageable.

To do so calls at the very beginning for a careful specification of our task. There are a number of important ways in which machines make history that will not concern us here. For example, one can study the impact of technology

Dr. Heilbroner, chairman of the Graduate Department of Economics, New School for Social Research (New York City), is best known for his ability to translate the esoteric language of economic theory and development into terms understandable to the educated layman. His books include The Worldly Philosophers, The Future as History, *and* The Limits of American Capitalism.

This essay is an outgrowth of Dr. Heilbroner's paper delivered at the Ninth Annual Meeting of the Society for the History of Technology, held in Washington, D.C., in December 1966. It was published in the July 1967 issue of Technology and Culture *(Vol. 8, No. 3): 335–45, and has been reprinted in Dr. Heilbroner's latest book,* Between Capitalism and Socialism: Essays in Political Economics *(New York, 1970).*

on the *political* course of history, evidenced most strikingly by the central role played by the technology of war. Or one can study the effect of machines on the *social* attitudes that underlie historical evolution: one thinks of the effect of radio or television on political behavior. Or one can study technology as one of the factors shaping the changeful content of life from one epoch to another: when we speak of "life" in the Middle Ages or today we define an existence much of whose texture and substance is intimately connected with the prevailing technological order.

None of these problems will form the focus of this essay. Instead, I propose to examine the impact of technology on history in another area—an area defined by the famous quotation from Marx that stands beneath our title. The question we are interested in, then, concerns the effect of technology in determining the nature of the *socioeconomic order*. In its simplest terms the question is: did medieval technology bring about feudalism? Is industrial technology the necessary and sufficient condition for capitalism? Or, by extension, will the technology of the computer and the atom constitute the ineluctable cause of a new social order?

Even in this restricted sense, our inquiry promises to be broad and sprawling. Hence, I shall not try to attack it head-on, but to examine it in two stages:

1. If we make the assumption that the hand-mill does "give" us feudalism and the steam-mill capitalism, this places technological change in the position of a prime mover of social history. Can we then explain the "laws of motion" of technology itself? Or to put the question less grandly, can we explain why technology evolves in the sequence it does?

2. Again, taking the Marxian paradigm at face value, exactly what do we mean when we assert that the hand-mill "gives us" society with the feudal lord? Precisely how does the mode of production affect the superstructure of social relationships?

These questions will enable us to test the empirical content—or at least to see if there *is* an empirical content—in the idea of technological determinism. I do not think it will come as a surprise if I announce now that we will find *some* content, and a great deal of missing evidence, in our investigation. What will remain then will be to see if we can place the salvageable elements of the theory in historical perspective—to see, in a word, if we can explain technological determinism historically as well as explain history by technological determinism.

I

We begin with a very difficult question hardly rendered easier by the fact that there exist, to the best of my knowledge, no empirical studies on which to base our speculations. It is the question of whether there is a fixed sequence to technological development and therefore a necessitous path over which technologically developing societies must travel.

I believe there is such a sequence—that the steam-mill follows the hand-mill not by chance but because it is the next "stage" in a technical conquest of nature that follows one and only one grand avenue of advance. To put it differently, I believe that it is impossible to proceed to the age of the steam-mill until one has passed through the age of the hand-mill, and that in turn one cannot move to the age of the hydroelectric plant before one has mastered the steam-mill, nor to the nuclear power age until one has lived through that of electricity.

Before I attempt to justify so sweeping an assertion, let me make a few reservations. To begin with, I am fully conscious that not all societies are interested in developing a technology of production or in channeling to it the same quota of social energy. I am very much aware of the different pressures that different societies exert on the direction in which technology unfolds. Lastly, I am not unmindful of the difference between the discovery of a given machine and its application as a technology—for example, the invention of a steam engine (the aeolipile) by Hero of Alexandria long before its incorporation into a steam-mill. All these problems, to which we will return in our last section, refer, however, to the way in which technology makes its peace with the social, political, and economic institutions of the society in which it appears. They do not directly affect the contention that there exists a determinate sequence of productive technology for those societies that are interested in originating and applying such a technology.

What evidence do we have for such a view? I would put forward three suggestive pieces of evidence:

1. The Simultaneity of Invention

The phenomenon of simultaneous discovery is well known.[1] From our view, it argues that the process of discovery takes place along a well-defined frontier of knowledge rather than in grab-bag fashion. Admittedly, the concept of "simultaneity" is impressionistic,[2] but the related phenomenon of technological "clustering" again suggests that technical evolution follows a sequential and determinate rather than random course.[3]

2. The Absence of Technological Leaps

All inventions and innovations, by definition, represent an advance of the art beyond existing base lines. Yet, most advances, particularly in retrospect, appear essentially incremental, evolutionary. If nature makes no sudden leaps, neither, it would appear, does technology. To make my point by exaggeration, we do not find experiments in electricity in the year 1500, or attempts to extract power from the atom in the year 1700. On the whole, the development of the technology of production presents a fairly smooth and continuous profile rather than one of jagged peaks and discontinuities.

3. The Predictability of Technology

There is a long history of technological prediction, some of it ludicrous and some not.[4] What is interesting is that the development of technical progress has always seemed *intrinsically* predictable. This does not mean that we can lay down future timetables of technical discovery, nor does it rule out the possibility of surprises. Yet I venture to state that many scientists would be willing to make *general* predictions as to the nature of technological capability twenty-five or even fifty years ahead. This, too, suggests that technology follows a developmental sequence rather than arriving in a more chancy fashion.

I am aware, needless to say, that these bits of evidence do not constitute anything like a "proof" of my hypothesis. At best they establish the grounds on which a *prima facie* case of plausibility may be rested. But I should like now to strengthen these grounds by suggesting two deeper-seated reasons why technology *should* display a "structured" history.

The first of these is that a major constraint always operates on the technological capacity of an age, the constraint of its accumulated stock of available knowledge. The application of this knowledge may lag behind its reach; the technology of the hand-mill, for example, was by no means at the frontier of medieval technical knowledge, but technical realization can hardly precede what men generally know (although experiment may incrementally advance both technology and knowledge concurrently). Particularly from the mid-nineteenth century to the present do we sense the loosening constraints on technology stemming from successively yielding barriers of scientific knowledge—loosening constraints that result in the successive arrival of the electrical, chemical, aeronautical, electronic, nuclear, and space stages of technology.[5]

The gradual expansion of knowledge is not, however, the only order-bestowing constraint on the development of technology. A second controlling factor is the material competence of the age, its level of technical expertise. To make a steam engine, for example, requires not only some knowledge of the elastic properties of steam but the ability to cast iron cylinders of considerable dimensions with tolerable accuracy. It is one thing to produce a single steam-machine as an expensive toy, such as the machine depicted by Hero, and another to produce a machine that will produce power economically and effectively. The difficulties experienced by Watt and Boulton in achieving a fit of piston to cylinder illustrate the problems of creating a technology, in contrast with a single machine.

Yet until a metal-working technology was established—indeed, until an embryonic machine-tool industry had taken root—an industrial technology was impossible to create. Furthermore, the competence required to create such a technology does not reside alone in the ability or inability to make a particular machine (one thinks of Babbage's ill-fated calculator as an example of a machine born too soon), but in the ability of many industries to change their products or processes to "fit" a change in one key product or process.

This necessary requirement of technological congruence[6] gives us an additional cause of sequencing. For the ability of many industries to cooperate in producing the equipment needed for a "higher" stage of technology depends not alone on knowledge or sheer skill but on the division of labor and the specialization of industry. And this in turn hinges to a considerable degree on the sheer size of the stock of capital itself. Thus the slow and painful accumulation of capital, from which springs the gradual diversification of industrial function, becomes an independent regulator of the reach of technical capability.

In making this general case for a determinate pattern of technological evolution—at least insofar as that technology is concerned with production—I do not want to claim too much. I am well aware that reasoning about technical sequences is easily faulted as *post hoc ergo propter hoc.* Hence, let me leave this phase of my inquiry by suggesting no more than that the idea of a roughly ordered progression of productive technology seems logical enough to warrant further empirical investigation. To put it as concretely as possible, I do not think it is just by happenstance that the steam-mill follows, and does not precede, the hand-mill, nor is it mere fantasy in our own day when we speak of the coming of the automatic factory. In the future as in the past, the development of the technology of production seems bounded by the constraints of knowledge and capability and thus, in principle at least, open to prediction as a determinable force of the historical process.

II

The second proposition to be investigated is no less difficult than the first. It relates, we will recall, to the explicit statement that a given technology imposes certain social and political characteristics upon the society in which it is found. Is it true that, as Marx wrote in *The German Ideology,* "A certain mode of production, or industrial stage, is always combined with a certain mode of cooperation, or social stage,"[7] or as he put it in the sentence immediately preceding our hand-mill, steam-mill paradigm, "In acquiring new productive forces men change their mode of production, and in changing their mode of production they change their way of living—they change all their social relations"?

As before, we must set aside for the moment certain "cultural" aspects of the question. But if we restrict ourselves to the functional relationships directly connected with the process of production itself, I think we can indeed state that the technology of a society imposes a determinate pattern of social relations on that society.

We can, as a matter of fact, distinguish at least two such modes of influence:

1. The Composition of the Labor Force

In order to function, a given technology must be attended by a labor force of a particular kind. Thus, the hand-mill (if we may take this as referring to late medieval technology in general) required a work force composed of skilled or semiskilled craftsmen, who were free to practice their occupations at home or in a small atelier, at times and seasons that varied considerably. By way of contrast, the steam-mill—that is, the technology of the nineteenth century—required a work force composed of semiskilled or unskilled operatives who could work only at the factory site and only at the strict time schedule enforced by turning the machinery on or off. Again, the technology of the electronic age has steadily required a higher proportion of skilled attendants; and the coming technology of automation will still further change the needed mix of skills and the locale of work, and may as well drastically lessen the requirements of labor time itself.

2. The Hierarchical Organization of Work

Different technological apparatuses not only require different labor forces but different orders of supervision and coordination. The internal organization of the eighteenth-century handicraft unit, with its typical man-master relationship, presents a social configuration of a wholly different kind from that of the nineteenth-century factory with its men-manager confrontation, and this in turn differs from the internal social structure of the continuous-flow, semi-automated plant of the present. As the intricacy of the production process increases, a much more complex system of internal controls is required to maintain the system in working order.

Does this add up to the proposition that the steam-mill gives us society with the industrial capitalist? Certainly the class characteristics of a particular society are strongly implied in its functional organization. Yet it would seem wise to be very cautious before relating political effects exclusively to functional economic causes. The Soviet Union, for example, proclaims itself to be a socialist society although its technical base resembles that of old-fashioned capitalism. Had Marx written that the steam-mill gives you society with the industrial *manager,* he would have been closer to the truth.

What is less easy to decide is the degree to which the technological infrastructure is responsible for some of the sociological features of society. Is anomie, for instance, a disease of capitalism or of all industrial societies? Is the organization man a creature of monopoly capital or of all bureaucratic industry wherever found? These questions tempt us to look into the problem of the impact of technology on the existential quality of life, an area we have

ruled out of bounds for this paper. Suffice it to say that
superficial evidence seems to imply that the similar technologies of Russia and America are indeed giving rise to
similar social phenomena of this sort.

As with the first portion of our inquiry, it seems advisable to end this section on a note of caution. There is a
danger, in discussing the structure of the labor force or
the nature of intrafirm organization, of assigning the sole
causal efficacy to the visible presence of machinery and
of overlooking the invisible influence of other factors at
work. Gilfillan, for instance, writes, "engineers have committed such blunders as saying the typewriter brought
women to work in offices, and with the typesetting machine
made possible the great modern newspaper, forgetting that
in Japan there are women office workers and great modern newspapers getting practically no help from typewriters
and typesetting machines."[8] In addition, even where technology seems unquestionably to play the critical role, an
independent "social" element unavoidably enters the scene
in the *design* of technology, which must take into account
such facts as the level of education of the work force or
its relative price. In this way the machine will reflect, as
much as mold, the social relationships of work.

These caveats urge us to practice what William James
called a "soft determinism" with regard to the influence
of the machine on social relations. Nevertheless, I would
say that our cautions qualify rather than invalidate the
thesis that the prevailing level of technology imposes itself
powerfully on the structural organization of the productive side of society. A foreknowledge of the shape of the
technical core of society fifty years hence may not allow
us to describe the political attributes of that society, and
may perhaps only hint at its sociological character, but
assuredly it presents us with a profile of requirements, both
in labor skills and in supervisory needs, that differ considerably from those of today. We cannot say whether the
society of the computer will give us the latter-day capitalist or the commissar, but it seems beyond question that it
will give us the technician and the bureaucrat.

III

Frequently, during our efforts thus far to demonstrate
what is valid and useful in the concept of technological
determinism, we have been forced to defer certain aspects
of the problem until later. It is time now to turn up the

rug and to examine what has been swept under it. Let us try to systematize our qualifications and objections to the basic Marxian paradigm:

1. Technological Progress Is Itself a Social Activity

A theory of technological determinism must contend with the fact that the very activity of invention and innovation is an attribute of some societies and not of others. The Kalahari bushmen or the tribesmen of New Guinea, for instance, have persisted in a neolithic technology to the present day; the Arabs reached a high degree of technical proficiency in the past and have since suffered a decline; the classical Chinese developed technical expertise in some fields while unaccountably neglecting it in the area of production. What factors serve to encourage or discourage this technical thrust is a problem about which we know extremely little at the present moment.[9]

2. The Course of Technological Advance Is Responsive to Social Direction

Whether technology advances in the area of war, the arts, agriculture, or industry depends in part on the rewards, inducements, and incentives offered by society. In this way the direction of technological advance is partially the result of social policy. For example, the system of interchangeable parts, first introduced into France and then independently into England, failed to take root in either country for lack of government interest or market stimulus. Its success in America is attributable mainly to government support and to its appeal in a society without guild traditions and with high labor costs.[10] The general *level* of technology may follow an independently determined sequential path, but its areas of application certainly reflect social influences.

3. Technological Change Must be Compatible with Existing Social Conditions

An advance in technology not only must be congruent with the surrounding technology but must also be compatible with the existing economic and other institutions of society. For example, labor-saving machinery will not find ready acceptance in a society where labor is abundant and cheap as a factor of production. Nor would a mass-production technique recommend itself to a society that did not have a mass market. Indeed, the presence of slave labor seems generally to inhibit the use of machinery and the presence of expensive labor to accelerate it.[11]

These reflections on the social forces bearing on technical progress tempt us to throw aside the whole notion of technological determinism as false or misleading.[12] Yet, to relegate technology from an undeserved position of *primum mobile* in history to that of a mediating factor, both acted upon by and acting on the body of society, is not to write off its influence but only to specify its mode of operation with greater precision. Similarly, to admit we understand very little of the cultural factors that give rise to technology does not depreciate its role but focuses our attention on that period of history when technology is clearly a major historical force, namely Western society since 1700.

IV

What is the mediating role played by technology within modern Western society? When we ask this much more modest question, the interaction of society and technology begins to clarify itself for us:

1. The Rise of Capitalism Provided a Major Stimulus for the Development of a Technology of Production

Not until the emergence of a market system organized around the principle of private property did there also emerge an institution capable of systematically guiding the inventive and innovative abilities of society to the problem of facilitating production. Hence the environment of the eighteenth and nineteenth centuries provided both a novel and an extremely effective encouragement for the development of an *industrial* technology. In addition, the slowly opening political and social framework of late mercantilist society gave rise to social aspirations for which the new technology offered the best chance of realization. It was not only the steam-mill that gave us the industrial capitalist but the rising inventor-manufacturer who gave us the steam-mill.

2. The Expansion of Technology within the Market System Took on a New "Automatic" Aspect

Under the burgeoning market system not alone the initiation of technical improvement but its subsequent adoption and repercussion through the economy was largely governed by market considerations. As a result, both the rise and the proliferation of technology assumed the attributes of an impersonal diffuse "force" bearing on social and economic life. This was all the more pronounced because the political control needed to buffer its disruptive consequences was seriously inhibited by the prevailing laissez-faire ideology.

3. The Rise of Science Gave a New Impetus to Technology

The period of early capitalism roughly coincided with and provided a congenial setting for the development of an independent source of technological encouragement—the rise of the self-conscious activity of science. The steady expansion of scientific research, dedicated to the exploration of nature's secrets and to their harnessing for social use, provided an increasingly important stimulus for technological advance from the middle of the nineteenth century. Indeed, as the twentieth century has progressed, science has become a major historical force in its own right and is now the indispensable precondition for an effective technology.

It is for these reasons that technology takes on a special significance in the context of capitalism—or, for that matter, of a socialism based on maximizing production or minimizing costs. For in these societies, both the continuous appearance of technical advance and its diffusion throughout the society assume the attributes of autonomous process, "mysteriously" generated by society and thrust upon its members in a manner as indifferent as it is imperious. This is why, I think, the problem of technological determinism—of how machines make history—comes to us with such insistence despite the ease with which we can disprove its more extreme contentions.

Technological determinism is thus peculiarly a problem of a certain historical epoch—specifically that of high capitalism and low socialism—*in which the forces of technical change have been unleashed, but when the agencies for the control or guidance of technology are still rudimentary.*

The point has relevance for the future. The surrender of society to the free play of market forces is now on the wane, but its subservience to the impetus of the scientific ethos is on the rise. The prospect before us is assuredly that of an undiminished and very likely accelerated pace of technical change. From what we can foretell about the direction of this technological advance and the structural alterations it implies, the pressures in the future will be toward a society marked by a much greater degree of organization and deliberate control. What other political, social, and existential changes the age of the computer will also bring we do not know. What seems certain, however, is that the problem of technological determinism—that is, of the impact of machines on history—will remain germane until there is forged a degree of public control over technology far greater than anything that now exists.

REFERENCES

[1] See Robert K. Merton, "Singletons and Multiples in Scientific Discovery: A Chapter in the Sociology of Science," *Proceedings* of the American Philosophical Society, CV (October 1961), 470–486.

[2] See John Jewkes, David Sawers, and Richard Stillerman, *The Sources of Invention* (New York, 1960 [paperback edition]), p. 227, for a skeptical view.

[3] "One can count 21 basically different means of flying, at least eight basic methods of geophysical prospecting; four ways to make uranium explosive; . . . 20 or 30 ways to control birth. . . . If each of these separate inventions were autonomous, i.e., without cause, how could one account for their arriving in these functional groups?" S. C. Gilfillan, "Social Implications of Technological Advance," *Current Sociology,* I (1952), 197. See also Jacob Schmookler, "Economic Sources of Inventive Activity," *Journal of Economic History* (March 1962), 1–20; and Richard Nelson, "The Economics of Invention: A Survey of the Literature," *Journal of Business,* XXXII (April 1959), 101–119.

[4] Jewkes *et al.* (see n. 2) present a catalogue of chastening mistakes (p. 230 f.). On the other hand, for a sober predictive effort, see Francis Bello, "The 1960s: A Forecast of Technology," *Fortune,* LIX (January 1959), 74–78; and Daniel Bell, "The Study of the Future," *Public Interest,* I (Fall 1965), 119–130. Modern attempts at prediction project likely avenues of scientific advance or technological function rather than the feasibility of specific machines.

[5] To be sure, the inquiry now regresses one step and forces us to ask whether there are inherent stages for the expansion of knowledge, at least insofar as it applies to nature. This is a very uncertain question. But having already risked so much, I will hazard the suggestion that the roughly parallel sequential development of scientific understanding in those few cultures that have cultivated it (mainly classical Greece, China, the high Arabian culture, and the West since the Renaissance) makes such a hypothesis possible, provided that one looks to broad outlines and not to inner detail.

[6] The phrase is Richard LaPiere's in *Social Change* (New York, 1965), p. 263 f.

[7] Karl Marx and Friedrich Engels, *The German Ideology* (London, 1942), p. 18.

[8] Gilfillan (see n. 3), p. 202.

[9] An interesting attempt to find a line of social causation is found in E. Hagen, *The Theory of Social Change* (Homewood, Ill., 1962).

[10] See K. R. Gilbert, "Machine-Tools," in Charles Singer, E. J. Holmyard, A. R. Hall, and Trevor I. Williams (eds.), *A History of Technology* (Oxford, 1958), IV, Chap. xiv.

[11] See LaPiere (see n. 6), p. 284; also H. J. Habbakuk, *British and American Technology in the 19th Century* (Cambridge, 1962), *passim*.

[12] As, for example, in A. Hansen, "The Technological Determination of History," *Quarterly Journal of Economics* (1921): 76–83.

The First Technological Revolution and Its Lessons

Peter F. Drucker

Aware that we are living in the midst of a technological revolution, we are becoming increasingly concerned with its meaning for the individual and its impact on freedom, on society, and on our political institutions. Side by side with messianic promises of utopia to be ushered in by technology, there are the most dire warnings of man's enslavement by technology, his alienation from himself and from society, and the destruction of all human and political values.

Tremendous though today's technological explosion is, it is hardly greater than the first great revolution technology wrought in human life seven thousand years ago when the first great civilization of man, the irrigation civilization, established itself. First in Mesopotamia, and then in Egypt and in the Indus Valley, and finally in China there

---

Dr. Drucker, of the Claremont Graduate School, is one of the world's leading economists and management consultants. Among his many books are The Future of Industrial Man, The New Society, Landmarks of Tomorrow, The Practice of Management, *and* The Age of Discontinuity. *His most recent book is a series of essays on the interrelationships among economic, political, and social thought and actions,* Men, Ideas and Politics *(New York, 1971).*

This article, which appeared in the Spring 1966 issue of Technology and Culture *(Vol. 7, No. 2): 143–151, was Dr. Drucker's presidential address to the Society for the History of Technology at its Eighth Annual Meeting in San Francisco in December 1965. The article has been republished in a series of essays by Dr. Drucker entitled* Technology, Management and Society *(New York, 1970).*

appeared a new society and a new polity: the irrigation city, which then rapidly became the irrigation empire. No other change in man's way of life and in his making a living, not even the changes under way today, so completely revolutionized human society and community. In fact, the irrigation civilizations were the beginning of history, if only because they brought writing.

The age of the irrigation civilization was pre-eminently an age of technological innovation. Not until a historical yesterday, the eighteenth century, did technological innovations emerge which were comparable in their scope and impact to those early changes in technology, tools, and processes. Indeed, the technology of man remained essentially unchanged until the eighteenth century insofar as its impact on human life and human society is concerned.

But the irrigation civilizations were not only one of the great ages of technology. They represent also mankind's greatest and most productive age of social and political innovation. The historian of ideas is prone to go back to ancient Greece, to the Old Testament prophets, or to the China of the early dynasties for the sources of the beliefs that still move men to action. But our fundamental social and political institutions antedate political philosophy by several thousand years. They all were conceived and established in the early dawn of the irrigation civilizations. Anyone interested in social and governmental institutions and in social and political processes will increasingly have to go back to those early irrigation cities. And, thanks to the work of archeologists and linguists during the last fifty years, we increasingly have the information, we increasingly know what the irrigation civilizations looked like, we increasingly can go back to them for our understanding both of antiquity and of modern society. For essentially our present-day social and political institutions, practically without exception, were then created and established. Here are a few examples.

1. The irrigation city first established government as a distinct and permanent institution. It established an impersonal government with a clear hierarchical structure in which very soon there arose a genuine bureaucracy— which is of course what enabled the irrigation cities to become irrigation empires.

Even more basic: the irrigation city first conceived of man as a citizen. It had to go beyond the narrow bounds of tribe and clan and had to weld people of very different origins and blood into one community. This required the first super-tribal deity, the god of the city. It also required the first clear distinction between custom and law and the development of an impersonal, abstract, codified legal system. Indeed, practically all legal concepts, whether of criminal or of civil law, go back to the irrigation city. The first great code of law, that of Hammurabi, almost four thousand years ago, would still be applicable to a good deal of legal business in today's highly developed, industrial society.

The irrigation city also first developed a standing army —it had to. For the farmer was defenseless and vulnerable and, above all, immobile. The irrigation city which, thanks to its technology, produced a surplus, for the first time in human affairs, was a most attractive target for the barbarian outside the gates, the tribal nomads of steppe and desert. And with the army came specific fighting technology and fighting equipment: the war horse and the chariot, the lance and the shield, armor and the catapult.

2. It was in the irrigation city that social classes first developed. It needed people permanently engaged in producing the farm products on which all the city lived; it needed farmers. It needed soldiers to defend them. And it needed a governing class with knowledge, that is, originally a priestly class. Down to the end of the nineteenth century these three "estates" were still considered basic in society.[1]

But at the same time the irrigation city went in for specialization of labor resulting in the emergence of artisans and craftsmen: potters, weavers, metal workers, and so on; and of professional people: scribes, lawyers, judges, physicians.

And because it produced a surplus it first engaged in organized trade which brought with it not only the merchant but money, credit, and a law that extended beyond the city to give protection, predictability, and justice to the stranger, the trader from far away. This, by the way, also made necessary international relations and international law. In fact, there is not very much difference between a nineteenth-century trade treaty and the trade treaties of the irrigation empires of antiquity.

3. The irrigation city first had knowledge, organized it, and institutionalized it. Both because it required considerable knowledge to construct and maintain the complex engineering works that regulated the vital water supply and because it had to manage complex economic transactions stretching over many years and over hundreds of

miles, the irrigation city needed records, and this of course meant writing. It needed astronomical data, as it depended on a calendar. It needed means of navigating across sea or desert. It therefore had to organize both the supply of the needed information and its processing into learnable and teachable knowledge. As a result, the irrigation city developed the first schools and the first teachers. It developed the first systematic observation of natural phenomena, indeed, the first approach to nature as something outside of and different from man and governed by its own rational and independent laws.

4. Finally, the irrigation city created the individual. Outside the city, as we can still see from those tribal communities that have survived to our days, only the tribe had existence. The individual as such was neither seen nor paid attention to. In the irrigation city of antiquity, however, the individual became, of necessity, the focal point. And with this emerged not only compassion and the concept of justice; with it emerged the arts as we know them, the poets, and eventually the world religions and the philosophers.

This is, of course, not even the barest sketch. All I wanted to suggest is the scope and magnitude of social and political innovation that underlay the rise of the irrigation civilizations. All I wanted to stress is that the irrigation city was essentially "modern," as we have understood the term, and that, until today, history largely consisted in building on the foundations laid five thousand or more years ago. In fact, one can argue that human history, in the last five thousand years, has largely been an extension of the social and political institutions of the irrigation city to larger and larger areas, that is, to all areas on the globe where water supply is adequate for the systematic tilling of the soil. In its beginnings, the irrigation city was the oasis in a tribal, nomadic world. By 1900 it was the tribal, nomadic world that had become the exception.

The irrigation civilization was based squarely upon a technological revolution. It can with justice be called a "technological polity." All its institutions were responses to opportunities and challenges that new technology offered. All its institutions were essentially aimed at making the new technology most productive.

I return to the question I posed at the beginning: what we can learn from the first technological revolution regarding the impacts likely to result on man, his society, and his government from the new industrial revolution, the one we are living in. Does the story of the irrigation civilization show man to be determined by his technical achievements, in thrall to them, coerced by them? Or does it show him capable of using technology to his own, to human ends, and of being the master of the tools of his own devising?

The answer which the irrigation civilizations give us to this question is threefold.

1. Without a shadow of doubt, major technological change creates the need for social and political innovation. It does make obsolete existing institutional arrangements. It does require new and very different institutions of community, society, and government. To this extent there can be no doubt: technological change of a revolutionary character coerces; it *demands innovation.*

2. The second answer also implies a strong necessity. There is little doubt, one would conclude from looking at the irrigation civilizations, that specific technological changes demand equally specific social and political innovations. That the basic institutions of the irrigation cities of the Old World, despite great cultural difference, all exhibited striking similarity may not prove much. After all, there probably was a great deal of cultural diffusion (though I refuse to get into the quicksand of debating whether Mesopotamia or China was the original innovator). But the fact that the irrigation civilizations of the New World around the Lake of Mexico and in Maya Yucatán, though culturally completely independent, millennia later evolved institutions which, in fundamentals, closely resemble those of the Old World (e.g., an organized government with social classes and a permanent military, and writing) would argue strongly that the solutions to specific conditions created by new technology have to be specific and are therefore limited in number and scope.

In other words, one lesson to be learned from the first technological revolution is that new technology creates what a philosopher of history might call "objective reality." And objective reality has to be dealt with on *its* terms. Such a reality would, for instance, be the conversion, in the course of the first technological revolution, of human space from "habitat" into "settlement," that is, into a permanent territorial unit always to be found in the same place—unlike the migrating herds of pastoral people or the hunting grounds of primitive tribes. This alone makes obsolete the tribe and demands a permanent, impersonal, and rather powerful government.

3. But the irrigation civilizations can teach us also that the new objective reality determines only the gross parameters of the solutions. It determines where, and in respect to what, new institutions are needed. It does not make anything "inevitable." It leaves wide open *how* the new problems are to be tackled, what the purposes and values of the new institutions are to be.

In the irrigation civilizations of the New World the individual, for instance, failed to make his appearance. Never, as far as we know, did these civilizations get around to separating law from custom nor, despite a highly developed trade, did they invent money, and so on.

Even within the Old World, where one irrigation civilization could learn from the others, there were very great differences. They were far from homogeneous even though all had similar tasks to accomplish and developed similar institutions for these tasks. The different specific answers expressed above all different views regarding man, his position in the universe, and his society—different purposes and greatly differing values.

Impersonal bureaucratic government had to arise in all these civilizations; without it they could not have functioned. But in the Near East it was seen at a very early stage that such a government could serve equally to exploit and hold down the common man and to establish justice for all and protection for the weak. From the beginning the Near East saw an ethical decision as crucial to government. In Egypt, however, this decision was never seen. The question of the purpose of government was never asked. And the central quest of government in China was not justice but harmony.

It was in Egypt that the individual first emerged, as witness the many statues, portraits, and writings of professional men, such as scribes and administrators, that have come down to us—most of them superbly aware of the uniqueness of the individual and clearly asserting his primacy. It is early Egypt, for instance, which records the names of architects who built the great pyramids. We have no names for the equally great architects of the castles and palaces of Assur or Babylon, let alone for the early architects of China. But Egypt suppressed the individual after a fairly short period during which he flowered (perhaps as part of the reaction against the dangerous heresies of Ikhnaton). There is no individual left in the records of the Middle and New Kingdoms, which perhaps explains their relative sterility.

In the other areas two entirely different basic approaches emerged. One, that of Mesopotamia and of the Taoists, we might call "personalism," the approach that found its greatest expression later in the Hebrew prophets and in the Greek dramatists. Here the stress is on developing to the fullest the capacities of the person. In the other approach—we might call it "rationalism," taught and exemplified above all by Confucius—the aim is the molding and shaping of the individual according to pre-established ideals of rightness and perfection. I need not tell you that both these approaches still permeate our thinking about education.

Or take the military. Organized defense was a necessity for the irrigation civilization. But three different approaches emerged: a separate military class supported through tribute by the producing class, the farmers; the citizen-army drafted from the peasantry itself; and mercenaries. There is very little doubt that from the beginning it was clearly understood that each of these three approaches had very real political consequences. It is hardly coincidence, I believe, that Egypt, originally unified by overthrowing local, petty chieftains, never developed afterwards a professional, permanent military class.

Even the class structure, though it characterizes all irrigation civilizations, showed great differences from culture to culture and within the same culture at different times. It was being used to create permanent castes and complete social immobility, but it was also used with great skill to create a very high degree of social mobility and a substantial measure of opportunities for the gifted and ambitious.

Or take science. We now know that no early civilization excelled over China in the quality and quantity of scientific observations. And yet we also know that early Chinese culture did not point toward anything we would call "science." Perhaps because of their rationalism the Chinese refrained from generalization. And though fanciful and speculative, it is the generalizations of the Near East or the mathematics of Egypt which point the way toward systematic science. The Chinese, with their superb gift for accurate observation, could obtain an enormous amount of information about nature. But their view of the universe remained totally unaffected thereby—in sharp contrast to what we know about the Middle Eastern developments out of which Europe arose.

In brief, the history of man's first technological revolution indicates the following:

1. Technological revolutions create an objective need for social and political innovations. They create a need also for identifying the areas in which new institutions are needed and old ones are becoming obsolete.

2. The new institutions have to be appropriate to specific new needs. There are right social and political responses to technology and wrong social and political responses. To the extent that only a right institutional response will do, society and government are largely circumscribed by new technology.

3. But the values these institutions attempt to realize, the human and social purposes to which they are applied, and, perhaps most important, the emphasis and stress laid on one purpose as against another, are largely within human control. The bony structure, the hard stuff of a society, is prescribed by the tasks it has to accomplish. But the ethos of the society is in man's hands and is largely a matter of the "how" rather than of the "what."

For the first time in thousands of years, we face again a situation that can be compared with what our remote ancestors faced at the time of the irrigation civilization. It is not only the speed of technological change that creates a "revolution," it is its scope as well. Above all, today, as seven thousand years ago, technological developments from a great many areas are growing together to create a new human environment. This has not been true of any period between the first technological revolution and the technological revolution that got under way two hundred years ago and has still clearly not run its course.

We therefore face a big task of identifying the areas in which social and political innovations are needed. We face a big task in developing the institutions for the new tasks, institutions adequate to the new needs and to the new capacities which technological change is casting up. And, finally, we face the biggest task of them all, the task of ensuring that the new institutions embody the values we believe in, aspire to the purposes we consider right, and serve human freedom, human dignity, and human ends.

If an educated man of those days of the first technological revolution—an educated Sumerian perhaps or an educated ancient Chinese—looked at us today, he would certainly be totally stumped by our technology. But he would, I am sure, find our existing social and political institutions reasonably familiar—they are after all, by and large, not fundamentally different from the institutions he and his contemporaries first fashioned. And, I am quite certain, he would have nothing but a wry smile for both those among us who predict a technological heaven and those who predict a technological hell of "alienation," of "technological unemployment," and so on. He might well mutter to himself, "This is where I came in." But to us he might well say, "A time such as was mine and such as is yours, a time of true technological revolution, is not a time for exultation. It is not a time for despair either. It is a time for work and for responsibility."

REFERENCE

[1] See the brilliant though one-sided book by Karl A. Wittvogel, *Oriental Despotism: A Comparative Study of Total Power* (New Haven, Conn., 1957).

Authoritarian and Democratic Technics

Lewis Mumford

"Democracy" is a term now confused and sophisticated by indiscriminate use, and often treated with patronizing contempt. Can we agree, no matter how far we might diverge at a later point, that the spinal principle of democracy is to place what is common to all men above that which any organization, institution, or group may claim for itself? This is not to deny the claims of superior natural

Mr. Mumford, of Amenia, New York, is one of the most honored and respected of America's social philosophers and humanistic thinkers. He has made major contributions to the history of technology, literary criticism, and architectural and city planning. His Technics and Civilization, *first published in 1934, was one of the pioneer works in the history of technology, emphasizing especially the human and social impact of technological development, and it remains a classic. In 1961 his book,* The City in History, *received the National Book Award; he has been awarded the President's Medal of Freedom and many other honors, including the Leonardo da Vinci Medal of the Society for the History of Technology.*

This paper is Mr. Mumford's speech at the Fund for the Republic Tenth Anniversary Convocation on "Challenges to Democracy in the Next Decade," held in New York City in January 1963. It was published in Technology and Culture *(Vol. 5, No. 1): 1–8, by permission of the Fund for the Republic and the Center for the Study of Democratic Institutions. Mr. Mumford has amplified the thesis presented in this article in his most recent work,* The Myth of the Machine, *consisting of two volumes:* Technics and Human Development *(New York, 1967) and* The Pentagon of Power *(New York, 1970).*

endowment, specialized knowledge, technical skill, or institutional organization: all these may, by democratic permission, play a useful role in the human economy. But democracy consists in giving final authority to the whole, rather than the part; and only living human beings, as such, are an authentic expression of the whole, whether acting alone or with the help of others.

Around this central principle clusters a group of related ideas and practices with a long foreground in history, though they are not always present, or present in equal amounts, in all societies. Among these items are communal self-government, free communication as between equals, unimpeded access to the common store of knowledge, protection against arbitrary external controls, and a sense of individual moral responsibility for behavior that affects the whole community. All living organisms are in some degree autonomous, in that they follow a life-pattern of their own; but in man this autonomy is an essential condition for his further development. We surrender some of our autonomy when ill or crippled: but to surrender it every day on every occasion would be to turn life itself into a chronic illness. The best life possible—and here I am consciously treading on contested ground—is one that calls for an ever greater degree of self-direction, self-expression, and self-realization. In this sense, personality, once the exclusive attribute of kings, belongs on democratic theory to every man. Life itself in its fullness and wholeness cannot be delegated.

In framing this provisional definition I trust that I have not, for the sake of agreement, left out anything important. Democracy, in the primal sense I shall use the term, is necessarily most visible in relatively small communities and groups, whose members meet frequently face to face, interact freely, and are known to each other as persons. As soon as large numbers are involved, democratic association must be supplemented by a more abstract, depersonalized form. Historical experience shows that it is much easier to wipe out democracy by an institutional arrangement that gives authority only to those at the apex of the social hierarchy than it is to incorporate democratic practices into a well-organized system under centralized direction, which achieves the highest degree of mechanical efficiency when those who work it have no mind or purpose of their own.

The tension between small-scale association and large-scale organization, between personal autonomy and insti-

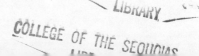

tutional regulation, between remote control and diffused local intervention, has now created a critical situation. If our eyes had been open, we might long ago have discovered this conflict deeply embedded in technology itself.

I wish it were possible to characterize technics with as much hope of getting assent, with whatever quizzical reserves you may still have, as in this description of democracy. But the very title of this paper is, I confess, a controversial one; and I cannot go far in my analysis without drawing on interpretations that have not yet been adequately published, still less widely discussed or rigorously criticized and evaluated. My thesis, to put it bluntly, is that from late neolithic times in the Near East, right down to our own day, two technologies have recurrently existed side by side: one authoritarian, the other democratic, the first system-centered, immensely powerful, but inherently unstable, the other man-centered, relatively weak, but resourceful and durable. If I am right, we are now rapidly approaching a point at which, unless we radically alter our present course, our surviving democratic technics will be completely suppressed or supplanted, so that every residual autonomy will be wiped out, or will be permitted only as a playful device of government, like national balloting for already chosen leaders in totalitarian countries.

The data on which this thesis is based are familiar; but their significance has, I believe, been overlooked. What I would call democratic technics is the small-scale method of production, resting mainly on human skill and animal energy but always, even when employing machines, remaining under the active direction of the craftsman or the farmer, each group developing its own gifts, through appropriate arts and social ceremonies, as well as making discreet use of the gifts of nature. This technology had limited horizons of achievement, but, just because of its wide diffusion and its modest demands, it had great powers of adaptation and recuperation. This democratic technics has underpinned and firmly supported every historical culture until our own day, and redeemed the constant tendency of authoritarian technics to misapply its powers. Even when paying tribute to the most oppressive authoritarian regimes, there yet remained within the workshop or the farmyard some degree of autonomy, selectivity, creativity. No royal mace, no slave-driver's whip, no bureaucratic directive left its imprint on the textiles of Damascus or the pottery of fifth-century Athens.

If this democratic technics goes back to the earliest use of tools, authoritarian technics is a much more recent achievement: it begins around the fourth millennium B.C. in a new configuration of technical invention, scientific observation, and centralized political control that gave rise to the peculiar mode of life we may now identify, without eulogy, as civilization. Under the new institution of kingship, activities that had been scattered, diversified, cut to the human measure, were united on a monumental scale into an entirely new kind of theological-technological mass organization. In the person of an absolute ruler, whose word was law, cosmic powers came down to earth, mobilizing and unifying the efforts of thousands of men, hitherto all-too-autonomous and too decentralized to act voluntarily in unison for purposes that lay beyond the village horizon.

The new authoritarian technology was not limited by village custom or human sentiment: its herculean feats of mechanical organization rested on ruthless physical coercion, forced labor and slavery, which brought into existence machines that were capable of exerting thousands of horsepower centuries before horses were harnessed or wheels invented. This centralized technics drew on inventions and scientific discoveries of a high order: the written record, mathematics and astronomy, irrigation and canalization: above all, it created complex human machines composed of specialized, standardized, replaceable, interdependent parts—the work army, the military army, the bureaucracy. These work armies and military armies raised the ceiling of human achievement: the first in mass construction, the second in mass destruction, both on a scale hitherto inconceivable. Despite its constant drive to destruction, this totalitarian technics was tolerated, perhaps even welcomed, in home territory, for it created the first economy of controlled abundance: notably, immense food crops that not merely supported a big urban population but released a large trained minority for purely religious, scientific, bureaucratic, or military activity. But the efficiency of the system was impaired by weaknesses that were never overcome until our own day.

To begin with, the democratic economy of the agricultural village resisted incorporation into the new authoritarian system. So even the Roman Empire found it expedient, once resistance was broken and taxes were collected, to consent to a large degree of local autonomy in religion and government. Moreover, as long as agriculture absorbed the labor of some 90 per cent of the population, mass technics were confined largely to the populous urban centers. Since authoritarian technics first took form in an age when metals were scarce and human raw material, captured in war, was easily convertible into machines, its directors never bothered to invent inorganic mechanical substitutes. But there were even greater weaknesses: the system had no inner coherence: a break in communication, a missing link in the chain of command, and the great human machines fell apart. Finally, the myths upon which the whole system was based—particularly the essential myth of kingship—were irrational, with their paranoid suspicions and animosities and their paranoid claims to unconditional obedience and absolute power. For all its redoubtable constructive achievements, authoritarian technics expressed a deep hostility to life.

By now you doubtless see the point of this brief historical excursus. That authoritarian technics has come back today in an immensely magnified and adroitly perfected form. Up to now, following the optimistic premises of nineteenth-century thinkers like Auguste Comte and Herbert Spencer, we have regarded the spread of experimental science and mechanical invention as the soundest guarantee of a peaceful, productive, above all democratic, industrial society. Many have even comfortably supposed that the revolt against arbitrary political power in the seventeenth century was causally connected with the industrial revolution that accompanied it. But what we have interpreted as the new freedom now turns out to be a much more sophisticated version of the old slavery: for the rise of political democracy during the last few centuries has been increasingly nullified by the successful resurrection of a centralized authoritarian technics—a technics that had in fact for long lapsed in many parts of the world.

Let us fool ourselves no longer. At the very moment Western nations threw off the ancient regime of absolute government, operating under a once-divine king, they were restoring this same system in a far more effective form in their technology, reintroducing coercions of a military character no less strict in the organization of a factory than in that of the new drilled, uniformed, and regimented army. During the transitional stages of the last two centuries, the ultimate tendency of this system might be in doubt, for in many areas there were strong democratic reactions; but with the knitting together of a scientific ideology, itself liberated from theological restrictions or humanistic purposes, authoritarian technics found an instrument at hand that has now given it absolute command of physical energies of cosmic dimensions. The inventors of nuclear bombs, space rockets, and computers are the pyramid builders of our own age: psychologically inflated by a similar myth of unqualified power, boasting through their science of their increasing omnipotence, if not omniscience, moved by obsessions and compulsions no less irrational than those of earlier absolute systems: particularly the notion that the system itself must be expanded, at whatever eventual cost to life.

Through mechanization, automation, cybernetic direction, this authoritarian technics has at last successfully overcome its most serious weakness: its original dependence upon resistant, sometimes actively disobedient servomechanisms, still human enough to harbor purposes that do not always coincide with those of the system.

Like the earliest form of authoritarian technics, this new technology is marvelously dynamic and productive: its power in every form tends to increase without limits, in quantities that defy assimilation and defeat control, whether we are thinking of the output of scientific knowledge or of industrial assembly lines. To maximize energy, speed, or automation, without reference to the complex conditions that sustain organic life, have become ends in themselves. As with the earliest forms of authoritarian technics, the weight of effort, if one is to judge by national budgets, is toward absolute instruments of destruction, designed for absolutely irrational purposes whose chief by-product would be the mutilation or extermination of the human race. Even Ashurbanipal and Genghis Khan performed their gory operations under normal human limits.

The center of authority in this new system is no longer a visible personality, an all-powerful king: even in totalitarian dictatorships the center now lies in the system itself, invisible but omnipresent: all its human components, even the technical and managerial elite, even the sacred priesthood of science, who alone have access to the secret knowledge by means of which total control is now swiftly being effected, are themselves trapped by the very perfection of the organization they have invented. Like the pharaohs of the Pyramid Age, these servants of the system identify its goods with their own kind of well-being: as with the divine king, their praise of the system is an act of self-worship;

and again like the king, they are in the grip of an irrational compulsion to extend their means of control and expand the scope of their authority. In this new systems-centered collective, this Pentagon of power, there is no visible presence who issues commands: unlike Job's God, the new deities cannot be confronted, still less defied. Under the pretext of saving labor, the ultimate end of this technics is to displace life, or rather, to transfer the attributes of life to the machine and the mechanical collective, allowing only so much of the organism to remain as may be controlled and manipulated.

Do not misunderstand this analysis. The danger to democracy does not spring from any specific scientific discoveries or electronic inventions. The human compulsions that dominate the authoritarian technics of our own day date back to a period before even the wheel had been invented. The danger springs from the fact that, since Francis Bacon and Galileo defined the new methods and objectives of technics, our great physical transformations have been effected by a system that deliberately eliminates the whole human personality, ignores the historical process, overplays the role of the abstract intelligence, and makes control over physical nature, ultimately control over man himself, the chief purpose of existence. This system has made its way so insidiously into Western society that my analysis of its derivation and its intentions may well seem more questionable—indeed more shocking—than the facts themselves.

Why has our age surrendered so easily to the controllers, the manipulators, the conditioners of an authoritarian technics? The answer to this question is both paradoxical and ironic. Present-day technics differs from that of the overtly brutal, half-baked authoritarian systems of the past in one highly favorable particular: it has accepted the basic principle of democracy, that every member of society should have a share in its goods. By progressively fulfilling this part of the democratic promise, our system has achieved a hold over the whole community that threatens to wipe out every other vestige of democracy.

The bargain we are being asked to ratify takes the form of a magnificent bribe. Under the democratic-authoritarian social contract, each member of the community may claim every material advantage, every intellectual and emotional stimulus he may desire, in quantities hardly available hitherto even for a restricted minority: food, housing, swift transportation, instantaneous communication, medical care, entertainment, education. But on one condition: that one must not merely ask for nothing that the system does not provide, but likewise agree to take everything offered, duly processed and fabricated, homogenized and equalized, in the precise quantities that the system, rather than the person, requires. Once one opts for the system no further choice remains. In a word, if one surrenders one's life at source, authoritarian technics will give back as much of it as can be mechanically graded, quantitatively multiplied, collectively manipulated and magnified.

"Is this not a fair bargain?" those who speak for the system will ask. "Are not the goods authoritarian technics promises real goods? Is this not the horn of plenty that mankind has long dreamed of, and that every ruling class has tried to secure, at whatever cost of brutality and injustice, for itself?" I would not belittle, still less deny, the many admirable products this technology has brought forth, products that a self-regulating economy would make good use of. I would only suggest that it is time to reckon up the human disadvantages and costs, to say nothing of the dangers, of our unqualified acceptance of the system itself. Even the immediate price is heavy; for the system is so far from being under effective human direction that it may poison us wholesale to provide us with food or exterminate us to provide national security, before we can enjoy its promised goods. Is it really humanly profitable to give up the possibility of living a few years at Walden Pond, so to say, for the privilege of spending a lifetime in *Walden Two?* Once our authoritarian technics consolidates its powers, with the aid of its new forms of mass control, its panoply of tranquilizers and sedatives and aphrodisiacs, could democracy in any form survive? That question is absurd: life itself will not survive, except what is funneled through the mechanical collective. The spread of a sterilized scientific intelligence over the planet would not, as Teilhard de Chardin so innocently imagined, be the happy consummation of divine purpose: it would rather ensure the final arrest of any further human development.

Again: do not mistake my meaning. This is not a prediction of what *will* happen, but a warning against what *may* happen.

What means must be taken to escape this fate? In characterizing the authoritarian technics that has begun to dominate us, I have not forgotten the great lesson of history: prepare for the unexpected! Nor do I overlook the immense reserves of vitality and creativity that a more humane democratic tradition still offers us. What I wish to do is to persuade those who are concerned with maintaining democratic institutions to see that their constructive efforts must include technology itself. There, too, we must return to the human center. We must challenge this authoritarian system that has given to an underdimensioned ideology and technology the authority that belongs to the human personality. I repeat: life cannot be delegated.

Curiously, the first words in support of this thesis came forth, with exquisite symbolic aptness, from a willing agent —but very nearly a classic victim!—of the new authoritarian technics. They came from the astronaut, John Glenn, whose life was endangered by the malfunctioning of his automatic controls, operated from a remote center. After he barely saved his life by personal intervention, he emerged from his space capsule with these ringing words: "Now let man take over!"

That command is easier to utter than obey. But if we are not to be driven to even more drastic measures than Samuel Butler suggested in *Erewhon,* we had better map out a more positive course: namely, the reconstitution of both our science and our technics in such a fashion as to insert the rejected parts of the human personality at every stage in the process. This means gladly sacrificing mere quantity in order to restore qualitative choice, shifting the seat of authority from the mechanical collective to the human personality and the autonomous group, favoring variety and ecological complexity, instead of stressing undue uniformity and standardization, above all, reducing the insensate drive to extend the system itself, instead of containing it within definite human limits and thus releasing man himself for other purposes. We must ask, not what is good for science or technology, still less what is good for General Motors or Union Carbide or IBM or the Pentagon, but what is good for man: not machine-conditioned, system-regulated, mass-man, but man in person, moving freely over every area of life.

There are large areas of technology that can be redeemed by the democratic process, once we have overcome the infantile compulsions and automatisms that now threaten to cancel out our real gains. The very leisure that the machine now gives in advanced countries can be profitably used, not for further commitment to still other kinds of machine, furnishing automatic recreation, but by doing significant forms of work, unprofitable or technically impossible under mass production: work dependent upon special skill, knowledge, aesthetic sense. The do-it-yourself movement prematurely got bogged down in an attempt to sell still more machines; but its slogan pointed in the right direction, provided we still have a self to do it with. The glut of motor cars that is now destroying our cities can be coped with only if we redesign our cities to make fuller use of a more efficient human agent: the walker. Even in childbirth, the emphasis is already happily shifting from an officious, often lethal, authoritarian procedure, centered in hospital routine, to a more human mode, which restores initiative to the mother and to the body's natural rhythms.

The replenishment of democratic technics is plainly too big a subject to be handled in a final sentence or two: but I trust I have made it clear that the genuine advantages our scientifically based technics has brought can be preserved only if we cut the whole system back to a point at which it will permit human alternatives, human interventions, and human destinations for entirely different purposes from those of the system itself. At the present juncture, if democracy did not exist, we would have to invent it, in order to save and recultivate the spirit of man.

Technology in Focus—
The Emerging Nations

Lord Ritchie-Calder

Science and technology could, without a further remit, solve all the material problems of the world today. It is not a question of "Whether" but "How," and the "How" is not a matter of knowledge but of intention.

The story of man's survival and, indeed, evolution has been one of his ability to master his environment. The one creature which was naked to the elements without fur, feather, or carapace, managed to migrate from the tropics into the temperate zone and into the Arctic. He contrived to clothe himself, build himself shelter, and master fire to provide menial heat. The most defenseless of creatures, without fang or claw, he protected himself against his natural enemies by the fire he had mastered, by the clubs he contrived to extend his forearm and give weight to his fist, by the spears he could throw from safe range, or by

Lord Ritchie-Calder, formerly professor of international relations at the University of Edinburgh, has written many books, including After the Seventh Day *and* The Evolution of the Machine. *He has been the United Kingdom delegate to UNESCO and has served the United Nations extensively as adviser to several agencies and as a member of missions to Asia and the Arctic.*

This paper, published in Technology and Culture *(Vol. 3, No. 4): 563–80, was part of an issue containing the proceedings of the Encyclopaedia Britannica Conference on the Technological Order, held in March 1962 at the Center for the Study of Democratic Institutions in Santa Barbara, California. The proceedings of this conference have also been published in Carl F. Stover (ed.),* The Technological Order *(Detroit, 1963).*

the stockades he could build, or by the lake dwellings which set water between him and his predators.

From being a beast of prey, a hunter, he domesticated animals and drove his herds, seeking pastures which would feed them so that they could feed him. From a food gatherer, he became a food grower. He noticed that edible seeds could themselves produce seeds and that they would grow more plentifully if he scratched the ground and better still when the ground was well watered. So he became a tiller of the alluvial lands of the river valleys. He found that he could be the victim of floods or the manager of the flood waters, so he developed the system of irrigation. But, while the individual could till, by irrigation, the land in the vicinity of rivers, if the area of cultivation was to be extended, it could only be by master canals, public utilities. So the individual became a hostage to water management, vested in pharaohs or kings.

The settled tiller could produce surpluses in excess of his, or his dependents', needs for sustenance. He could barter food for better skills. He could make his own implements or domestic utensils, but there were others who could make them better, so they became the artisans maintained by the productivity of the tiller. His knowledge was mixed up with the unknown, and he needed intercessors who could intercede with or appease the gods. Thus a priesthood emerged, and his family altar became the temple or the ziggurat. But a priesthood has a continuity beyond the span, or experience, of a man. The tiller might observe the configuration of the stars, or the lengthening shadows which divided the day, and the passage of the seasons; or the woman, from her own biological rhythm, would understand the living cycle. They would know when to plant or to harvest or when to expect calves or lambs or kids. The priests, however, had a longer sweep. Their observations could extend beyond a generation or even the centuries. Thus we find the priests of Chaldea turned astronomers, giving a length to the year only 26 minutes 26 seconds longer than that determined by modern precision instruments. We find that they discovered the *saros,* the 18 years and 11½ days, intervening between the eclipses of the sun. The priestly ledger clerks of the Nile, a river on the rise and fall of which the whole livelihood of the valley settlers depended, had the Nilometer, a simple gauge. With the records derived from that they could predict not only the seasonal but the cyclical changes. From such natural observations they could assert supernatural powers. The Chaldean priests could predict to the moment when the sun would be blotted out, and the watergaugers of the temple were obviously the confidants of Isis and Osiris. Thus from awe of science, as well as innate piety, the tillers endowed the temples with the fruits of their toil.

This had further consequences because the tithebarns of the temples became the warehouses of surplus products, which could be traded beyond the limits of the community in exchange for alien products or crafts to enhance the temples. Ledger-keeping needed ciphers, the clay jars of ancient packaging needed seals, and trade needed credit tokens. In this we have the beginning of the scholarly clerks, of men of commerce, and of financiers.

With the growth of the priestly hierarchy, the emergence of the priest-king was inevitable. Either by his learning or by his inheritance of temple lands an individual was bound at some stage to exercise authority over his fellows. He would reinforce that authority by identifying himself with a god and become the living embodiment of a deity to which the citizens already acknowledged allegiance. And as the cities grew and their external trade increased, the hostility which a nomadic people felt toward the tillers, whose settlements in the river valleys intruded upon their summer pastures, turned to envy and to a predatory urge. The cities and the accessible wealth which they represented were likely to be raided by marauders, and they, worshiping alien gods, would not be deterred by the supernatural sanctions of a priest-king. So the cities had to have walls and soldiers to man them, and the priest-kings became, or gave way to, warrior-kings. Defense became aggression, and any pretext was sufficient for aggrandizement by conquest. There was another encouragement toward wars of conquest: the need for slaves. The peasant and the husbanded soil were now supporting a social superstructure as massive as the ziggurats which had been built over the modest shrines and peasant altars—the craftsmen, the priesthood, the tax-collectors, the dynastic households, the tradesmen, the money-changers, and the soldiers. The produce of the soil had to be increased to satisfy the stomachs and the avarice of those who no longer worked the land. The pious tithe-payer of the primitive temple, paying his dues for the land the gods had made and the crops which they gave him, had now become a serf of a feudal system, but the prisoner of war could, in turn, become his slave.

Among the public works for which the slaves were most needed was the construction of water channels. Once, however, irrigation systems were constructed, then, as now, competition for water rights led to violent conflicts. (We have only to remind ourselves that the word "rivalry" comes from the Latin *rivus,* a stream, to realize that the struggle for water has always been a cause for feuding.) It called for an overlord who would not only make himself responsible for the building of the master-channels but could command the labor force to maintain them, the laws to regulate them, and the soldiers to protect them. The story of ancient Mesopotamia is that of quarrels over competing canal systems. The systematic ruin of an irrigation system was the method of punishing a defeated enemy, and the dead cities of the desert are those which died of thirst through a water blockade, or in some instances, those which died through the malaria which came with collapse of neglected canal systems into marshes.

Conversely, ancient systems of water management and soil conservation are reminders, more significant than pyramids or ziggurats which remain as gravestones of lost civilizations, of survival skills of thousands of years ago. It would be difficult to suggest or devise today systems more ingenious or rational than those of the past: the exploitation of the gravitational flow between the Tigris and Euphrates; the flash-flooding of the delta fields of the Nile; contour farming and terracing in middle east China and Peru; Persian *wanaats* or horizontal wells; Nabatean desert farming which from 100 millimeters of rain could produce crops in the Biblical wilderness; plant drainage of Sumer and Aral civilizations which without engineering enabled farmers to cope with salt aggregations beyond modern tolerance. We can provide bulldozers, mechanical shovels, gelignite, steel, and concrete, but we can scarcely improve on their principles.

Mastering his environment, man created his own problems. Contagion and infections of crowded cities are not present in small wandering groups of pastoral peoples. Urban hazards included polluted waters, community sewage, and proximity diseases, such as leprosy, or insect-borne diseases such as typhus, bubonic plague, and malaria, and the miasmic diseases of a fetid atmosphere. But we can find a recognition of all these. Nergal, the god of disease of the Babylonians, and Beelzebub, chief god of the Assyro-Phoenicians, had as their symbol the fly, a recognition of insect-borne disease; and Rhazes, the lute player of Baghdad, anticipated Pasteur by a thousand years when, ordered by the caliph to build a hospital, he hung pieces of meat around the city, and where the meat putrefied least he sited the hospital. In Mohenjo-Daro, in the Indus civilization of five thousand years ago, there was an impressive system of drains, and around the remains of water booths, where drinking water was sold, have been found spoil heaps of broken cups, far in excess of the casualties of the most careless dishwashers and strongly suggestive of sanitary laws which compelled the breaking of those rough cups; if someone bought a drink of water, the cup was thrown away, like the paper cups of today.

In the arts of industry, the ancients have also a great deal to teach. For example, the ancient Hindus were the supreme metal workers. They produced steel, which was imported into the West, to make the Damascene swords, the Toledo blades, and possibly the Excalibur of King Arthur, and which remained a craft secret until Réaumur discovered the metallurgical chemistry of steel in the mid-nineteenth century. And the metallurgists are still trying to discover the secret of the Delhi Pillar and the girders of the temples of Konarak (exposed to the salt winds from the sea) which are made of iron which, throughout the centuries, has remained rustless. And we must admire the combination of craft skills and medical knowledge which, at least two thousand years ago, produced the 121 surgical instruments of the *Susruta*—scalpels, lancets, saws, scissors, needles, hooks, probes, obstetric forceps, catheters, syringes, bougies, and so forth.

We should recall that the ancient Chinese anticipated the spinnerets of the rayon and nylon industries by employing the natural spinnerets of the silk worm; they anticipated block printing, rockets, and the mechanical timepiece, as well as so much else. And that Hero of Alexandria, in 140 A.D., forestalled Savery, Newcomen, James Watt, Parsons, and Whittle by inventing a steam engine which combined jet propulsion. This was a sphere into which steam was fed, to escape through right-angled jets forcing the sphere to spin. He also had a system by which hot air expanded and drove water out of a container into a bucket which, in descending with the weight of water, turned a spindle. By this, when fires were lit upon an altar, the doors of a shrine were mysteriously opened, disclosing the god. But Hero's ingenuity remained no more than such conjuring tricks because his was the age of slavery when muscle power was abundant and mechanical power unnecessary; Hero's ideas were "uneconomic."

If excuse is necessary for this strip-cartoon approach to "Technology in Focus," it is that invariably we look at technology in fine focus, a close-up in function and immediacy; but, if we are examining the effects, good or bad, on contemporary circumstance, it can only be by resolution of modern detail out of the panorama of history. Otherwise we may be only magnifying our own mistakes. After all, the ancients achieved their enduring results by trial and error over centuries. Just as mistaken as the conservatism of "We tried it once and it didn't work" is the forgetting that "They tried it once and it did work." Experience is a good substitute for experiment. For example, the modern Israelis, when they are attempting desert recovery, start not in the laboratory but in the field; their archaeologists and plant palaeontologists go to find out how, by techniques that have been forgotten, the ancient Israelites and the Nabateans and the Byzantines contrived to feed themselves out of what might seem a hopeless, inhospitable desert.

There is the further excuse that, in the arrogance of modern achievement, we tend to regard "Point Four" and other forms of technical assistance to underdeveloped and developing countries as a charitable endowment, an act of generosity by the highly advanced countries to the "backward" countries, forgetting that many of those countries had advanced civilization and what we would now call "technology" when the ancestors of the Industrial Revolution were running around painted in woad. If we take into account the heritage of science and technology from which our present achievements, and perhaps excesses, evolved, we may be disposed to think of technical assistance as a kind of reverse lend-lease. Moreover it is important to the self-esteem of countries conscious of their own traditions that they should not regard themselves as technological beggars.

To ignore the traditional is to assume a complacency which is dangerous. The United States discovered this when the U.S.S.R. put the first Sputnik into orbit. In the context of the International Geophysical Year, in which the U.S.A. and the U.S.S.R. were both scheduled to put up satellites, it was taken for granted that the Americans would be first. When the bleeps of Sputnik were heard they caused a consternation which the *Times* of London estimated cost the U.S. economy $13 billion. American technology had been forestalled. That itself was not traumatic; it had already been conceded that the Russians were no longer moujiks who tore the innards out of tractors copied from America and that they had pretty competent engineers. What was shattering was that they put it into an orbit which, at that moment, America was not going to attempt. This was science and not just technology, which would not have been surprising if it had not been forgotten that the Russians, long before the revolution, had a tradition of pre-eminent mathematicians, astronomers, astrophysicists, and ballistic scientists.

But in a wider context, the past has salutary reminders for the present. A succession of civilizations, empires, and cultures have flourished and foundered, but those were localized civilizations and when they became effete—reached, if you like, the limits of a dynamic technology and became the victims, not the masters, of their environment—others less improvident or more dynamic took over. Today, as a consequence of our technological advances, ours is a global civilization. It is an earth around which a man-made piece of hardware can circle sixteen times a day; on which a broadcast voice can be heard faster 13,000 miles away than it can be heard at the back of the room in which it is speaking—because, of course, radio waves travel at 186,000 miles per second, and sound waves amble along at 700 miles an hour—in which the fallout of an H-bomb, exploded in mid-Pacific, on Christmas Island, or in the Arctic, can return from the stratosphere, be distributed by the jet-stream, and be deposited in rain throughout the world; on which three billion people today and a certain four billion twenty years from now will have to contrive to live and work together.

We may talk about "Our Way of Life" as distinguishing us from the political or cultural values of others, but in terms of the material civilization or the problem of survival the whole world is interdependent and no longer bounded by the Tigris and the Euphrates or the valley of the Nile.

For one thing, man is now technologically capable of vetoing the continued evolution of his species. He can exercise that veto by the nuclear destruction of the race and the radioactive poisoning of his total environment. Or he can, by neglecting the development of food production and material resources, produce starvation. His ingenuity and capacity are such that he may reach out for the other planets, but his species as such must survive on the surface of this earth, subsist from the nine inches of topsoil which feeds, shelters, and clothes him or from the seas, which cover seven-tenths of this small planet and from which he emerged all those hundreds of millions of years ago.

Even if we are thinking of maintaining or extending the standards of living of countries which are highly advanced, it will be found that no unit, except the globe itself, will be self-sufficient. The Paley Report (U.S. President's Material Policy Commission, 1952) estimated that if the United States maintains full employment and continues to expand her manufacturing capacity, she will have to import 50 per cent of her raw materials from abroad by the year 1975. The European economy is even more dependent. Britain already has to import half its food and most of the raw materials of its factories.

The United States, which in fifty years has taken more minerals out of the crust of the States than has been taken out of the whole earth in the whole previous history of all mankind, will become dependent on countries which we now call "underdeveloped" and to which President Truman made his Point Four promise: "We must embark on a bold new program for making the benefits of our scientific advances and technical progress available for the improvement and growth of underdeveloped areas."

Apart from the merits of good intention and the political considerations involved, the development of countries on which the advanced nations are going to depend for material resources is enlightened and intelligent self-interest. Point Four said: "It must be a world-wide effort for the achievement of peace, plenty and freedom." The Paley Report expressed it: "The over-all objective of a national materials policy for the United States should be to insure an adequate and dependable flow of materials at the lowest cost consistent with national security and the welfare of friendly nations." But it also said that "the habit of regarding diversified economic growth for the underdeveloped countries as an alternative to materials development is erroneous." It pointed out that "development should involve a balanced growth of agriculture, manufacturing, and other industries directed towards maximizing the real income and improving the standard of living." In other words, it is not enough to quarry the wealth of another country. Something ought to be done to fill up the morass of poverty. With the emergence of new nations and the bargaining of the conflicting ideologies, there is not much option. Even the military know that you cannot build secure bases in countries of disaffected poverty. But a reappraisal of bilateral aid in the 1950's will probably show that it was a mistake to confuse military aid with economic aid. Apart from introducing skepticism and, politically, encouraging neutralism, it must introduce technological inconsistency. Proper technical aid ought always to involve "counterparts"—the nationals who, with training, ought to be left competent to carry on and reproduce the demonstration programs when the outside help has gone. But the essence of military technology is to get things done—on a pretty lavish engineering scale—as quickly as possible. It is usually a "crash" program, which can be most efficiently done by massive (and alien) equipment and by army engineers, who can drill unskilled local labor but will not want to waste their time on anything but a minimal training program.

Once, in Thailand, I was driving with a Thai forester over an excellent road through the northern teak forests. I congratulated him. "Oh," he said, "this is an American road. They built it." I remarked that it was very thoughtful of them to give the Thais a road on which to haul their teak. "This is not our road," he said, "it is the strategical road to the Mekong, and Laos." Never mind, I said, the road would still be there when the emergency was over. His comment was revealing: "They built it. They taught no one how it was built. They taught no one how to maintain it. To us, building a road meant mechanical shovels, bull-dozers, concrete mixers—all things we have not got."

The essence of technical cooperation, or mutual aid, on the other hand, is to allow people to grow up with changes, learn as they go, and to be discriminating about the modern methods they want to adopt—not just to go into what Professor P. M. S. Blackett has called the "supermarket of science" and pick things off the shelves; things they never knew they wanted and things which, in terms of their way of life, they may not need. Genuine mutual aid must never be the huckstering of gadgets or glib methods.

Another lesson of the 1950's and indeed of today is the ignoring of the statement in the Paley Report which called for a dependable flow of materials at the lowest cost *consistent with national security and with the welfare of friendly nations.* That means a fair price for the products of the underdeveloped, or developing countries. As the late Dag Hammarskjold, as Secretary General of the United Nations, pointed out: "A fall of only five per cent in the average of their export prices is approximately equivalent to the entire flow of capital which they receive not only from the International Bank loans but from all other public and private and government loans."

The fluctuations in their price-return on their exports have been 45 to 55 per cent greater than in the industrialized countries. The net loss in any recent year in the primary producer countries has been of the order of $1.5 billion—an amount which offsets the financial assistance they have received from all sources. At the same time, in the 100 countries and territories which need development, the net increase in population was as high as 2.2 to 2.4 per cent, as compared with 0.7 to 1.7 in the industrialized countries. At the same time, with all the incentives as well as their natural urge toward development, those countries had incurred the service on huge debts assumed at financial rates of interest and for normal periods of repayment. They had been told what to do and were borrowing to do it. According to the studies of the World Bank, during the years 1956–58 external medium- and long-term debt nearly doubled in the low-income countries of Asia, the Middle East, and Africa and increased by 40 per cent in Latin America. In both groups of countries it had reached $5 billion by the end of 1958, and it has increased since. They were selling cheap and buying dear.

When Prime Minister Mr. Harold Macmillan claimed that the British people had "never had it so good," it was because largely, as a country dependent on primary products from those countries, Britain had had the advantage of the decline in prices.

This is like trying to fill a bath with the plug out. A level could be maintained if the faucets were turned well on but if the inflow is less than the outflow, the level drops or drains away. And, in spite of all the zeroes behind the dollar signs of mutual aid, the inflow, in relation to the real needs, has been a trickle.

The advantages have been to the already highly developed countries. As George Orwell said in *Animal Farm:* "All animals are equal but some animals are more equal than others." So, it seems, "All aid is mutual but some aid is more mutual than other."

In the years which have intervened since Truman's "bold new program," the gulf between the prosperous countries and the impoverished ones has not narrowed; it has widened. The rich countries are richer and the poor countries are poorer.

As near as can be estimated, during the 1950's about $2 billion was spent on pre-investment activities such as technical assistance, fellowships, surveys, and so on, and about $28 billion in investment—$30 billion in all, for all countries, developed, developing, and underdeveloped. The average *per capita* income of the 100 underdeveloped territories rose from $90 to $100 in the ten years. Their annual income grew at the rate of 3 per cent, and their population increased at the rate of 2 per cent. This is Sisyphean.

This would seem to be the realistic background against which to consider the technological order and the impingement of science and technology on world affairs. We could, of course, start at the other end, with high development, and consider—as though we needed to be reminded!—the political convulsions due to nuclear energy and space research. Those, like the $110 billion a year which the world spends on armaments, represent an enormous diversion of human ingenuity, human effort, and human wealth from the historical purpose of man's survival and the justification of his evolution—the control of his material environment. This is not the occasion to discuss the political circumstances which have induced this competition of military skills, but it is obvious that, if we are trying to get technology into focus, this is a distortion.

On the chicken-wire fencing of Stagg Field, at the University of Chicago, there is a plaque which reads:

> On December 2, 1942, man achieved here the first self-sustaining chain reaction and thereby initiated the controlled release of atomic energy.

For the small group of scientists around Fermi in the squash court that winter's afternoon, the Atomic Age began as impatient clickings and a reassuring hum. Few, if any, thought that the reactor would explode; the drama and suspense for them was whether it would work. The prodigious energy of the atom, a thousand times greater than chemical energy, had been harmlessly tamed. If the story had been told as casually as the circumstances themselves, and if the atomic engine had been presented to the world as something as innocuous and as easily controlled as the internal combustion engine, the emotional problems would not have arisen. Instead, the world's imagination was stunned by the cataclysmic violence of the bombs on Hiroshima and Nagasaki. Military secrecy had compelled the suppression of the story of the squash court and the testing of the bomb at Alamogordo, in New Mexico, on July 16, 1945, and the world's first introduction to atomic energy was the death and destruction it could cause in one awful moment.

We know now that the safebreakers broke open the lock

of the atom before the locksmiths knew how it worked, and multimillions are being spent on cyclotrons, synchrotrons, and all the other "-trons" and other kinds of accelerators to discover how the wards of the lock fit together.

When the UN International Conference on the Peaceful Uses of Atomic Energy met in Geneva in August 1955, there was a second dawn less lurid than the man-made "sunrise" over the New Mexican desert. In that roseate moment, it seemed that atomic energy could, in terms of harnessed power, offer to the power-hungry underdeveloped countries their short cut to the second Industrial Revolution. By 1958, at the second Conference, it was obvious that that was hope deferred and that the promise of reactors of the type and size commensurate to their needs—using enriched fuels—were not to be forthcoming in any time which would redress the power-balance sheet between those who had industrial power and those who desperately needed it. The prospect of H-energy, fusion energy—putting the H-bomb into overalls and putting it to work—was even more remote. There were all kinds of reasons, but leaving out the possibility of "sneaking" fissile fuels and turning them into bombs, and the lack of skilled personnel (who could have been supplied internationally, anyway) to run the power plants, the argument could be summed up as: You need power to run your industries, but you have not the industries to absorb the power; and, since you have not got industries, you cannot afford power to give you the industries you have not got.

Space research, now escalating not only to the planets but in the budgets, began, so far as the world was aware, as an ambitious but innocent contribution to the I. G. Y. But, in the temper of our times, Sputnik (it might have been the U.S. satellite) became a potential mine-in-orbit and its rocket a potential intercontinental ballistic missile. A televising satellite capable of photographing the back of the moon or relaying information about our weather from outside becomes a means of espionage. A capsule capable of carrying a man into orbit becomes a potential warhead.

Man's adventure beyond the confines of his planet thus became the competition between contending defense departments. It could be argued that the defense departments are also the trucking companies carrying scientific research instruments into space to acquire scientific knowledge—man and his electronic senses reaching out for the stars. As a by-product of military urgencies, and the corresponding budgets, they could find out more about our terrestrial climate; about the nature of radio communications, with the possibilities of global television; about cosmic dust which might tell us about the prebiotic elements and "jump" many of the questions about DNA (deoxyribonucleic acid) and the secrets of life. It can be argued, and demonstrated, that the same urgencies, and unlimited means, can produce "quantum leaps" in the advancing of electronics, of the use of solar batteries for the direct conversion of the sun's rays into electricity, and in the development of computers.

It will seem grudging and unimaginative, if one raises philosophical doubts. There could be none if science and technology were ends in themselves, but, if we are here considering their relevance to the problems of mankind, doubts may be expressed even if they are rejected. Will, for instance, commensurable attention be given to the "secret of life" when we have got it? Or are we going to confront it as ill-prepared as we were for the discovery of the "secret of matter"? Where are those "quantum leaps" in electronics going to land us? Solar batteries? Maybe. They have already, on the space account, been made more efficient, and research and technology have shown how they might eventually be made cheap enough to meet terrestrial need for solar power.

And computers? At the Computer Conference, organized by UNESCO in Paris two years ago, we had 2,000 delegates representing 100,000 scientists, technologists, and technicians, in an industry which had barely existed twelve years before. We were told then that there were already $2.5 billion worth of orders "on the books" for 1965. They discussed calculating machines; translation machines; machines which could be "programed" to make machines; pattern-identification machines with Pavlovian reactions; machines into which could be built the experience which previous machines had "acquired" as electronic-apprentices; machines which would make "value judgments," or at least which produced judgments not predicated by the programmers, and Professor Edward Teller said that "now we had machines which could make value judgments, he could produce the mathematical model for machine emotions." We heard about giant-midget memories, or midget-giant memories, which by cryogenics, could predictably store "all the knowledge of all the libraries of all the world" in a casket no bigger than a cigar box. (They admitted that they could "put it in, but did not yet know how to get it out.") Only one muted voice in all the deliberations of all those computer people was raised to ask whether they had ever got or could ever get together to consider the social implications of all this.

There are scientific implications as well. For the last twenty years nuclear physics has been able to command the big money and the big battalions of scientists; then it was the turn, with transistors, masers, and computers, of the solid-state physicists; now it is the turn of the space engineers; presently it will be the cult of DNA. Nothing derogatory is intended toward those who are engaged in any of these projects, but all of them represent an imbalance in the ledger of problems of human survival and of human well-being.

For instance, it is possible to clear the earth of malaria at a cost of about $50 million spread over ten years. That probably is less than one space sortie or a bigger and better synchrotron. But if we do not eradicate malaria in a measurable period, the consequences may be dire. This is an example of what we can see in the sands of time and the story of the lost civilizations—when man interferes with nature, he must go on interfering, otherwise nature revolts. With DDT and insecticides it became possible to control malaria, which afflicted 300 million people and caused 3 million deaths a year, throughout the world. By killing off the adult mosquitoes, the transfer of malaria from the blood of a sick person to that of a healthy person could be prevented. Whole areas where malaria had been endemic were dealt with in this way. Then insects developed DDT resistance. This was not Mithridatic—taking small doses of poisons to resist lethal doses—the insects had not just "acquired the habit." This was in fact evolution before our eyes. A tiny minority of mosquitoes have a genetic resistance to chemical insecticides. They survived, and without even the competition of their kin, they multiply and will presently replenish the earth with mosquitoes. It is possible to "buy time" by switching insecticides and, possibly, finding new ones, but the genetic resistance reasserts itself. There is, however, a complementary way of getting rid of malaria. There are prophylactic and therapeutic drugs which can purge the parasites from the blood of human beings. If by insecticide control the extent of malaria can be restricted to public health or clinical proportions, individual cases can be treated. If there is no malarial blood to be transferred, the resurgence of the mosquitoes will not be a menace. This means eradication, as proposed by the World

Health Organization. Without it, malaria may come back in epidemic proportions without the brakes which exist in chronic, or endemic, malaria. The response of governments has not been commensurate to the threat nor to their allocations for the incidentals of a defense program.

Next to the threat of nuclear destruction of the human race or, at the least, of civilized living, the biggest threat is the rising tide of population. It has taken *Homo sapiens* two hundred thousand years to reach the present figure of three billion; it will take less than forty years to double it. The figure by 1980 will be, inescapably, four billion—it may be more, but, short of a man-made cataclysm, it cannot be less. This is not due to the fact that women have suddenly become more fecund or that couples are having more children than they had in the past. It is because science and technology have introduced death control (more lives have been saved by antibiotics in the past two decades than have, in the aggregate, been lost in all the wars of all history). Fewer mothers and babies are dying at birth; fewer infants are dying. They are surviving the diseases of childhood, to marry and to multiply. The span of life has been increased. It is not the birth rate which has risen but the survival rate. The result is that the world population is increasing at the rate of 120,000 extra mouths to feed every day.

Already, more than half the existing population is inadequately fed. Actual starvation afflicts millions, but, apart from walking corpses, vastly many more do not get enough food for well-being. If they are sick and hungry, they have not the energy to produce food or the means, or perception, to improve their methods. Today, even the global figures for food calories are, for the first time since the war, not matching the population increase. In any event, the global figures have been misleading for years because they have included the rising standards, and yields, and farm price-support surpluses of the advanced countries, particularly in North America. The measure is not the global *per capita* but the local *per stomach*.

To the present shortages have to be added the needs of that extra one billion mouths before 1980. The answer, except in famines or acute emergencies, cannot be to disburse, as an act of charity, the surpluses of the United States or Canada. The present accumulation of grain would be enough to provide the full calorie needs of the Indian subcontinent for one year or give an extra 200 calories a day for three years to the peoples of Asia.

The only effective answer, however, is to increase the capacity of those peoples to feed themselves. If, for instance, an effective hybrid of the *Japonica* rice of the high latitudes and the *Indica* rice of tropical Asia could be established, it would, theoretically, be possible to double the yields from the existing acreages of Southeast Asia. If the impoverished peasant of India, with an average income of less than $100 a year and unable to afford artificial fertilizers even if they were available, were to use his cow dung as manure instead of as cooking fuel, he could nourish his sickly earth. An effective, cheap, and acceptable solar cooker could increase crop yields. If in those areas where the population density is mainly riverine, the floods which drown or smother the lands could be controlled, multipurpose dams could help to irrigate the thirsty uplands and spread the population on yielding soil. With malaria control jungle regions could be recovered for cultivation. In Africa, the tsetse fly prevents man and his domesticated animals from using nearly four million square miles of what might be productive land. But here one enters the caveat, on which intelligent Africans themselves insist. Before the "protecting" tsetse is banished, the Africans must learn better how to husband this land which the insect has preserved from the ravages of shifting cultivation and overgrazing.

As a global contribution there are deserts, hot and cold, to be regenerated. The population is sparse and will increase slowly, and any yields will be surplus to their needs. This, of course, is complicated by politics and economics. Throughout North Africa and the Middle East there are few indications that others are prepared to convert their deserts as intensively and as effectively as the Israelis have done the Biblical wilderness, and there is little encouragement for the Canadians with their embarrassing surpluses to extend, as might be feasible, their crop-growing northwards.

Seven-tenths of the globe is covered by oceans. As far as the plants and creatures of the seas are concerned, we are still at the cave-man stage; we hunt our seafood. By and large we neglect the fish of the deep oceans and concentrate on those of the continental shelves. For thousands of years fish have been cultivated in inland fisheries, in the fishponds and the rice terraces in the East and in the carp ponds of the medieval monasteries, but, apart from oyster beds and some attempts to increase the growth and yield of inshore fish, practically nothing has been done to husband or herd the sea creatures. It is estimated that the nutrient material produced annually by the sea amounts to 100,000 million tons, of which only 30 million tons the world over are recovered as edible fish. This contrasts with 1,000 million tons of vegetable produce and 100 million tons of animal produce from the land surface. Only about 2 per cent of the fish food available in the seas is eaten by the food fish. The invertebrates, like starfish, eat about four times as much nutrient as do a quantity of edible fish of the same total weight.

Sir Alister Hardy (British Association for the Advancement of Science, 1960) pointed out that the potential fish catches from any fishing ground could be increased many times by merely getting rid of the pests. This could be done by the submarine raking of the fishing banks, and the starfish would be useful as poultry food. In the discussions of the use of artificial fertilizers to increase the fish pastures and hence the fish crops, he pointed out that the bottom of the sea was itself a vast compost heap and that all that was needed was mechanical methods of stirring the nutrients up and getting them into the food fish layers.

The idea of sea ranching—The Riders of the Purple Kelp—and of range managing the sea bottom is no more fantastic than the idea of reaching out to the planets. Developing the sea pastures will be much more rewarding than exploring H. G. Wells' Moon Pastures. Sea farming is entirely within the range of feasibility. There are the already proven devices, like the fish hatcheries. The infantile mortality in the fish population is very high. The British Fisheries Research Laboratory at Lowestoft has shown that by rearing young plaice in tanks until they are past the stage of greatest mortality risk and then liberating them in the North Sea, the catches substantially increased. Such projects, however, would have to be done on an international basis, for the reason that no one nation, or national fishing industry, would put up the capital for this kind of stocking of fishing grounds which are free for all.

There have been schemes for fish farming in arms of the sea. The trouble is that one cannot clip the fins of fish, as one might clip the wings of poultry to keep them from migrating, and the problem is how to keep them within the confines. In experiments in a Scottish sea loch, the results of fertilizing the inlet to intensify the growth of fish food and consequently of fish were encouraging. It was found that the inlet was colder in winter than the open sea and the fish migrated. One proposal to overcome this was that one of the atomic-energy electricity-generating stations should be sited on such a loch and that the waste heat be used to warm the water. Just as it is practical to put a single wire around a field with a harmless electric charge to teach the animals not to stray, it should be possible to "fence" a sea inlet by passing an electric charge through the water—an invisible gate—to discourage the fish from leaving.

Just as the introduction of the scythe and the hoe into Afghanistan in the 1950's for winter fodder saved the flocks and the only dollar earning trade (in karakul pelts), so comparatively simple diffusion of technological experience may produce extensive benefits. "For the sake of a nail the shoe was lost; for the sake of a shoe, the horse was lost; for the sake of a horse, the kingdom was lost." Or, to mechanize the metaphor, a missing piston ring or the absence of the ability to make one may be holding up an industry.

There is a tendency for those in advanced countries looking at the problems of underdeveloped countries, and for those in those countries who are looking for prestige, to think of spectacular projects. Multipurpose dams and giant steel plants may be necessary for Walt W. Rostow's "take-off" into self-stabilizing prosperity, but in the pre-investment stage (before investment flows by gravitation), there are a host of technological problems to which the answers are not spectacular—just pile-driving into a morass of poverty.

Just as the Panama Canal could not be built (and $200 million had been lost in the ill-fated de Lesseps attempt) until Gorgas cleared the isthmus of yellow fever, so countries have to be cleared of diseases. Medical technology and the chemical and pharmaceutical industries can provide the answer. Hungry people represent a heavy labor turnover, so people have to be better fed as a precondition of industrial development. Sick and hungry people are ignorant because lassitude blunts perception and the capacity to learn to help themselves. But, given better health and food, they need teachers, not only to provide literates but to create the climate in which new knowledge is acceptable.

It is not difficult to decide what they *need,* but it is necessary to find out what they *want* ("felt need"), because only through that can they be taught what they need. And there is a dangerous tendency, in the huckstering of "mutual aid," to make them want things they never knew they needed and to feel deprived when they find they cannot afford them.

By way of illustration: At the World Power Congress in

Belgrade, the advanced countries paraded for the edifica-
tion of the delegates of underdeveloped countries the wonders of 500-megawatt atomic power stations and multipurpose dams costing hundreds of millions. A Malayan was
asked "What are you shopping for?" His reply was "A
little buoyant generator which would float up and down in
our streams, just enough to give light to read in our villages.
Then we could read and learn about all those wonderful
things they are talking about here." His next wish was for a
cinema projector which could be used to teach better methods of cultivation and crop preparation and preservation.
Then he wanted a road to the next village and to the nearest market town, because then they could market the surpluses which they would thus learn to produce. Then, with
the village earnings, they might be able to afford a diesel
generator and have local industries. "Then," he said, "we
will come and buy all those wonderful things we will read
and hear about."

Basically, it all comes back to the land. In underdeveloped
countries with rising populations, the land is too congested
for proper cultivation and improved methods. It is sub-
subsistence farming, with underemployment on the land.
The little that is not enough, from the plots which are not
big enough, can be got in the few man-hours of scraping
the soil, planting, and harvesting. The units must be made
bigger. Systems of land tenure come in but so does the
fractionating of patrimonial land among the members of a
family. With bigger units, better methods of farming can be
introduced, but that means that a large proportion of those
on the land must, beneficially, leave it. If it is just displace-
ment (like the dispossession from the common lands at the
time of the enclosures in Britain), the landless will drift
into the industrial conurbations. Instead there should be
local industries to absorb them and still keep them available, at least in a transition stage, for seasonal labor on the
land. This means small factories and industrial units far
removed in size and geography from the massive, high-
capital power sources. It also means making the most of
local materials and developing local craft skills, with whatever help in terms of money, equipment, or experience they
can get from outside.

Sometimes it is difficult to reconcile the three—local materials, craft skills, and outside help. For example, in India
there is the problem of the sacred cow. A cow cannot be
killed. It must be allowed to die. The farming community,

however impoverished, is of a higher caste than the Untouchables, and by definition a skinner is an Untouchable. So a peasant will not readily become a practitioner in leather. But, as experience has shown, the expert brought in from some other country, however highly qualified, will have been accustomed to handling "live" hides, from freshly killed animals, and may be baffled by the chemistry of "dead leather."

As Professor P. M. S. Blackett warned the leaders at the Rehovoth Conference in *Science and the New Nations* (Basic Books), there are risks in going shopping in the "supermarket of science." They ought to compile a cautious shopping list, embodying a rational choice of priorities, otherwise they might "come home with plenty of sweetmeats but no bread and butter."

More than embarking upon ambitious programs of advanced research, they need people who are capable of understanding and selecting "know-how." And that "know-how" cannot just be "ordered off the shelf" because what is valid to the needs and experience of advanced countries may not be suitable to their circumstances.

Again the sacred cow: in the Ganges Delta 300 tube wells were sunk and pumps brought in from abroad. But the fresh water was left unused, and the people preferred their slimy ponds—because they had found that the pumps contained leather valves and they would not drink water that came through the hide of the cow. This was not the mistake of a sacrilegious outsider but of Hindu water engineers who had flicked through a mail-order catalogue.

Perhaps this contribution, which has been cautionary rather than constructive, can be best rounded off by recalling the United Nations International Conference on Science and New Sources of Energy held in Rome in August 1961. In a triennial progression—1955, 1958, 1961—this might have been the occasion of the Third International Conference on the Peaceful Uses of Atomic Energy. The sixteen volumes of the first conference which became the thirty-six volumes of the second might have become the eighty volumes of the third. But it would, for the underdeveloped countries, have followed the law of diminishing returns. There would have been plenty of new detailed knowledge but no dividend. At the first conference they had been bewitched by the new magic—the reactors which were going to be their short cut to the second Industrial Revolution. At the second conference, they had been told that they were not ready for this new benefaction of advanced science. At the third conference (judging from the papers at the Rehovoth Conference) they would have been told that they would be "burning their own rocks," in breeder reactors—forty years from now.

The United Nations promoted instead "New Sources of Energy." Ironically, they were the oldest sources of all—the sun, the wind, and geothermal energy, heat from the crust of the earth. Atomic energy was dethroned. There were no nuclear physicists present.

It was a do-it-yourself conference. It was about solar stoves, refrigerators, solar ponds (with prospects of generating electricity from them at unit cost less than from the atom), windmills, which they could contrive themselves, and, with some surprising disclosures, the possibilities of finding geothermal energy and of "custom tailoring" it to their own needs. This was the first time, in any international science conference, that geothermal energy had been seriously considered.

As a generalization it may be said that a large number of underdeveloped countries, geologically deprived of fossil fuels, have been compensated by the existence of geothermal energy. This had been thought of previously only as the obvious "showings"—the geysers, the bubbling mud-pools, and the hot gases escaping from rock clefts. It has been shown, however, that the geothermal structures usually, but not always, associated with volcanic formations could yield "economic" power. Tapped sources could produce hot steam or hot gases at as much as twenty times the temperatures of the open sources. This energy has the advantage that it could be tapped to meet local requirements and provide power units with which industries could grow up.

Perhaps the sacred cow might remain the cautionary symbol for vaunting technology: Methane, from cow dung, with the residue returned to the soil as useful fertilizer, may be more useful in the short term than the civil atomic-energy electricity-generating stations on which the Indian government is embarking.

The "Know-How" Tradition: Technology in American History

John B. Rae

The emergence of the History of Technology as a discipline in its own right is an event to be hailed by all those who believe that the role of technology in the story of mankind needs more thorough study and more careful evaluation than it has so far received. It also provides an appropriate opportunity to consider the relationship of this subject to other areas of history. There is a great deal of valuable work being done and still to be done in the History of Technology per se. No field of history, however, can usefully be dealt with in isolation; unless it is related to the whole current of historical change, it is likely to deteriorate into futile scholasticism.

Having thus taken a firm stand for the "seamless web" concept of history I now propose to tear the web apart and

Professor of history and head of the Department of Humanities and Social Sciences at Harvey Mudd College, Dr. Rae is the author of American Automobile Manufacturers: The First Forty Years *and* Climb to Greatness: A History of the Aircraft Industry. *Dr. Rae was one of the founders of the Society for the History of Technology and is now first vice-president of that organization.*

This paper, published in Technology and Culture *(Vol. 1, No. 2): 139–50, was originally presented at a joint meeting of the Society for the History of Technology, the History of Science Society, and Section L (History and Philosophy of Science) of the American Association for the Advancement of Science, held in Washington, D.C., in December 1958. The research on which this paper is based was supported by the Sloan Research Fund of the School of Industrial Management at Massachusetts Institute of Technology and the Earhart Foundation.*

deal with only a portion of it: to wit, the portion compris-
ing the history of the United States. For one thing, this is
the area I know most about; for another, American society
has been influenced to a unique degree by the forces of
technology.

As a people we have given ample lip service to technology. Until a year or so ago it was an article of faith that
Americans had an ingrained superiority in technical "know-
how," and if we have lost a little of our self-confidence, it
has been only a little. We have a completely justified pride
in the achievement of American civilization in applying
technology to the material advancement of its people, although our pride in this achievement has not been matched
by understanding of how it came about. For many Americans, information on the growth of their country's technology is about on the level of the conversation in a Pullman
smoking room which Frederick L. Smith, one-time president of the Olds Motor Works, reported overhearing in the
1920's.[1]

"Who invented the automobile anyway?"

"Henry Ford. Started as a racer by beating Barney Old-
field on the ice at Detroit. Right away after that he built a
plant to turn out the same kind of car in 50,000 lots."

"Doesn't he own the Lincoln now?"

"Yeah, owns the Lincoln and the Packard, Cadillac,
Buick—all the big ones and a lot of the little ones besides."

"Is it true about his taking over the Detroit City Hospital?"

"I'll say it's true. Bought it and runs it for his employees.
Charges everybody a fixed rate for every job and makes it
pay."

I wish to suggest here that a clearer and more complete
knowledge of the role of technology in American history is
not merely desirable but necessary to a full understanding
of the evolution of American civilization. In particular, unless adequate weight is given to the technological factor, it
is completely impossible to give an accurate picture of the
growth and character of American business enterprise.

Before we go any further, we should perhaps try to define what we mean by technology. It is not easy to do so.
Technology includes engineering, but the two are not synonymous, and no one who is concerned with curricular
problems in present-day engineering colleges is likely to be
bold enough to attempt to explain exactly where technology
stops and science begins. I like the definition of engineering

formulated by President R. E. Doherty of the Carnegie Institute of Technology:

"Engineering is the art, based primarily upon training in mathematical and physical sciences, of utilizing economically the forces and materials of nature for the benefit of man."[2]

Technology may be regarded as comprehending trial-and-error and rule-of-thumb as well as the systematic application of scientific principles, which brings us close enough for practical purposes to the definition given by V. Gordon Childe in Singer's *History of Technology:*

"Technology should mean the study of those activities, directed to the satisfaction of human needs, which produce alterations in the material world."[3]

The technological aspect of American history comes to us mostly in unrelated fragments—the principal inventors and inventions, with at best casual reference to the circumstances which stimulated them; something of early industrial development, emphasizing its political repercussions; railroad expansion, generally in terms of the organization of large systems; the growth of big business, again with political overtones.

To cite one conspicuous example, every student of American history learns about the economic importance of the Erie Canal, but reference to the fact that the canal was a monumental engineering achievement is likely to be incidental. It would not detract from the credit to which DeWitt Clinton is rightfully entitled if we recognized the work of Benjamin Wright and Canvass White or pointed out that the Erie Canal was the training school for a substantial group of brilliant civil engineers.

However, merely enlarging the bits and pieces is not enough. A society which has been as profoundly influenced by technology as ours needs to be able to see how technology has been woven into the fabric of its national life. I do not propose to accomplish that in this paper. Even if space permitted, I do not possess any such encyclopedic information. All I can do is point out some of the factors which I consider to be of major importance.

The first of these is the overwhelming pragmatic character of American technology. It is most appropriate that our favorite synonym for technical skill should be "know-how," because as a people we have placed a far higher premium on knowing how than on knowing why. The folk heroes of American technology are the Edisons and the Fords, men with a minimum of formal training, dedicated to cut-and-try, lacking in scientific background, and inclined to be scornful of scientific method. Until well into the twentieth century, Americans have been content to let most of the basic discoveries in science and technology originate in Europe while they themselves have followed a policy of "adapt, improve, and apply."

The reason for this phenomenon is obvious enough. The United States began as an undeveloped area, without benefit of a fairy godmother distributing largesse in the form of economic aid or technical assistance. There was chronically more work to be done than there were people to do it, so that labor, particularly skilled labor, was likely to be in short supply and expensive. Securing trained ironworkers for the ironworks at Saugus, Massachusetts, in the 1640's was one of the company's most difficult problems,[4] and a century and a half later the high cost of craftsmen in the United States was a primary factor in stimulating Eli Whitney and Simeon North to turn their attention to the manufacture of firearms by a method which would permit the substitution of mechanical for human skill.[5] Similarly, while machine production of textiles began in the United States later than in Great Britain, the American industry mechanized more rapidly because of the need to employ technics in which skill would be a minor factor.[6] For example, although Cartwright's power loom appeared in England in the 1780's, a good half century elapsed before mechanized weaving displaced the crafts technique in the British textile industry, in part because the power loom needed considerable refinement before it could compete in quality with the work of the hand weaver. As a result, the weaving process in Great Britain was generally organized separately from the other operations of textile manufacturing. On the other hand, when Francis Cabot Lowell designed his power loom in 1813, he and his associates developed it into a large-scale business enterprise in a little more than ten years. In characteristic fashion, cotton manufacturing by the "Lowell system" was organized on an integrated basis, with all the operations conducted in a single factory, a maximum of mechanization, and even a labor policy devised to meet the conditions of early nineteenth-century New England.[7]

American conditions, in other words, placed a high valuation on getting things done, preferably in the shortest possible time and with the minimum of human labor. The man who could devise a gadget or a technique which would work was making a recognizable contribution to the growth of the country, whether he understood the fundamental principles he was using or not. One consequence was that to the end of the nineteenth century the relationship between science and technology in the United States was somewhat casual. Where science had a practical application, it was invoked. The invention of the telegraph depended heavily on Joseph Henry's research in magnetism—Morse, indeed, could hardly have succeeded without Henry's assistance—and George H. Bissell invoked the skill of Benjamin Silliman, Jr., the leading American chemist of his day, for an evaluation of the commercial possibilities of petroleum when the Pennsylvania oil fields began to attract attention. These contacts, however, were spasmodic and rare. There was little interest in accumulating scientific knowledge as a foundation for future technological advance and still less in accumulating such knowledge for its own sake. It was completely in accord with the pattern of American thought that when Congress undertook to promote higher technical education by the Morrill Act of 1862, it singled out agriculture and the mechanic arts as the areas to be given particular attention.

This attitude has been subjected to substantial modification as our technology has become increasingly complex and increasingly dependent on science. Nevertheless, the emphasis on the practical still dominates American thinking, and its existence has to be recognized if we are to evaluate properly the place of technology in American history. Moreover, while we can concede that we have unduly neglected "pure" or "basic" research in the past and now need to give it more vigorous encouragement, it does not follow that the American emphasis on the practical has been wrong. It was determined in the first place by circumstances; beyond that, if we recall our definition that technology is an activity directed to the satisfaction of human needs, then American technology has been performing its function with phenomenal success. "Adapt, improve, and apply" may have less glamour than original creativity, but the technique of application may in itself be more significantly creative than the original idea or invention. In the reminiscences from which the story on Henry Ford came, the author also remarked—this in the late 1920's—that since Karl Benz introduced spark ignition in 1886, nothing has been added to the gasoline automobile but the assembling of known parts.[8] Maybe so—but what has happened to the process of assembly, and the economic and social consequences thereof, makes quite a story. Nor does it follow that the pragmatic approach necessarily precludes research in fundamentals. Kendall Birr's scholarly study of the General Electric Research Laboratory makes it clear that a good deal of "basic" research grew out of efforts to find a solution to a specific problem.[9]

The second major factor which could profitably be given more attention is the interrelationship between technological development and industrial application. While some work has been done on the problem of invention and innovation, this whole area needs more intensive research in order to untangle some of its complex of causes and effects. Technological change may and frequently does originate in an isolated act of genius, but its effectiveness in an economic sense is a matter of time and circumstances. A classic American illustration is George B. Selden's patent on a motor vehicle powered by an internal combustion engine, for which he filed his application in 1879. This story cannot be told in detail here. It is sufficient to point out that Selden had worked out the essential technical features of the gasoline automobile but that he was unable to exploit his idea because he was ahead of his time. The highway system of the United States was totally inadequate to the demands of travel by motor vehicle: as late as 1900 there was not enough paved road outside the big cities to make a continuous route from New York to Boston[10]—a matter of just over two hundred miles. Moreover, considerable development in manufacturing technics and machine tools was needed before a practical automobile could be successfully put into commercial production. So Selden, who might have been the father of the American automobile industry, became instead an obscure figure in an elaborate and much misunderstood lawsuit.

The story of aluminum offers an even clearer illustration. The properties of the metal were known and some experimental work had been done in reducing aluminum ore before Charles Martin Hall and Paul L. T. Heroult independently and almost simultaneously discovered the electrolytic process.[11] There had even been experiments with electrolysis, but a practical method of producing aluminum at a cost low enough to remove it from the list of precious metals had to wait for Charles S. Bradley's electric furnace, patented in 1885,[12] and that in turn depended on the development of a dynamo capable of generating electric power in large quantities. Here indeed is an excellent example of the distinction between science and technology.

Hans Christian Oersted isolated aluminum in his laboratory in 1825; industrial utilization had to wait for over sixty years and the development of other separate technologies.

My final point in this paper is that we must give more attention to the role of technology in the growth and organization of industry. Our failure to do so in the past has resulted in misconceptions and distortions, particularly in the area of big business. Despite the very substantial amount of work which has been done in reappraising business and businessmen, the predominant attitude is still suspicion of bigness. We may acknowledge that large-scale organization has advantages in economy and efficiency of operation, but in our hearts we continue to take it for granted that the creators of these organizations invariably had conscious monopolistic intent or wished to manipulate security issues. We seldom inquire into the possible relationship between the growth and structure of the business and the technological processes in which it was engaged.

It is not merely a question of reinterpreting past events in the light of new information. Much of the information has been available all along and has simply been disregarded. As a prime illustration, let me go back to the old familiar story of the Standard Oil Company. I have no intention of rewriting this story or of suggesting that railroad rate rebates no longer form part of the record—although I could argue that they have been badly overworked. I simply wish to point out certain features which do not appear in the conventional picture handed down from the era of Henry D. Lloyd and Ida M. Tarbell.

Historians agree that John D. Rockefeller owed much to his strategic position as the leading refiner in Cleveland— but how did he get to be the leading refiner in Cleveland? Lloyd sees a deep-seated, nefarious plot here, saying, "He (Rockefeller) started a little refinery in Cleveland, hundreds of miles from the oil wells."[13] Since the distance from Cleveland to Oil City, Pennsylvania, is about 115 miles, and to Titusville only a little more, one might question the author's standard of accuracy, but this is perhaps a digression. In the same passage Lloyd goes on to say: "With him were his brother and an English mechanic. The mechanic was bought out later, as all the expert skill needed could be bought for wages, which were cheaper than dividends."

Now this "English mechanic" which Lloyd dismisses so cavalierly was none other than Samuel Andrews, a Rockefeller partner for fifteen years and a key figure in the early history of Standard Oil. He is described by Ida M. Tarbell as "a mechanical genius. He devised new processes, made a better quality of oil, got larger and larger percentages of refined from his crude,"[14] and by Allan Nevins as "the best superintending refiner in Cleveland."[15] Miss Tarbell is a little more lyrical; Dr. Nevins, on the other hand, makes the essential point that Rockefeller's success was in no small part due to his ability to select gifted associates. In this case, Rockefeller seems entitled to full credit not only for appreciating Andrews' skill but also for recognizing at the outset the necessity for being technically superior to his competitors.

Miss Tarbell's acknowledgment of Andrews' merits does not extend to approval of Rockefeller and his company. While she is more aware of the technological side of the oil business than Lloyd, her emphasis is on the achievements of Standard's opponents like the Tidewater Pipe Line. There is no reference whatever to Rockefeller's support of Herman Frasch's experiments in the late 1880's with the heavily sulfurated petroleum known as Lima-Indiana crude, experiments which represented the first large-scale application of chemical and engineering research to a refining problem in the United States. It was a gamble which could just as well have been a total loss and which no small firm could have risked, and while the outcome was highly profitable to Standard, the whole oil industry benefited as well— for one thing because in trying to find markets for the Indiana crude, Standard's engineers and salesmen succeeded in giving a vigorous boost to the use of oil as an industrial fuel.[16]

If I seem to have spent undue effort criticizing two admittedly outdated works, it is because these books and others like them had a pronounced influence on public opinion and public policy. The evidence marshaled by the authors pointed to the conclusion that big business was inherently bad. There is no suggestion, for example, that the nature of the refining operation gave the large-scale organization an advantage which unavoidably meant that the lesser firms could not hope to compete successfully, although this fact was brought out in a contemporary analysis of Standard Oil.[17]

The fact needs to be faced that it was the popular image of Standard Oil which was prosecuted and dissolved under the Sherman Anti-Trust Act, not the reality as it actually was in the early years of this century. Few people knew or cared what specific acts the company was charged with; Standard Oil had simply become the pre-eminent symbol of monopoly. Yet whatever prospect Standard had ever had of monopolizing the oil industry in the United States had disappeared irrevocably ten years before the Supreme Court's decree of dissolution—specifically, on January 10, 1901, when the gusher at Spindletop in Beaumont, Texas, started a new era in the history of the industry,[18] and the Spindletop achievement was primarily a technological triumph. It combined Anthony F. Lucas's informed guess about the domes on the East Texas coastal plain, based on his experience as a mining engineer, and the ingenuity of his drilling crew in devising new technics to meet unfamiliar conditions.

The automobile industry provides another illustration of a growth pattern largely determined by technology. So far the disappearance of the lesser automobile manufacturers has been a source of regret to those who cherish the memory of the Pierce-Arrow and the Marmon rather than a cause of public and governmental indignation, but this situation could change. There is a definite suggestion of trust-busting crusade in the recent decision of the Supreme Court holding that the DuPont Company's stock ownership in General Motors was a violation of the Anti-Trust Laws.[19] The company was adjudged to have violated Section 7 of the Clayton Act forty years before, on the ground that its stock purchase might have restricted competition in selling paints and finishes to General Motors. The evidence that any such restriction occurred was not such as a historian would consider conclusive; the court, indeed, made such evidence irrelevant when it held that "The section (7 of the Clayton Act) is violated whether or not actual restraints or monopolies, or the substantial lessening of competition, have occurred or are intended." The fact that DuPont, working in cooperation with Charles F. Kettering, was first in the field with the quick-drying finishes which broke a serious bottleneck in automobile production seems to have carried no weight whatsoever.

I do not wish to suggest that the present organization of the industry is necessarily the best possible; we ought, however, to appreciate the fact that the prospect for the small producer became gloomy as early as 1913, when Ford's moving assembly line went into full-scale operation. This was a superior production process, but it was feasible only for a big organization. To put it another way, the widespread ownership of automobiles which is a distinctive feature of American society has been made possible by a technology which can be most effectively employed in what the economists like to call "oligopoly." The manufacture of a car at a price within the reach of what Henry Ford referred to as "the multitudes" can be achieved only by mass-production techniques. These require a tremendous investment in plant and equipment, and this investment in turn demands a high volume of sales. To cite a recent example, Ford Motor Company is reputed to have spent a quarter of a billion dollars in putting the Edsel on the market, with something less than unqualified success in spite of Ford's production and marketing facilities. If an established firm can encounter difficulty in introducing a new make of automobile, then the prospect of a complete newcomer getting a foothold in a fiercely competitive market is remote, except in the unlikely event of a radical change in automotive technology which the established companies chose to disregard.[20]

The technology of production has also had a direct influence on the organizational structure of specific firms. Alfred D. Chandler, Jr., has pointed out that since 1900 the major innovations in the techniques of management and control have come in the industries most affected by the newer technologies, i.e., the application of chemistry and physics and the coming of electricity and the internal combustion engine.[21] To put it another way, the industrial processes which lend themselves most readily to diversification of product also stimulate decentralization of management.

In recent years there has been a considerable volume of scholarly work in industrial history in which technology has been given its proper weight, but much more needs to be done. Apart from the numerous other areas of economic activity which could use similar studies,[22] there are some general problems on which we have barely scratched the surface. For example, such research as has been done on innovation indicates that while the technology of production generally favors the large firm, significant innovations are more likely to originate in the small ones,[23] but we need to know much more than we do about the process of innovation. We cannot yet make adequate comparisons of situations in which technological developments have come in response to immediate economic pressures with those in which a technological advance has stimulated a new industrial growth. In addition, it cannot be said that the recent work in industrial and technological history has as yet appreciably made its way into the mainstream of historical thought.

Historians of American technology have therefore a dual

task to perform: first, to write the history of the technology itself; second, to have their findings incorporated into the total picture of American life. The attempt is worth making. We can all agree that science and technology occupy a vital place in our present-day civilization; we will agree equally that their importance is likely to increase in the future, unless they prevent us from having a future. A more thorough and accurate understanding of how they have influenced and been influenced by the American environment is a desirable step toward an intelligent appreciation of what they may hold for us now.

REFERENCES

[1] F. L. Smith, *Motoring Down a Quarter Century* (Detroit, 1929), p. 7.

[2] C. R. Young, *Engineering and Society* (Toronto, Ontario, 1946), pt. 1.

[3] Charles Singer *et al.* (ed.), *A History of Technology* (London and New York, 1954), Vol. I, p. 38.

[4] For a full discussion of this problem, see E. N. Hartley, *Ironworks on the Saugus* (Norman, Oklahoma, 1957), p. 187.

[5] C. McL. Greene, *Eli Whitney and the Birth of American Technology* (Boston, 1956), pp. 101–102.

[6] M. W. Copeland, *The Cotton Manufacturing Industry in the United States* (Cambridge, Mass., 1917), pp. 9, 10, 15.

[7] See Nathan Appleton, *The Introduction of the Power Loom and the Origin of Lowell* (Lowell, Mass., 1858).

[8] Smith, *op. cit.*, p. 10.

[9] Kendall Birr, *Pioneering in Industrial Research* (Washington, D.C., 1957), Chap. IV.

[10] Christy Borth, "The Automobile; Power Plant and Transport Tool of a Free People," *Centennial of Engineering, 1852–1952* (Chicago, 1953), p. 440.

[11] Alfred Cowles, *The True Story of Aluminum* (Chicago, 1958), pp. 7, 8.

[12] *Ibid.*, p. 47.

[13] Henry D. Lloyd, *Wealth Against Commonwealth* (New York, 1894), p. 44.

[14] Ida M. Tarbell, *History of the Standard Oil Company* (New York, 1904), Vol. I, p. 42.

[15] Allan Nevins, *Study in Power* (New York, 1953), Vol. I, p. 42.

[16] This episode is described in R. W. and M. E. Hidy, *Pioneering in Big Business* (New York, 1955), pp. 165–168.

[17] G. H. Montague, *Rise and Progress of the Standard Oil Company* (New York, 1903), p. 8.

[18] J. A. Clarke and M. T. Halbouty, *Spindletop* (New York, 1952) is an accurate if highly dramatized account of this event.

[19] 353 *U. S.*, 586 (June 3, 1957).

[20] I am referring here to production for the mass market. The manufacture of specialized vehicles is another matter altogether.

[21] A. D. Chandler, Jr., "The Beginnings of 'Big Business' in American Industry," *Business History Review,* XXXIII, 1 (Spring, 1959), 25. See also his "Management Decentralization; An Historical Analysis," *ibid.,* XXI, 2 (June, 1956), 115 ff.

[22] Agriculture, iron and steel, mining, food processing, and the use of electric power in the twentieth century are conspicuous fields to which this statement applies.

[23] The best studies of this phenomenon are W. R. MacLaurin, *Invention and Innovation in the Radio Industry* (New York, 1949); Harold C. Passer, *The Electrical Manufacturers, 1875–1900* (Cambridge, Mass., 1953); W. Paul Strassman, *Risk and Technological Innovation* (Ithaca, N. Y., 1959); and John Jewkes, David Sawers, and Richard Stillerman, *The Sources of Invention* (London and New York, 1958).

Bursting Boilers
and the Federal Power

John G. Burke

I

When the United States Food and Drug Administration removes thousands of tins of tuna from supermarket shelves to prevent possible food poisoning, when the Civil Aeronautics Board restricts the speed of certain jets until modifications are completed, or when the Interstate Commerce Commission institutes safety checks of interstate motor carriers, the federal government is expressing its power to regulate dangerous processes or products in interstate commerce. Although particular interests may take issue with a regulatory agency about restrictions placed upon certain products or seek to alleviate what they consider to be unjust directives, few citizens would argue that government regulation of this type constitutes a serious invasion of private property rights.[1]

Though federal regulatory agencies may contribute to the general welfare, they are not expressly sanctioned by any provisions of the U.S. Constitution. In fact, their genesis was due to a marked change in the attitude of many early nineteenth-century Americans who insisted that the federal government exercise its power in a positive way in an area that was nonexistent when the Constitution was enacted. At the

Dr. Burke is an associate dean and professor of history at the University of California, Los Angeles. He is the author and editor of books and articles on the history of science and technology, including Origins of the Science of Crystals *and* The New Technology and Human Values. *This article, which appeared in the Winter 1966 issue of* Technology and Culture *(Vol. 7, No. 1): 1–23, was awarded the Abbott Payson Usher Prize of the Society for the History of Technology.*

time, commercial, manufacturing, and business interests were willing to seek the aid of government in such matters as patent rights, land grants, or protective tariffs, but they opposed any action that might smack of governmental interference or control of their internal affairs. The government might act benevolently but never restrictively.

The innovation responsible for the changed attitude toward government regulation was the steam engine. The introduction of steam power was transforming American culture, and while Thoreau despised the belching locomotives that fouled his nest at Walden, the majority of Americans were delighted with the improved modes of transportation and the other benefits accompanying the expanding use of steam. However, while Americans rejoiced over this awesome power that was harnessed in the service of man, tragic events that were apparently concomitant to its use alarmed them—the growing frequency of disastrous boiler explosions, primarily in marine service. At the time, there was not even a governmental agency that could institute a proper investigation of the accidents. Legal definitions of the responsibility or negligence of manufacturers or owners of potentially dangerous equipment were in an embryonic state. The belief existed that the enlightened self-interest of an entrepreneur sufficed to guarantee the public safety. This theory militated against the enactment of any legislation restricting the actions of the manufacturers or users of steam equipment.

Although the Constitution empowered Congress to regulate interstate commerce, there was still some disagreement about the extent of this power even after the decision in *Gibbons* v. *Ogden,* which ruled that the only limitations on this power were those prescribed in the Constitution. In the early years of the republic, Congress passed legislation under the commerce clause designating ports of entry for customs collections, requiring sailing licenses, and specifying procedures for filing cargo manifests. The intent of additional legislation in this area, other than to provide for these normal concomitants of trade, was to promote commerce by building roads, dredging canals, erecting lighthouses, and improving harbors. Congress limited its power under the commerce clause until the toll of death and destruction wrought by bursting steamboat boilers mounted, and some positive regulations concerning the application of steam power seemed necessary. Thomas Jefferson's recommendation that we should have "a wise and frugal Government, which shall restrain men from injuring one another, shall leave them otherwise free to regulate their own pursuits of industry and improvement" took on a new meaning.[2]

Although several historians have noted the steamboat explosions and the resulting federal regulations, the wider significance of the explosions as an important factor in altering the premises concerning the role of government vis à vis private enterprise has been slighted.[3] Further, there has been no analysis of the role of the informed public in this matter. The scientific and technically knowledgeable members of society were—in the absence of a vested interest— from the outset firmly committed to the necessity of federal intervention and regulation. They conducted investigations of the accidents; they proposed detailed legislation which they believed would prevent the disasters. For more than a generation, however, successive Congresses hesitated to take forceful action, weighing the admitted danger to the public safety against the unwanted alternative, the regulation of private enterprise.

The regulatory power of the federal government, then, was not expanded in any authoritarian manner. Rather, it evolved in response to novel conditions emanating from the new machine age, which was clearly seen by that community whose educations or careers encompassed the new technology. In eventually reacting to this danger, Congress passed the first positive regulatory legislation and created the first agency empowered to supervise and direct the internal affairs of a sector of private enterprise in detail. Further, certain congressmen used this precedent later in efforts to protect the public in other areas, notably in proposing legislation that in time created the Interstate Commerce Commission. Marine boiler explosions, then, provoked a crisis in the safe application of steam power, which led to a marked change in American political attitudes. The change, however, was not abrupt but evolved between 1816 and 1852.

II

Throughout most of the eighteenth century, steam engines worked on the atmospheric principle. Steam was piped to the engine cylinder at atmospheric pressure, and a jet of cold water introduced into the cylinder at the top of the stroke created a partial vacuum in the cylinder. The atmospheric pressure on the exterior of the piston caused the power stroke. The central problem in boiler construction, then, was to prevent leakage. Consequently, most eighteenth-century boilers were little more than large wood, copper, or cast-iron containers placed over a hearth and encased with firebrick. In the late eighteenth century, Watt's utilization of the expansive force of steam compelled more careful boiler design. Using a separate condenser in conjunction with steam pressure, Watt operated his engines at about 7 p.s.i. above that of the atmosphere. Riveted wrought-iron boilers were introduced, and safety valves were employed to discharge steam if the boiler pressure exceeded the designed working pressure.

Oliver Evans in the United States and Richard Trevithick in England introduced the relatively high-pressure noncondensing steam engine almost simultaneously at the turn of the nineteenth century. This development led to the vast extension in the use of steam power. The high-pressure engines competed in efficiency with the low-pressure type, while their compactness made them more suitable for land and water vehicular transport. But, simultaneously, the scope of the problem faced even by Watt was increased, that is, the construction of boilers that would safely contain the dangerous expansive force of steam. Evans thoroughly respected the potential destructive force of steam. He relied chiefly on safety valves with ample relieving capacity but encouraged sound boiler design by publishing the first formula for computing the thickness of wrought iron to be used in boilers of various diameters carrying different working pressures.[4]

Despite Evans' prudence, hindsight makes it clear that the rash of boiler explosions from 1816 onward was almost inevitable. Evans' design rules were not heeded. Shell thickness and diameter depended upon available material, which was often of inferior quality.[5] In fabrication, no provision was made for the weakening of the shell occasioned by the rivet holes. The danger inherent in the employment of wrought-iron shells with cast-iron heads affixed because of the different coefficients of expansion was not recognized, and the design of internal stays was often inadequate. The openings in the safety valves were not properly proportioned to give sufficient relieving capacity. Gauge cocks and floats intended to ensure adequate water levels were inaccurate and subject to malfunction by fouling with sediment or rust.

In addition, there were also problems connected with

boiler operation and maintenance.[6] The rolling and pitching of steamboats caused alternate expansion and contraction of the internal flues as they were covered and uncovered by the water, a condition that contributed to their weakening. The boiler feedwater for steamboats was pumped directly from the surroundings without treatment or filtration, which accelerated corrosion of the shell and fittings. The sediment was frequently allowed to accumulate, thus requiring a hotter fire to develop the required steam pressure, which led, in turn, to a rapid weakening of the shell. Feed pumps were shut down at intermediate stops without damping the fires, which aggravated the danger of low water and excessive steam pressure. With the rapid increase in the number of steam engines, there was a concomitant shortage of competent engineers who understood the necessary safety precautions. Sometimes masters employed mere stokers who had only a rudimentary grasp of the operation of steam equipment. Increased competition also led to attempts to gain prestige by arriving first at the destination. The usual practice during a race was to overload or tie down the safety valve so that excessive steam pressure would not be relieved.

III

The first major boiler disasters occurred on steamboats, and, in fact, the majority of explosions throughout the first half of the nineteenth century took place on board ship.[7] By mid-1817, four explosions had taken five lives in the Eastern waters, and twenty-five people had been killed in three accidents on the Ohio and Mississippi rivers.[8] The city council of Philadelphia appears to have been the first legislative body in the United States to take cognizance of the disasters and attempt an investigation. A joint committee was appointed to determine the causes of the accidents and recommend measures that would prevent similar occurrences on steamboats serving Philadelphia. The question was referred to a group of practical engineers who recommended that all boilers should be subjected to an initial hydraulic proof test at twice the intended working pressure and additional monthly proof tests to be conducted by a competent inspector. Also, appreciating the fact that marine engineers were known to overload the safety-valve levers, they advocated placing the valve in a locked box. The report of the joint committee incorporated these recommendations, but it stated that the subject of regulation was outside the competence of municipalities. Any municipal enactment would be inadequate for complete regulation. The matter was referred, therefore, to the state legislature, and there it rested.[9]

Similar studies were being undertaken abroad. In England, a fatal explosion aboard a steamboat near Norwich prompted Parliament to constitute a Select Committee in May 1817 to investigate the conditions surrounding the design, construction, and operation of steam boilers. In its report, the committee noted its aversion to the enactment of any legislation but stated that where the public safety might be endangered by ignorance, avarice, or inattention, it was the duty of Parliament to interpose. Precedents for legislation included laws covering the construction of party walls in buildings, the qualification of physicians, and the regulation of stagecoaches. The committee recommended that passenger-carrying steam vessels should be registered, that boiler construction and testing should be supervised, and that two safety valves should be employed with severe penalties for tampering with the weights.[10]

No legislation followed this report, nor were any laws enacted after subsequent reports on the same subject in 1831, 1839, and 1843.[11] The attitude of the British steamboat owners and boiler manufacturers was summarized in a statement that the prominent manufacturer, Sir John Rennie, made to the Select Committee in 1843. There should be, he said, no impediments in the application of steam power. Coroners' juries made such complete investigations of boiler explosions that no respectable manufacturer would risk his reputation in constructing a defective boiler. Constant examination of boilers, he argued, would cause serious inconvenience and would give no guarantee that the public safety would be assured. Admittedly, it would be desirable for steam equipment to be perfect, but with so many varied boiler and engine designs, it would be next to impossible to agree on methods of examination. Besides, he concluded, there were really few accidents.[12]

In this latter remark, Sir John was partially correct. In England, from 1817 to 1839, only 77 deaths resulted from twenty-three explosions.[13] This record was relatively unblemished compared to the slaughter in the United States, where in 1838 alone, 496 lives had been lost as a result of fourteen explosions.[14] The continued use of low-pressure engines by the British; the fact that by 1836 the total number of U.S. steamboats—approximately 750—was greater than the total afloat in all of Europe; and the fact that the average tonnage of U.S. steamboats was twice that of British vessels, implying the use of larger engines and boilers and more numerous passengers, accounted for the large difference in the casualty figures.[15]

In France, the reaction to the boiler hazard was entirely different than in Great Britain and the United States. Acting under the authority of Napoleonic legislation, the government issued a Royal Ordinance on October 29, 1823, relative to stationary and marine steam engines and boilers.[16] A committee of engineers of mines and civil engineers prepared the regulations, but the scientific talent of such men as Arago, Dulong, and Biot was enlisted to prepare accurate steam tables.[17] By 1830, amendments resulted in the establishment of a comprehensive boiler code. It incorporated stress values for iron and copper and design formulas for these materials. It required the use of hemispherical heads on all boilers operating above 7 p.s.i. and the employment of two safety valves, one of which was enclosed in a locked grating. Boiler shells had to be fabricated with fusible metal plates made of a lead-tin-bismuth alloy and covered with a cast-iron grating to prevent swelling when close to the fusing point. Boilers had to be tested initially at three times the designed working pressure and yearly thereafter. The French engineers of mines and government civil engineers were given detailed instructions on the conduct of the tests and were empowered to remove any apparently defective boiler from service. The proprietors of steamboats or factories employing boilers were liable to criminal prosecution for evasion of the regulations, and the entire hierarchy of French officialdom was enjoined to report any infractions.[18]

Proper statistics proving that this code had a salutary effect in the prevention of boiler explosions are not available. It is certain that some explosions occurred despite the tight regulations. Arago, writing in 1830, reported that a fatal explosion on the "Rhone" resulted from the tampering with a safety valve and pointed out that fusion of the fusible metal plates could be prevented by directing a stream of water on them.[19] Undoubtedly, in some instances the laws were evaded, but Thomas P. Haldeman, an experienced Cincinnati steamboat captain, said in 1848 that the code had been effective. He wrote: "Since those laws were enforced we have scarcely heard of an explosion in that country. . . . What a misfortune our government did not follow the example of France twenty years ago."[20] Significantly, both Belgium and Holland promulgated boiler laws that were in all essentials duplicates of the French regulations.[21]

IV

From 1818 to 1824 in the United States, the casualty figures in boiler disasters rose, about forty-seven lives being lost in fifteen explosions. In May 1824 the "Aetna," built in 1816 to Evans' specifications, burst one of her three wrought-iron boilers in New York harbor, killing about thirteen persons and causing many injuries. Some experts attributed the accident to a stoppage of feedwater due to incrustations in the inlet pipes, while others believed that the rupture in the shell had started from an old fracture in a riveted joint.[22] The accident had two consequences. Because the majority of steamboats plying New York waters operated at relatively low pressures with copper boilers, the public became convinced that wrought-iron boilers were unsafe. This prejudice forced New York boat builders who were gradually recognizing the superiority of wrought iron to revert to the use of copper even in high-pressure boilers. Some owners recognized the danger of this step, but the outcry was too insistent. One is reported to have said: "We have concluded therefore to give them [the public] a copper boiler, the strongest of its class, and have made up our minds that they have a perfect right to be scalded by copper boilers if they insist upon it."[23] His forecast was correct, for within the next decade, the explosion of copper boilers employing moderate steam pressures became common in Eastern waters.[24]

The second consequence of the "Aetna" disaster was that it caught the attention of Congress. A resolution was introduced in the House of Representatives in May 1824 calling for an inquiry into the expediency of enacting legislation barring the issuance of a certificate of navigation to any boat operating at high steam pressures. Although a bill was reported out of committee, it was not passed due to lack of time for mature consideration.[25]

In the same year, the Franklin Institute was founded in Philadelphia for the study and promotion of the mechanical arts and applied science.[26] The institute soon issued its *Journal,* and, from the start, much space was devoted to the subject of boiler explosions. The necessity of regulatory legislation dealing with the construction and operation of boilers was discussed, but there was a diversity of opinion as to what should be done. Within a few years, it became apparent that only a complete and careful investigation of the causes of explosions would give sufficient knowledge for suggesting satisfactory regulatory legislation. In June 1830, therefore, the Institute empowered a committee of its members to conduct such an investigation and later authorized it to perform any necessary experiments.

The statement of the purpose of the committee reflects clearly the nature of the problem created by the frequent explosions. The public, it said, would continue to use steamboats, but if there were no regulations, the needless waste of property and life would continue. The committee believed that these were avoidable consequences; the accidents resulted from defective boilers, improper design, or carelessness. The causes, the committee thought, could be removed by salutary regulations, and it affirmed: "That there must be a power in the community lodged somewhere, to protect the people at large against any evil of serious and frequent recurrence, is self-evident. But that such power is to be used with extreme caution, and only when the evil is great, and the remedy certain of success, seems to be equally indisputable."[27]

Here is a statement by a responsible group of technically oriented citizens that public safety should not be endangered by private negligence. It demonstrates the recognition that private enterprise was considered sacrosanct, but it calls for a reassessment of societal values in the light of events. It proposes restrictions while still professing unwillingness to fetter private industry. It illustrates a change in attitude that was taking place with respect to the role of government in the affairs of industry, a change that was necessitated by technological innovation. The committee noted that boiler regulation proposals had been before Congress twice without any final action. Congressional committees, it said, appeared unwilling to institute inquiries and elicit evidence from practical men, and therefore they could hardly determine facts based upon twenty years of experience with the use of steam in boats. Since Congress was apparently avoiding action, the committee asserted, it was of paramount importance that a competent body whose motives were above suspicion should shoulder the burden.[28] Thus, the Franklin Institute committee began a six-year investigation of boiler explosions.

From 1825 to 1830, there had been forty-two explosions killing about 273 persons, and in 1830 a particularly serious one aboard the "Helen McGregor" near Memphis, which killed 50 or 60 persons, again disturbed Congress. The House requested the secretary of the treasury, Samuel D. Ingham of Pennsylvania, to investigate the boiler accidents and submit a report.[29] Ingham had served in Congress from

1813 to 1818, and again from 1822 to 1829. He was a successful manufacturer who owned several paper mills; he was acquainted with the activities of the Franklin Institute and had written to the *Journal* about steam boiler problems.[30] Ingham was thus in a unique position to aid the Franklin Institute committee which had begun its inquiries. Before his resignation from Jackson's cabinet over the Peggy O'Neill Eaton affair, Ingham committed government funds to the Institute to defray the cost of apparatus necessary for the experiments.[31] This was the first research grant of a technological nature made by the federal government.[32]

Ingham attempted to make his own investigation while still secretary of the treasury. His interim report to the House in 1831 revealed that two investigators, one on the Atlantic seaboard and the other in the Mississippi basin, had been employed to gather information on the boiler explosions. They complained that owners and masters of boats seemed unwilling to aid the inquiry. They were told repeatedly that the problem was purely individual, a matter beyond the government's right to interfere.[33] In the following year, the new secretary, Louis B. McLane, circulated a questionnaire among the collectors of customs, who furnished information and solicited opinions about the explosions. Their answers formed the basis of McLane's report to Congress. They mentioned the many causes of boiler explosions. One letter noted that steamboat trips from New Orleans to Louisville had been shortened from twenty-five to twelve days since 1818 without increasing the strength of the boilers. A frequent remark was that the engineers in charge of the boilers were ignorant, careless, and usually drunk.[34]

This report prompted a bill proposed in the House in May 1832. It provided for the appointment of inspectors at convenient locations to test the strength of the boilers every three months at three times their working pressure, and the issuance of a license to navigate was made contingent upon this inspection. To avoid possible objections on the score of expense, inspection costs were to be borne by the government. To prevent explosions caused by low water supply, the bill provided that masters and engineers be required under threat of heavy penalties to supply water to the boilers while the boat was not in motion.

The half-hearted tone of the House committee's report on the bill hardly promised positive legislative action. The Constitution gave Congress the power to regulate commerce, the report noted, but the right of Congress to prescribe the mode, manner, or form of construction of the vehicles of conveyance could not be perceived. Whether boats should be propelled by wind, paddles, or steam, and if by steam, whether by low or high pressure, were questions that were not the business of Congress. No legislation was competent to remove the causes of boiler explosions, so that steam and its application must be left to the control of intellect and practical science. The intelligent conduct of those engaged in its use would be the best safeguard against the dangers incident to negligence. Besides, the report concluded, the destruction was much less than had been thought; the whole number of explosions in the United States was only fifty-two, with total casualties of 256 killed and 104 injured.[35] Supporters of the bill could not undo the damage of the watered-down committee report, however. The bill died, and the disasters continued.

In his State of the Union message in December 1833, President Jackson noted that the distressing accidents on steamboats were increasing. He suggested that the disasters often resulted from criminal negligence by masters of the boats and operators of the engines. He urged Congress to pass precautionary and penal legislation to reduce the accidents.[36] A few days later, Senator Daniel Webster proposed that the Committee on Naval Affairs study the problem. He suggested that all boilers be tested at three times their working pressure and that any steamboat found racing be forfeited to the government. Thomas Hart Benton followed Webster, stating that the matter properly was the concern of the Judiciary Committee. The private waters of states were involved, Benton said; interference with their sovereignty might result. In passing, Benton remarked that the masters and owners of steamboats were, with few exceptions, men of the highest integrity. Further, Benton said, *he* had never met with any accident on a steamboat despite the fact that he traveled widely; upon boarding he was always careful to inquire whether the machinery was in good order. Webster still carried the day, since the matter went to the Committee on Naval Affairs; however, Benton's attitude prevailed in the session, for the reported Senate bill failed to pass.[37]

V

A program of experiments carried out by the Franklin Institute from 1831 to 1836 was based largely upon the reports of circumstances surrounding previous boiler explosions, the contemporary design and construction of boilers and their accessories, and methods of ensuring an adequate water supply. The work was done by a committee of volunteers led by Alexander Dallas Bache, later superintendent of the U.S. Coast Survey, who, at the time, was a young professor of natural philosophy at the University of Pennsylvania. A small boiler, one foot in diameter and about three feet long, with heavy glass viewing ports at each end, was used in most of the experiments. In others, the zeal of the workers led them to cause larger boilers to burst at a quarry on the outskirts of Philadelphia.

The group's findings overturned a current myth, proving conclusively that water did not decompose into hydrogen and oxygen inside the boiler, with the former gas exploding at some high temperature. The experimenters demonstrated that an explosion could occur without a sudden increase of pressure. Another widely held theory they disproved was that when water was injected into a boiler filled with hot and unsaturated steam, it flashed into an extremely high-pressure vapor, which caused the boiler to rupture. The group proved that the reverse was true: the larger the quantity of water thus introduced, the greater the decrease in the steam pressure.

The Franklin Institute workers also produced some positive findings. They determined that the gauge cocks, commonly used to ascertain the level of water inside the boilers, did not in fact show the true level, and that a glass tube gauge was much more reliable, if kept free from sediment. They found the fusing points of alloys of lead, tin, and bismuth, and recommended that fusible plates be employed with caution, because the more fluid portion of an alloy might be forced out prior to the designated fusion temperature, thus leaving the remainder with a higher temperature of fusion.[38] They investigated the effect of the surface condition of the shell on the temperature and time of vaporization, and they determined that properly weighted safety valves opened at calculated pressures within a small margin of error. The results of their experiments on the relationship of the pressure and temperature of steam showed close correspondence with those of the French, although, at this time, values of the specific heat of steam were erroneous due to the inability to differentiate between constant volume and constant temperature conditions.[39]

Simultaneously, another committee, also headed by Bache, investigated the strength of boiler materials. In these experiments, a sophisticated tensile testing machine was constructed, and corrections were made for friction and stresses produced during the tests. The investigators tested numerous specimens of rolled copper and wrought iron, not only at ambient temperatures but up to 1,300° F. They showed conclusively that there were substantial differences in the quality of domestic wrought irons by the differences in yield and tensile strengths. Of major importance was their finding that there was a rapid decrease in the ultimate strength of copper and wrought iron with increasing temperature. Further, they determined that the strength of iron parallel to the direction of rolling was about 6 per cent greater than in the direction at right angles to it. They proved that the laminated structure in "piled" iron, forged from separate pieces, yielded much lower tensile values than plate produced from single blooms. Their tests also showed that special precautions should be taken in the design of riveted joints.[40]

Taken as a whole, the Franklin Institute reports demonstrate remarkable experimental technique as well as a thorough methodological approach. They exposed errors and myths in popular theories on the nature of steam and the causes of explosions. They laid down sound guidelines on the choice of materials, on the design and construction of boilers, and on the design and arrangement of appurtenances added for their operation and safety. Further, the reports included sufficient information to emphasize the necessity for good maintenance procedures and frequent proof tests, pointing out that the strength of boilers diminished as the length of service increased.

VI

The Franklin Institute report on steam-boiler explosions was presented to the House through the secretary of the treasury in March 1836, and the report on boiler materials was available in 1837. The Franklin Institute committee also made detailed recommendations on provisions that any regulatory legislation should incorporate. It proposed that inspectors be appointed to test all boilers hydraulically every six months; it prohibited the licensing of ships using boilers whose design had proved to be unsafe; and it recommended penalties in cases of explosions resulting from improper maintenance, from the incompetence or negligence of the master or engineer, or from racing. It placed responsibility for injury to life or property on owners who neglected to have the required inspections made, and it recommended that engineers meet certain standards of experience, knowledge, and character. The committee had no doubt of the right of Congress to legislate on these matters.[41]

Congress did not act immediately. In December 1836 the House appointed a committee to investigate the explosions, but there was no action until after President Van Buren urged the passage of legislation in December 1837.[42] That year witnessed a succession of marine disasters. Not all were attributable to boiler explosions, although the loss of 140 persons in a new ship, the "Pulaski," out of Charleston, was widely publicized. The Senate responded quickly to Van Buren's appeal, passing a measure on January 24, 1838. The House moved less rapidly. An explosion aboard the "Moselle" at Cincinnati in April 1838, which killed 151 persons,[43] caused several Congressmen to request suspension of the rules so that the bill could be brought to the floor, but in the face of more pressing business the motion was defeated.[44] The legislation was almost caught in the logjam in the House at the end of the session, but on June 16 the bill was brought to the floor. Debate centered principally upon whether the interstate commerce clause in the Constitution empowered Congress to pass such legislation. Its proponents argued affirmatively, and the bill was finally approved and became law on July 7, 1838.[45]

The law incorporated several sections relating to the prevention of collisions, the control of fires, the inspection of hulls, and the carrying of lifeboats. It provided for the immediate appointment by each federal judge of a competent boiler inspector having no financial interest in their manufacture. The inspector was to examine every steamboat boiler in his area semiannually, ascertain its age and soundness, and certify it with a recommended working pressure. For this service the owner paid the inspector $5.00 —his sole remuneration—and a license to navigate was contingent upon the receipt of this certificate. The law specified no inspection criteria. It enjoined the owners to employ a sufficient number of competent and experienced engineers, holding the owners responsible for loss of life or property damage in the event of a boiler explosion for their failure to do so. Further, any steamboat employee whose negligence resulted in the loss of life was to be considered guilty of manslaughter, and upon conviction could be sentenced to not more than ten years' imprisonment. Finally, it provided that in suits against owners for damage to persons or property, the fact of the bursting of the boilers should be considered *prima facie* evidence of negligence until the defendant proved otherwise.[46]

This law raises several questions, because the elimination of inspection criteria and the qualification of engineers rendered the measure ineffectual. Why was this done? Did Congress show restraint because it had insufficient information? Did it yield to the pressure of steamboat interests who feared government interference? Such questions cannot be definitely answered, but there are clues for some tentative conclusions.

The bill, as originally introduced, was similar to the Franklin Institute proposals, so that the Senate committee to which it was referred possessed the most recent informed conclusions as to the causes of boiler explosions and the means of their prevention. The President's plea to frame legislation in the face of the mounting fatalities undoubtedly persuaded the Democratic majority to act. They were unmoved by a memorial from steamboat interests urging the defeat of the bill.[47] But the majority was not as yet prepared to pass such detailed regulations as had originally been proposed. In response to a question as to why the provision for the qualification of engineers had been eliminated, the Senate committee chairman stated that the committee had considered this requirement desirable but foresaw too much difficulty in putting it into effect. Further, the Senate rejected an amendment to levy heavy penalties for racing, as proposed by the Whig, Oliver Smith of Indiana. The Whigs appeared to have seen the situation as one in which the federal government should use its powers and interpose firmly. Henry Clay, R. H. Bayard of Delaware, and Samuel Prentiss of Vermont supported Smith's amendment, and John Davis of Massachusetts declared that he would support the strongest measures to make the bill effective. Those who had urged rapid action of the bill in the House were William B. Calhoun and Caleb Cushman of Massachusetts and Elisha Whittlesey of Ohio, all Whigs. But at this time the majority hewed to the doctrine that enlightened self-interest should motivate owners to provide safe operation. The final clause, specifying that the bursting of boilers should be taken as *prima facie* evidence of negligence until proved otherwise, stressed this idea.

The disappointment of the informed public concerning

the law was voiced immediately in letters solicited by the secretary of the treasury, contained in a report that he submitted to Congress in December 1838.[48] There were predictions that the system of appointment and inspection would encourage corruption and graft. There were complaints about the omission of inspection criteria and a provision for the licensing of engineers. One correspondent pointed out that it was impossible legally to determine the experience and skill of an engineer, so that the section of the law that provided penalties for owners who failed to employ experienced and skilful engineers was worthless. One critic who believed that business interests had undue influence upon the government wrote: "We are mostly ruled by corporations and joint-stock companies. . . . If half the citizens of this country should get blown up, and it should be likely to affect injuriously the trade and commerce of the other half by bringing to justice the guilty, no elective officer would risk his popularity by executing the law."[49]

But there also was a pained reaction from the owners of steamboats. A memorial in January 1841 from steamboat interests on the Atlantic seaboard stressed that appropriate remedies for the disasters had not been afforded by the 1838 law as evidenced by casualty figures for 1839 and 1840. They provided statistics to prove that in *their* geographical area the loss of life per number of lives exposed had decreased by a factor of sixteen from 1828 to 1838, indicating that the troubles centered chiefly in the Western waters. But at the same time the memorial emphasized that the 1838 law acted as a deterrent for prudent men to continue in the steamboat business, objecting particularly to the clause that construed a fatal disaster as *prima facie* evidence of negligence. They argued that if Congress considered steam navigation too hazardous for the public safety, it would be more just and honorable to prohibit it entirely.[50]

However, it not only was the Congress that was reconsidering the concepts of negligence and responsibility in boiler explosions. The common law also searched for precedents to meet the new conditions, to establish guidelines by which to judge legal actions resulting from technological innovation. A key decision, made in Pennsylvania in 1845, involved a boiler explosion at the defendant's flour mill that killed the plaintiff's horse. The defense pleaded that any negligence was on the part of the boiler manufacturer. The court, however, ruled otherwise, stating that the owner of a public trade or business which required the use of a steam engine was responsible for any injury resulting from its deficiency.[51] This case was used as a precedent in future lawsuits involving boiler explosions.

VII

Experience proved that the 1838 law was not preventing explosions or loss of life. In the period 1841–1848, there were some 70 marine explosions that killed about 625 persons. In December 1848 the commissioner of patents, to whom Congress now turned for data, estimated that in the period 1816–1848 a total of 233 steamboat explosions had occurred in which 2,563 persons had been killed and 2,097 injured, with property losses in excess of $3 million.[52]

In addition to the former complaints about the lack of proof tests and licenses for engineers, the commissioner's report included testimony that the inspection methods were a mockery. Unqualified inspectors were being appointed by district judges through the agency of highly placed friends. The inspectors regarded the position as a lifetime office. Few even looked at the boilers but merely collected their fees. The inspector at New York City complained that his strict inspection caused many boats to go elsewhere for inspections. He cited the case of the "Niagara," plying between New York City and Albany, whose master declined to take out a certificate from his office because it recommended a working pressure of only 25 p.s.i. on the boiler. A few months later the boiler of the "Niagara," which had been certified in northern New York, exploded while carrying a pressure of 44 p.s.i. and killed two persons.[53]

Only eighteen prosecutions had been made in ten years under the manslaughter section of the 1838 law. In these cases there had been nine convictions, but the penalties had, for the most part, been fines which were remitted. It was difficult to assemble witnesses for a trial, and juries could not be persuaded to convict a man for manslaughter for an act of negligence, to which it seemed impossible to attach this degree of guilt. Also, the commissioner's report pointed out that damages were given in cases of bodily injury but that none were awarded for loss of life in negligence suits. It appeared that exemplary damages might be effective in curbing rashness and negligence.[54]

The toll of life in 1850 was 277 dead from explosions, and in 1851 it rose to 407.[55] By this time Great Britain had joined France in regulatory action, which the Congress noted.[56] As a consequence of legislation passed in 1846 and

1851, a rejuvenated Board of Trade was authorized to inspect steamboats semiannually, to issue or deny certificates of adequacy, and to investigate and report on accidents.[57] The time had come for the Congress to take forceful action, and in 1852 it did.

John Davis, senator from Massachusetts, who had favored stricter legislation in 1838, was the driving force behind the 1852 law. In prefacing his remarks on the general provisions of the bill, he said: "A very extensive correspondence has been carried on with all parts of the country . . . there have been laid before the committee a great multitude of memorials, doings of chambers of commerce, of boards of trade, of conventions, of bodies of engineers; and to a considerable extent of all persons interested, in one form or another, in steamers . . . in one thing . . . they are all . . . agreed—that is, that the present system is erroneous and needs correction."[58]

Thus again, the informed public submitted recommendations on the detailed content of the measure. An outstanding proponent who helped shape the bill was Alfred Guthrie, a practical engineer from Illinois. With personal funds, Guthrie had inspected some two hundred steamboats in the Mississippi valley to ascertain the causes of boiler explosions. Early in the session, Senator Shields of Illinois succeeded in having Guthrie's report printed, distributed, and included in the Senate documents.[59] Guthrie's recommendations were substantially those made by the Franklin Institute in 1836. His reward was the post as first supervisor of the regulatory agency which the law created.

After the bill reached the Senate floor, dozens of amendments were proposed, meticulously scrutinized, and disposed of. The measure had been, remarked one senator, "examined and elaborated . . . more patiently, thoroughly, and faithfully than any other bill before in the Senate of the United States."[60] As a result, in place of the 1838 law which embodied thirteen sections and covered barely three pages, there was passed such stringent and restrictive legislation that forty-three sections and fourteen pages were necessary.[61]

The maximum allowable working pressure for any boiler was set at 110 p.s.i., and every boiler had to be tested yearly at one and one-half times its working pressure. Boilers had to be fabricated from suitable quality iron plates, on which the manufacturer's name was stamped. At least two ample safety valves—one in a locked grating—were required, as well as fusible plates. There were provisions relating to adequate supply of boiler feedwater and outlawing designs that might prove dangerous. Inspectors were authorized to order repairs at any time. All engineers had to be licensed by inspectors, and the inspectors themselves issued certificates only under oath. There were stiff monetary penalties for any infractions. The penalty for loading a safety valve excessively was a two hundred dollar fine and eighteen months' imprisonment. The fine for manufacturing or using a boiler of unstamped material was five hundred dollars. Fraudulent stamping carried a penalty of five hundred dollars and two years' imprisonment. Inspectors falsifying certificates were subject to a five hundred dollar fine and six months' imprisonment, and the law expressly prohibited their accepting bribes.

A new feature of the law, which was indicative of the future, was the establishment of boards of inspectors empowered to investigate infractions or accidents, with the right to summon witnesses, to compel their attendance, and to examine them under oath. Above the local inspectors were nine supervisors appointed by the President. Their duties included the compilation of evidence for the prosecution of those failing to comply with the regulations and the preparation of reports to the secretary of the treasury on the effectiveness of the regulations. Nor did these detailed regulations serve to lift the burden of presumptive negligence from the shoulders of owners in cases of explosion. The explosion of boilers was not made *prima facie* evidence as in the 1838 law, but owners still bore a legal responsibility. This was made clear in several court decisions which held that proof of strict compliance with the 1852 law was not a sufficient defense to the allegations of loss by an explosion caused by negligence.[62]

The final Senate debate and the vote on this bill shows how, in thirty years, the public attitude and, in turn, the attitudes of its elected representatives had changed toward the problem of unrestricted private enterprise, mainly as a result of the boiler explosions. The opponents of the bill still argued that the self-interest of the steamboat companies was the best insurance of the safety of the traveling public.[63] But their major argument against passage was the threat to private property rights which they considered the measure entailed. Senator Robert F. Stockton of New Jersey was most emphatic:

It is this—how far the Federal Government . . . shall be permitted to interfere with the rights of personal

property—or the private business of any citizen . . .
under the influence of recent calamities, too much sensibility is displayed on this subject . . . I hold it to be
my imperative duty not to permit my feelings of humanity and kindness to interfere with the protection
which I am bound, as a Senator of the United States,
to throw around the liberty of the citizen, and the investment of his property, or the management of his
own business . . . what will be left of human liberty
if we progress on this course much further? What will
be, by and by, the difference between citizens of this
far-famed Republic and the serfs of Russia? Can a
man's property be said to be his own, when you take
it out of his own control and put it into the hands of
another, though he may be a Federal officer?[64]

This expression of a belief that Congress should in no circumstances interfere with private enterprise was now supported by only a small minority. One proponent of the bill
replied: "I consider that the only question involved in the
bill is this: Whether we shall permit a legalized, unques-
tioned, and peculiar class in the community to go on com-
mitting murder at will, or whether we shall make such
enactments as will compel them to pay some attention to
the value of life."[65] It was, then, a question of the sanctity
of private property rights against the duty of government
to act in the public weal. On this question the Senate voted
overwhelmingly that the latter course should prevail.[66]

Though not completely successful, the act of 1852 had
the desired corrective effects. During the next eight years
prior to the outbreak of the Civil War, the loss of life on
steamboats from all types of accidents dropped to 65 per
cent of the total in the corresponding period preceding its
passage.[67] A decade after the law became effective, John
C. Merriam, editor and proprietor of the *American Engineer,* wrote: "Since the passage of this law steamboat explosions on the Atlantic have become almost unknown, and
have greatly decreased in the west. With competent inspectors, this law is invaluable, and we hope to hail the day
when a similar act is passed in every legislature, touching
locomotive and stationary boilers."[68]

There was, of course, hostility and opposition to the law
immediately after its passage, particularly among the owners
and masters of steamboats.[69] It checked the steady rise in
the construction of new boats, which had been character-

istic of the earlier years.[70] The effect, however, was chastening rather than emasculating. Associations for the prevention of steam-boiler explosions were formed; later, insurance companies were organized to insure steam equipment that was manufactured and operated with the utmost regard for safety. In time, through the agency of the American Society of Mechanical Engineers, uniform boiler codes were promulgated and adopted by states and municipalities.[71]

Thus, the reaction of the informed public, expressed by Congress, to boiler explosions caused the initiation of positive regulation of a sector of private enterprise through a governmental agency. The legislation reflected a definite change of attitude concerning the responsibility of the government to interfere in those affairs of private enterprise where the welfare and safety of the general public was concerned. The implications of this change for the future can be seen by reference to the Windom Committee report of 1874, which was the first exhaustive study of the conditions in the railroad industry that led ultimately to the passage of legislation creating the Interstate Commerce Commission. One section of this report was entitled: "The Constitutional Power of Congress to Regulate Commerce among the Several States." The committee cited the judicial interpretation of the Constitution in *Gibbons* v. *Ogden,* that it was the prerogative of Congress solely to regulate interstate commerce, and also referred to the decision of Chief Justice Taney in *Genesee Chief* v. *Fitzhugh,* wherein it was held that this power was as extensive upon land as upon water. The report pointed out that no decision of the Supreme Court had ever countenanced the view that the power of Congress was purely negative, that it could be constitutionally exercised only by disburdening commerce, by preventing duties and imposts on the trade between the states. In fact, the report argued, Congress had already asserted its power positively. Referring to the acts of 1838 and 1852, it stated that "Congress has passed statutes defining how steamboats shall be constructed and equipped."[72] Thus, the legislation that was provoked by bursting boilers was used as a precedent to justify regulatory legislation in another area where the public interest was threatened.

Bursting steamboat boilers, then, should be viewed not merely as unfortunate and perhaps inevitable consequences of the early age of steam, as occurrences which plagued nineteenth-century engineers and which finally, to a large degree, they were successful in preventing. They should be seen also as creating a dilemma as to how far the lives and property of the general public might be endangered by unrestricted private enterprise. The solution was an important step toward the inauguration of the regulatory and investigative agencies in the federal government.

REFERENCES

[1] See, e.g., *Report* on Practices and Procedures of Governmental Control, Sept. 18, 1944 (House of Representatives document 678, ser. 10873 [Washington: 78th Congress, 2d session]), p. 3, where it is stated: "Regulation, seen through modern eyes is not a violent departure from the ways of business to which the nation is both habituated and strongly attached. . . . regulation . . . enjoys, as a system, in large measure the confidence and approval of the parties concerned."

[2] Thomas Jefferson, "Inaugural Address," *Journal of the Executive Proceedings of the U.S. Senate,* I (Washington, 1828), 393.

[3] The most authoritative work is Louis C. Hunter, *Steamboats on the Western Rivers* (Cambridge, Mass., 1949), pp. 122–133, 271–304, 520–546.

[4] Greville and Dorothy Bathe, *Oliver Evans* (Philadelphia, 1935), pp. 151, 253. Also, see Walter F. Johnson, "On the Strength of Cylindrical Steam Boilers," *Journal of the Franklin Institute* (hereinafter cited as "JFI"), X, N.S. (1832), 149. Evans' formula reveals that he considered that a safe design tensile strength for good-quality wrought iron was about 42,000 p.s.i. and that a factor of safety of 10 should be used to arrive at a safe shell thickness.

[5] For reports of defective design and poor quality material see: Charles F. Partington, *An Historical and Descriptive Account of the Steam Engine* (London, 1822), p. 85; Committee on Steamboats *Report,* May 18, 1832 (House of Representatives document 478, ser. 228 [Washington: 22d Congress, 1st session]), pp. 44, 170 (hereinafter cited as "Doc. 478"). Also, *JFI,* VI, N.S. (1830), 44–51; VIII, N.S. (1831), 382; IX, N.S. (1832), 28, 100, 363; X, N.S. (1832), 226–32; XVII, N.S. (1836), 298–302; XX, N.S. (1837), 100, 103.

[6] For operating difficulties see: Partington, *op. cit.,* p. 118; *JFI,* V, N.S. (1830), 402; VI, N.S. (1830), 9; VIII, N.S. (1831), 277, 289–92; VIII, N.S. (1831), 309, 313, 382; IX, N.S. (1832), 20–22. Also, Secretary of the Treasury, *Report on Steam Engines,* Dec. 13, 1838 (House of Representatives document 21, ser. 345 [Washington: 25th Congress, 3d session]), p. 3 (hereinafter cited as "Doc. 21"). The whole number of steam engines in the United States in 1838 was estimated at 3,010: 800 on steamboats, 350 in locomotives, and 1,860 in manufacturing establishments. The majority of these engines were put into service after 1830. The term "practical engineer" was reserved

for a designer or builder of engines, while engine-room operatives were called "engineers." The complaints about the incompetence of the latter are very frequent in the literature.

[7] *Doc. 21,* p. 3.

[8] Bathe, *op. cit.,* p. 250.

[9] *Ibid.,* p. 255; *JFI,* VIII, N.S. (1831), 235–243.

[10] Parliamentary Sessional Papers, *Report* (1817), VI, 223.

[11] *Ibid.* (1831), VIII, 1; (1839), XLVII, 1; (1843), IX, 1.

[12] *Ibid.* (1842), IX, 383–384.

[13] *Ibid.* (1839), XLVII, 10.

[14] The number of explosions and the loss of life occasioned thereby, listed throughout this paper, were obtained by a comparison and tabulation of the figures listed in *Doc. 21,* pp. 399–403, and in the Commissioner of Patents, *Report,* Dec. 30, 1848 (Senate document 18, ser. 529 [Washington: 30th Congress, 2d session]), pp. 36–48 (hereinafter cited as "Doc. 18").

[15] Department of the Interior, Census Office, *10th Census* (Washington, 1883), IV, 6–7; *JFI,* IX, N.S. (1832), 350.

[16] *Archives Parlementaires* (Paris, 1864), III, ser. 2, 732; *JFI,* VII, N.S. (1831), 272.

[17] *JFI,* X, N.S. (1832), 106; *Doc. 478,* p. 145.

[18] *JFI,* VII, N.S. (1831), 272, 323, 399; VIII, N.S. (1831), 32; X, N.S. (1832), 105, 181.

[19] *JFI,* V, N.S. (1830), 399, 411.

[20] *Doc. 18,* p. 180.

[21] Parliamentary Sessional Papers, *Report* (1839), XLVII, 180.

[22] Bathe, *op. cit.,* p. 237; *JFI,* II (1826), 147.

[23] *Doc. 21,* p. 425.

[24] *Ibid.,* pp. 105, 424; *JFI,* XIII, N.S. (1834), 55, 126, 289.

[25] *Annals of Congress* (Washington: 18th Congress, 1st session), pp. 2670, 2694, 2707, 2708, 2765.

[26] For the history of the Franklin Institute, see S. L. Wright, *The Story of the Franklin Institute* (Philadelphia, 1938).

[27] *JFI,* VI, N.S. (1830), 33.

[28] *Ibid.,* 34.

[29] *Congressional Debates* (Washington: 21st Congress, 1st session), VI, Part 2, 739.

[30] *Dictionary of American Biography* (New York, 1932), IX, 473; *JFI,* IX, N.S. (1832), 12 (communicated Oct. 21, 1830).

[31] *JFI,* VII, N.S. (1831), 42.

[32] Arthur V. Greene, "The A.S.M.E. Boiler Code," *Mechanical Engineer,* LXXIV (1952), 555; A. Hunter Dupree, *Science in the Federal Government* (Cambridge, Mass., 1957), p. 50.

[33] Secretary of the Treasury, *Report,* March 3, 1831 (House of Representatives document 131, ser. 209 [Washington: 21st Congress, 2d session]), p. 1.

[34] *Doc. 478,* p. 44.

[35] *Ibid.,* pp. 1–7. Actually, the committee depended for its statistics upon the estimate of William C. Redfield, agent for

the Steam Navigation Company of New York, who could hardly have been expected to be impartial. Comparing Redfield's figures with those listed in *JFI*, IX, N.S. (1832), 24–30 and with the sources listed in n. 14, it is clear that he omitted many minor accidents; where the number of casualties were unknown, they were not counted; where they were estimated, Redfield took the lowest estimate.

[36] *Congressional Globe*, I (Washington: 23d Congress, 1st session), 7.

[37] *Ibid.*, I, 49, 442.

[38] In this series of experiments, the committee was actually investigating the solid solutions of these metals and determining points on what would later be called equilibrium diagrams.

[39] Franklin Institute, *Report*, March 1, 1836 (House of Representatives document 162, ser. 289 [Washington]).

[40] *JFI*, XVIII, N.S. (1836), 217, 289; XIX, N.S. (1837), 73, 157, 241, 325, 409; XX, N.S. (1837), 1, 73.

[41] *JFI*, XVIII, N.S. (1836), 369–75.

[42] *Congressional Globe*, IV (Washington: 24th Congress, 2d session), 29; VI (Washington: 25th Congress, 2d session), 7–9.

[43] The *Moselle* disaster was important because of its effect upon marine insurance policies. The estate of the captain and part owner, Isaac Perrin, sued for recovery under the policy (*The Administrators of Isaac Perrin* v. *The Protection Insurance Co.*, 11 Ohio [1842], 160). The defense gave evidence that Perrin was determined to outstrip another boat and that when passengers expostulated with him concerning the dangerous appearance of the boiler fires, he swore that he would be "that night in Louisville or hell." Despite proof of negligence on the part of the captain, the court ruled against the insurance company, stating that the explosion of boilers was a risk insured against. The insurance companies, thereafter, moved to exclude boiler explosions as a covered risk. See *Citizens Insurance Co.* v. *Glasgow, Shaw, and Larkin*, 9 Missouri (1852), 411 and *Roe and Kercheval* v. *Columbus Insurance Co.*, 17 Missouri (1852), 301.

[44] *Congressional Globe*, VI, 342.

[45] *Ibid.*, VI, 455.

[46] *U.S. Statutes at Large* (Washington: 25th Congress, 2d session, July 7, 1838), V, 304–6.

[47] *Congressional Globe*, VI, 265.

[48] *Doc. 21.*

[49] *Ibid.*, p. 396.

[50] *Memorial*, Jan. 23, 1841 (House of Representatives document 113, ser. 377 [Washington: 26th Congress, 2d session]).

[51] *Spencer* v. *Campbell*, 9 Watts & Sergeants (1845), 32.

[52] *Doc. 18*, p. 2.

[53] *Ibid.*, pp. 18, 78, 80.

[54] *Ibid.*, pp. 29, 52–53.

[55] *Congressional Globe* (Washington: 32d Congress, 1st session), Appendix, 287.

[56] *Ibid.*, p. 2426.

[57] Public and General Acts, 9, 10 Victoria (1846), chap. 1; 14, 15 Victoria (1851), chap. lxxix.

[58] *Congressional Globe* (32d Congress, 1st session), p. 1669. Organizations of experienced steamboat engineers were formed in many cities during the 1840's to promote safe operation and had attempted on previous occasions to influence Congress to improve the 1838 law, particularly with respect to providing for proof tests, better inspection methods, and the establishment of boards to qualify engineers. See *Relative to Steamboat Explosions* (House of Representatives document 68, ser. 441 [Washington: 28th Congress, 1st session]), which is a petition from a body in the city of Cincinnati.

[59] *Memorial of Alfred Guthrie, a Practical Engineer,* Feb. 6, 1852 (Senate miscellaneous document 32, ser. 629 [Washington: 32d Congress, 1st session]).

[60] *Congressional Globe* (32d Congress 1st session), p. 1742.

[61] *U.S. Statutes at Large* (Washington: 32d Congress, 1st session, Aug. 30, 1852), X, 61–75.

[62] *Curran* v. *Cheeseman,* 1 Cincinnati Rep. (1870), 52.

[63] *Congressional Globe* (32d Congress, 1st session), pp. 1741, 2425.

[64] *Ibid.*, pp. 2426, 2427.

[65] *Ibid.*, p. 2427.

[66] The strength of the vote can be gauged by the defeat, forty-three to eight, of a motion to table the bill by Senator Stockton just prior to its passage. The eight were: Bayard (D., Del.), Butler (States Rights D., S.C.), Clemens (D., Ala.); Hale (Antislavery D., N.H.), Hunter (D., Va.), James (Protective Tariff D., R.I.), Pratt (Whig, Md.), and Stockton (D., N.J.). Although these senators represented only states along the Eastern seaboard and in the South, it would be difficult to interpret their vote on a geographical basis, since eighteen senators from the same group of states voted against the motion. One might be tempted to ascribe some partisan basis to the vote, since only one Whig joined seven Democrats in supporting the motion. On the other hand, twenty-six Democrats and seventeen Whigs constituted the majority. Of those not voting—seven Democrats and four Whigs—by their comments during prior debates on the measure, Brodhead (D., Pa.) and De Saussure (D., S.C.) appear to have favored the bill, while Gwin (D., Calif.) was against it. The conclusion seems justified that the movement and final step toward positive regulation found support from congressmen of all political postures and from all geographical areas, that it was prompted by the recognition of the inadequacy of the 1838 law as evidenced by the continued severe loss of life, and that congressmen were urged to pass the legislation by constituents who were able to recognize how the problem could be solved.

[67] *10th Census,* IV, 5.

[68] L. Stebbins, pub., *Eighty Years Progress of the United States* (New York, 1864), p. 243.

[69] Lloyd M. Short, *Steamboat Inspection Service* (New York, 1922), p. 5.

[70] Department of the Interior, *op. cit.,* IV, 5.

[71] Greene, *op. cit.*

[72] *Report of the Select Committee on Transportation to the Seaboard* (Senate Report No. 307, Ser. 1588 [Washington: 43d Congress, 1st session]), pp. 79–92.

Part II
Technology and the Humanities

Introduction

There was a time when the word "humanistic" was used to distinguish what was human from what was brutal and coarse, or animal. But now we find the humanities increasingly contrasted with technology, and "humanistic" is used to denote everything which is not scientific or technical. While the ancient philosophers, such as Aristotle, held that the distinctive quality of human beings, which distinguished them from the rest of the animal world, was rational intelligence, the humanities in today's world tend to concentrate on the spiritual, emotional, and irrational aspects of man and ignore his rational attempts to cope with nature as embodied in his technological activities. Yet, technology is a very human activity; anthropologists tell us that it is perhaps the most basic of human cultural activities, for it helps to distinguish our human species from the rest of the animal kingdom on the basis of tool dependency.

In "Achieving a Perspective on the Technological Order," Aldous Huxley explains why men of letters tend to ignore or decry the technological aspects of human activities: literature deals with the lyrical and emotional aspects of life, while technology deals with reason. Hence, the rational side of man, expressed in his science and technology, finds little expression in the literary arts because human beings "are interested in a much livelier way in the passions than in reason."

As befits someone of his cultural inheritance, Huxley shows his concern for ecology, which, when he wrote this piece in 1962, was first coming into the public consciousness. As the heir to a great naturalist tradition, Huxley is concerned lest man's symbiosis with nature be destroyed by technology. He suggests that it is the duty of modern writers to express this danger clearly.

Huxley's humanistic concern with the relationships between technology and nature is shared by Scott Buchanan. Among the ancient Greeks, Buchanan points out, the arts included what we today call the fine arts and also the crafts, and in their concern with *technologia* (the giving of rules to the arts), they believed that art imitates nature. But the arts have passed from the individual craftsman to corporate organizations in today's world, and Buchanan claims we must find ways to bring human artists and human ends into the technological system in order to rediscover and maintain the true balance between man and nature.

If the question of the relationships of technology to human goals remains murky as a result of such discussions, it is perhaps because we have not yet thought through the implications of technological advance in terms of its meaning for the humanities.

The final two selections in this section provide specific examples of the use of technology in humanistic studies. The discovery of the Dead Sea Scrolls in the 1950's opened up many questions in biblical studies regarding the nature of the Jewish community at about the time of Christ, the sources of early Christian thought, and the like. Archaeological and biblical textual studies were first used to determine the authenticity of the Dead Sea Scrolls and to place them in relationship to the Jewish community and tradition of the times; J. B. Poole and R. Reed then looked at the problem from the standpoint of the history of technology. By their analysis of the leather and parchment technology of the Scrolls, they were able to demonstrate that these Dead Sea Scrolls were the product of a highly orthodox Jewish sect and not of any heretical group which turned away from the rabbinic regulations of the past.

The contributions of technology to the art of architecture in the twentieth century are viewed through Carl W. Condit's study of the skyscrapers of Louis Sullivan. This "first great modern architect," as Condit describes him, used the scientific technology of his time to transmute architecture "into an effective and valid artistic statement." Dr. Condit has pioneered in studying the relations of technological developments to artistic expressions in the field of structures, and other scholars have recently begun to study those relationships in other fields. Yet much further study is necessary if we are to understand the manifestations of human creativity embodied in the interdependence of technological and artistic developments.

Achieving a Perspective on the Technological Order

Aldous Huxley

I would like to talk about some aspects of technology's effects on human individuals and society which have interested me particularly. Dealing with the effect of the development of modern technological methods on the old arts and crafts, A. Rupert Hall has compared the Chinese porcelain cup with the plastic cup of today and pointed out that a plastic cup can be produced only as a result of the most elaborate interlocking technology. A huge volume of applied science has gone into the making of these curiously repulsive objects. The question is: what have we gained and what have we lost?

The cost of plastic containers is smaller than the cost of baked clay and porcelain containers. The people carrying water jugs are no doubt happy to have a lighter jug. Certainly we have gained. But there is one disturbing difference between the technology of mass production and the technology of the empirical handicrafts. The handicraftsman produces complex effects with incredibly simple tools simply by skill of hand and mind, whereas modern technology uses immensely complicated tools which require little skill of hand or mind. The technological systems of production and organization are virtually foolproof. But if anything is foolproof it is also spontaneity-proof, inspiration-proof, and

The late Aldous Huxley, novelist, essayist, and poet, is perhaps best known for his Brave New World, *a novel depicting the nightmare of a possible "technological future." This paper consists of Mr. Huxley's comments at the Encyclopaedia Britannica Conference on the Technological Order (March 1962) and appeared in the proceedings of that conference published in* Technology and Culture *(Vol. 3, No. 4): 636–42.*

even skill-proof. Against the advantages of the plastic cup we must set the grave disadvantages to the people deprived of the opportunity of craft expression, of artistic expression. Probably you will agree that it is a question of the greatest happiness for the greatest number. The number of jug carriers is greater than the number of plastic makers. On a voting basis there is a net gain. But in this universe nothing is free. We pay in one way or another for our gains. We must consider how much we pay, and whether the price is worth it.

The relationship of technology to nature has deeply concerned me. It is, in fact, immensely important to everyone. We often act as if we were not animals and did not have to live in a symbiotic relationship with our surroundings. We behave as though we were not part of the total ecology, as though in some way we were privileged and could throw our weight around *ad libitum*. This, of course, is perfectly untrue. We are part of the natural order and must conform to the rules of that order. Ecology, the youthful science, is beginning to discover this. The term "ecology" was invented by Haeckel not more than seventy-five years ago, so recent is the concept of the interwoven nature of reality and our own close involvement in the pattern. A moral tenet emerges from this study of the relationship of parts within the natural world. We are getting away from the infernal doctrine of the Middle Ages that animals have no soul and therefore may be treated as things. Now it is perceived that even things should not be treated as things. Things should be treated as though they were parts of a living organism. Our natural world is like a vast living organism. Indeed, a great deal of it actually is living. For instance, we now know that a large percentage of soil is composed of living bacteria. Bits cannot be cut out of a living organism without endangering the organism as a whole. We should be intensely aware of this.

We must approach nature with a good deal less bumptiousness than we have in the past. Modesty and absence of arrogance are essential ethical correlates to the enormous powers we have achieved. We can't commit acts of overweening pride against inanimate nature or we will suffer disastrous consequences. This is the Greek idea of hubris. In *The Persians* of Aeschylus, Xerxes is guilty of hubris—not only because he has attacked the Greeks, but because he has done something outrageous to nature. To us his action seems harmless enough—he built a bridge across the Hellespont. But to Aeschylus this seemed an outrage. The realization that nature cannot be recklessly outraged exists in the minds of good applied scientists and good technologists. It deeply affects their thinking. But caution is not a property of the advertisers of technology—popular science magazine writers, for example, who keep talking about conquering nature. The people who profit by technology have been boastful and hubristic. They have had to pipe down when consequences they failed to anticipate have shown up. But these alarming surprises have not cured their hubristic illusion that nature can be dominated.

Man has always thrown his weight around and upset the natural order. Now, with the enormous resources of modern technology he can tilt the balance disastrously. Even when his powers were small he was remarkably successful. It is astonishing how rapidly human beings can destroy their surroundings even in primitive countries. What mankind has made of its environment is a depressing spectacle. In Lebanon, for example, the magnificent mountain terrain was once covered from one end to the other with cedars. Now perhaps 1,200 trees are left. The land has been eroded down to the naked rock. Reforestation is no longer possible. Serious changes have come about in Africa where the desert is spreading, the savannahs getting drier. I remember the terrifying dust storms in Oklahoma in the thirties and again in the late forties. Every time the price of wheat went up more of the top cover went up in dust. The same thoughtlessness can be seen everywhere. It seems very difficult to drive the lesson home. Evidently we have to have a great many tremendous kicks in the pants before we learn anything.

We can only live in symbiosis with nature. If we treat the relationship intelligently we shall benefit. But thinking we can push nature around is absolutely wrong. It is absurd to attempt—to use that dreadful old-fashioned phrase—to conquer nature. We must take care before embarking on our grandiose technological schemes. The natural balance is easily upset, and we quickly get responses we have not anticipated.

Last summer in England I was saddened by changes that were unforeseen side-effects of technological progress. Walking in the countryside I was first struck with the extraordinary fact that there were virtually no butterflies and no cuckoos. I did hear one nightingale, but there were practically no chaffinches. The English hedgerows have been sprayed with weed killers. The weeds have gone and so have the caterpillars. Cuckoos, nightingales, and chaffinches live on caterpillars; caterpillars grow into butterflies. Perhaps this is not disastrous from an economic point of view, but to me it is very sad. These creatures have played an important part in English literature. They belong to the English countryside. Are we not going to see the butterflies any more, not hear the nightingales? Even in Switzerland, in high Alpine meadows where there used to be the most magnificent butterflies there are very few—again because of weed killers.

Another disheartening change had occurred in a valley near where I used to live as a boy. It was a long valley down to the river Wey, with smooth grass sides and chestnut woods at the top on either side. I entered this valley by footpath and I didn't recognize it. The footpath now winds through brush about twenty feet high. I couldn't imagine what had happened, then it suddenly struck me. Later friends confirmed that this was a result of myxomatosis introduced to combat the rabbits. The rabbits had kept down the brush. Their extermination has changed the face of the countryside. The rabbits were mischievous in many ways, but again, what have we gained by their extermination, and what have we lost?

We have rushed into nature's domain as if we knew exactly what we were doing. But we did not. We never could predict what we were going to bring about. The ecological structure is so delicate and so complex that it is impossible to imagine all the effects of a given action.

We don't yet know the consequences for human health of the insecticides being sprayed on all our food plants. A friend of mine in The Conservation Foundation has been collecting information about this and none of it is very reassuring. Nobody knows what the long-range consequences are of artificially immunizing the entire population. In the short run obviously it is a good thing. But if a population is created which genetically has no natural immunity to diseases, the slightest mutation in the cause of the disease means serious trouble.

DDT has spawned an unforeseen problem. Paeans of praise went up when DDT was first made available. Now there are large populations of insects that are not susceptible to the drug. In fact they seem to enjoy it as though it were a delicacy. The same thing has happened with antibiotics. When man tries to eradicate a species with a very rapid reproduction rate, he is up against natural selection and is almost bound to lose the fight. If 1 per cent of a rapidly reproducing species is genetically immune to a drug, it is clear that in a very short time the entire population will be immune.

One of the most serious unexpected consequences of advancing technology has been caused by the reduction of death rates without a balancing reduction of birth rates. Parts of the world are increasing their population at a rate of 3 per cent, even of 4 per cent per annum. The United Nations recently stepped up its estimate of the increase in India and Southeast Asia. Before the recent Indian census it was estimated to be at the rate of 1.7 per cent per annum. It is now known to be about 2.3 per cent per annum. On the basis of 450 million people this means nearly 10 million more people every year.

Last autumn I talked with a number of Indian politicians and officials. They all said the same thing: "We have just started on our third five-year plan. It is an ambitious plan, and will be difficult for us to fulfill even with a good deal of foreign aid. If we do succeed we will create 10 million new industrial jobs. When the plan is fulfilled there will be 15 million young people asking for the jobs. There are already between 50 and 100 million underemployed people with an annual income of less than $100 each who will hardly be touched by the plan." The best image one can make of this situation, it seems to me, is Lewis Carroll's parable in *Through the Looking Glass*. Alice and the Red Queen run at full speed for a very long time, until they are completely out of breath. When they stop, Alice is amazed to see that they are at the same place. She says, "In my country if we ran like this we should have gone a very long way." The Red Queen replies, "Well, yours must be a very slow country. Here we have to run as fast as we can in order to stay in the same place. If we want to get somewhere else we have to run twice as fast as we can." This is a comic statement of the tragic situation in which much of the world now finds itself. A 3 per cent increase doubles the population in twenty-four years, 2.3 per cent doubles it in a little over thirty-two years. How on earth are these people going to keep up?

There is a tendency to discuss the population problem solely in terms of food, as though man lived by bread alone. Certainly he can't live without bread. The solution to the food situation alone is difficult to see. But how is a country without educated personnel, without capital, without technical devices going to produce not only the additional food supply but also the additional housing, roads, schools? Even a rich country like this has trouble keeping up with the school population. India and Southeast Asian countries are desperately poor. Capital is that which remains when needs have been satisfied. In a country where 70 per cent of the people never satisfy their needs there isn't much left over.

The headlong increase in population presses heavily upon our resources. Preventing a mining of the soil and general ruin of the surface of the planet is going to be more difficult as the population increases. Our mineral resources are going to last longer than we thought twenty years ago, but they are not inexhaustible. In forty years America has consumed more than the rest of the world consumed since the Ice Age. Increasing amounts of energy and capital will have to be put into producing raw materials if anything like the present standard of industrial production is to be kept up. We will be forced to make use of poorer and poorer ores, finally getting down to sea water, the poorest ore of all. This won't happen for another one or two hundred years, but it will certainly happen. If there is a general industrialization, resources will be eaten up even faster. When the rest of the world is consuming as fast as America the acceleration in the destruction of raw materials will be enormous.

These tremendous problems have developed because technology accomplished something intrinsically good. They illustrate one of the greatest ethical questions of our time. Every one agrees that good ends do not justify bad means. The tragedy arises when good means result in catastrophic ends. This is a dilemma that requires a whole new type of ethical thinking.

The urgency of the need for answers has been stressed several times during this Conference on the Technological Order. We don't have much time. In fifteen years India will have added about another 150 million people, and China will have increased 255 million. The high expectations of a better life now prevalent in the underdeveloped countries will find themselves dreadfully frustrated. I think it will not take more than fifteen years for the ensuing deep social unrest to break down into various kinds of dictatorship. In this country the fifteen-year period will probably coincide with a maximum consequence of wholesale automation. This will not be easy to adapt to, and will produce peculiar results. For example, if the proletarianized middle executives are going to be pushed down, they may form a potentially dangerous class like the unemployed intellectuals in Eastern countries.

In the fifteen or twenty years before us a great many important decisions must be taken, decisions resulting in immediate action. I don't think disaster is inevitable. We can, if we choose to, use our intelligence and good will. As the Pope remarked the other day, if people would only use their heads and their hearts a little, a lot could be done. But if we don't choose to use our heads and our hearts a lot, we shall be in serious trouble in spite of technology, and in some sense because of it. We cannot retreat from technology. Gandhi's prescription is absurd. If we went back to the spinning wheel, four-fifths of the human race would die in about two years. We must go on. But we have somehow to see that we don't destroy our planet. We must not eat ourselves out of house and home, into misery and ignorance by simply overbreeding. The problems are enormous, and we have just the next few years to meet them.

This leads me finally to the question, what should we demand of letters? What should the attitude of the man of letters be toward problems of science and technology? What should he do?

I feel strongly that the man of letters should be intensely aware of the problems which surround him, of which technological and scientific problems are among the most urgent. It is his business to communicate his awareness and concern. Literature sets up a vision of man which guides people to a better understanding of themselves and their world.

But we are up against a curious paradox. Something of immense importance to all of us does not find expression in the literary arts. The rational side of man, with its scientific and technological expressions, gets little literary space. It is curious that science and technology have always occupied so small a place in literature. What important literary figure, except Diderot, seriously occupied himself with the problems of technology? This is all the more extraordinary when one considers that literature is supposed to hold the mirror up to life. In life people spend a great deal of time involved in the technology of the period in which they live. They work, and their jobs are connected with technology and the organizations technology engenders. Yet one sees little evidence of this in literature.

The explanation is, I think, quite simple. Human beings are interested in a much livelier way in the passions than in reason. Great works of literature, drama, and narrative have always had as their theme the passional and violent aspects of human nature. We see, after all, the same thing in the modern newspaper. "The news" is by definition bad news, something disastrous, unpleasant, something sinful or violent. Ordinary rational behavior is never news. And "news" in this sense has been the subject matter of literature from Aeschylus to Dostoevsky. Human beings, although they reluctantly conform to social usages, do their duty, and earn their living, are considerably more excited by their passional side. Literature has, in consequence, held the mirror up to this side. For example, I recently saw three plays by the most gifted of American dramatists. In the first the hero was eaten by dogs. In the second the hero was eaten by children. In the third (it had a happy ending) the hero was only castrated. These plays aren't mirroring reality. If they reflected the state of a society, that society wouldn't last a week. But this exceptionally violent side of man has more attraction in the sphere of art.

Science and technology are the theoretical and physical embodiments of logic and reason; and reason comes into literature only as that which creates conflict. The battle is between passion and the dictates of reason and sense. Writers have not described at length the rational side of man. It is taken for granted; it is merely the obstacle to passion.

It is difficult to see how a concern with the scientific and technological in life can be introduced into literature. Wordsworth, in *Preface to Lyrical Ballads,* laid it down that the remotest discoveries of the botanist and geologist would become fit subject matter for poetry when and if they should be of direct emotional concern to people. But they have never been of direct emotional concern to most people. As a consequence, they have not got into poetry, except that of Erasmus Darwin, which is not very good poetry.

How can adequate expression be found for these things in terms which shall be moving to people? People are concerned with love and hatred, with the seven deadly sins, the four cardinal virtues and the conflicts in human nature. For the ordinary person a good deal of technology can certainly be described as humdrum. It is extraordinarily difficult, for example, to see how the subject matter of this Conference could be rendered in persuasive literary terms. I wonder how I could find an adequate fictional expression for the problems of the population explosion. These massive events, correlated with technological advance, are peculiarly difficult to put across in the penetrating way characteristic of good literature. Good literature has what may be called an "X-ray" quality that penetrates into the mind.

The man of letters has a real duty to seek powerful means of expressing the nature of technology and the crises it has generated. Perhaps something can be done in the more mundane forms of literature, such as the article or the popularization of science. Many men of letters are so occupied with the lyrical and passional sides of life that this would be out of the question. But for those whose talents are less focused there is an obligation to try to make our world's dangerous and confusing state clear.

Technology As a
System of Exploitation

Scott Buchanan

The current discussion of technology in books and journals, both learned and popular, can be heard as a desperate clamor for a definition of terms. The reader or listener would like to call a moratorium on argument until the authors come to terms with each other and hopefully with their common subject matter, or less hopefully with their separate subject matters. Each new author is tempted to respond to the clamor and legislate clear and distinct definitions. There may be wisdom at the present stage in refusing to yield to the clamor. There have been many definitions in the past that simply do not comprehend the problematic phenomena of the present. There are always attempts to make loose and general definitions that do not penetrate and articulate the problems. There perhaps is need at present of a more patient ruminating discussion that will identify and arrange the materials for a later definition. I would like to offer one strand of such a discussion. Hopefully, it may throw some light on the cause of the confusion that delays definition.

Theories of the Arts

The discussion of technology is very old, and it is almost continuous with the Western intellectual tradition. I suspect

Mr. Buchanan was formerly dean of St. John's College in Annapolis, Maryland. His books have covered a broad range of topics, including Poetry and Mathematics, Symbolic Distance, *and* Essay in Politics. *This paper, presented at the Encyclopaedia Britannica Conference on the Technological Order (March 1962), was published in* Technology and Culture *(Vol. 3, No. 4): 535–43.*

there is a similar continuous discussion in the Oriental tradition. Technology is a Greek word. Unlike many apparently similar scientific terms in modern languages, it is not just stolen from the Greeks and recoined to fit a scientific novelty. It was a part of their discussion of the human arts. It meant the prescription of rules for the arts, and the context for it was a rich, subtle, and technical discussion of all the arts.

Plato is famous for his theory of ideas. He should be equally if not more famous for his technology, his theory of the arts. He is fascinated throughout his dialogues with the origin of the sciences from the arts. The modern notions about technology are elaborations of parts of the ancient discussion torn loose from the ancient context and developed independently because of the novelties of modern industry. I would like to trace very briefly and diagrammatically this separation and development.

The Greek word for art was *techné*, and it signified the power or capacity, the habit or skill, and the intellectual virtue of a man to make a product or an artifact. This formula is a highly condensed and for us oversimplified conclusion of the Greek discussion which continues on through the Roman period in somewhat degraded form and finally reaches its most subtle elaboration in scholastic thought. It belongs to the tradition of rational and moral psychology that we have honored more by neglect than cultivation since the eighteenth century. It emphasizes three distinct things: man the agent-artist, the end product, and the ordering of means by rational rules. A man realizes his natural capacity by acquiring a second nature in habits which are ordered rationally to the ends in the things he makes. Art is an intellectual virtue whose function it is to deal with contingent and empirical things by reasoned opinions or rules.

The Greek discussion begins and develops in a context wider than this compact humanism. The preceding analytic formula applies clearly to what the ancients called the intellectual or liberal arts and to what we call the fine arts, but it also applies to the useful arts or crafts, those arts that make useful things. The habits and skills of the formula are involved in the selection and acquisition of natural materials, what we would call raw materials, and in transforming them into the product.

It would seem that the scientific distinction between matter and form had its origin in the analysis of the arts. Matter is that which is fitted to receive forms which exist first in the artist's mind. The artist's or artisan's act is essentially putting such a form on this malleable matter. The means by which this is done are first the hands, but soon the extension of hands in tools or elementary machines. The typical example of the artistic act is the imprinting of the seal on wax, then the fashioning of the statue out of marble by the chisel, and then the making of a chair according to a pattern. These archetypes are easily extended to agriculture, carpentry, cooking, and building, some with many tools, and an order of applications, and some subordinated to others as wood-working to flute-making to flute-playing. There is also the art of tool-making to serve the arts which use the tools.

The tool or the hand operates between the form in the artist's mind and the raw material, but form also demands the training of the human capacity in skills that are learned. Some of these skills seem to be learned by rules that can be taught, some only by repetitive practice, some by the maturation of instinct, some apparently by inspiration, some by close attention to the material and its fitness to the end product.

This matter of skills and learning was the mysterious subject matter of great fascination and thought for the Greeks, and it led to one of their more familiar contributions to the tradition. They concluded that art imitates nature, that nature is the great teacher of men. This conclusion came at the end of a long period in which they believed wonderingly in the many legends in which the gods were said to have taught men the arts and still presided over the practice. All the wonder was finally precipitated in the doctrine of the four causes, first applied to the arts and then to nature. The causes of any end product were first in the matter which could be fashioned or formed; then in the form in the artist's mind which could be impressed on the matter; third in the hands, tools, skills, and energies of the agent-artist; and finally in the end itself, the product, which ruled over all the other causes in the actual making. These were the four causes, material, formal, efficient, and final.

The recognition that the artist was imitating nature almost forced a reversal of the insight. The four causes also operate in nature so that their discovery and formulation make science. Art or technology thus becomes the midwife of science. If you want to understand something, make a similar object or artifact; then impute that artistic process of making to nature. We are not far from this in our current use of models in science.

There is no doubt that the Greeks used the analogy of art to penetrate nature and its secrets. Travelers brought back technical lore from Egypt which they turned into geometry. This is apparently the origin of the Pythagorean development of number theory and geometry. The three, four, five triangle became the universal device for squaring the corners of buildings and for surveying lots of land property. It also led to the proof of the Pythagorean theorem in geometry.

This fragment of mathematics in its two aspects, applied and theoretical, precipitated a crisis in the Pythagorean society. The theorem proved too much—that any right triangle would have the sum of the squares of its legs equal to the square of the hypotenuse. In many right triangles this would mean that the hypotenuse was incommensurable with the legs. This would not have worried a builder, or an engineer. He could work with rough instruments and approximations. It would worry an incipient mathematician who had glimpsed the possibility of theoretical understanding. It is said that a member of the Pythagorean brotherhood who divulged the secret scandal about incommensurables was deliberately drowned at sea. The Pythagorean theorem had to be kept a professional secret until the problem of the diagonal of the square and the square root of two could be solved. Of course the problem was later solved, supposedly by Eudoxus who was a member of the Platonic Academy, and the solution appears in Euclid's *Elements* as highly sophisticated arithmetic and geometry. But the crisis in the Pythagorean society has been repeated many times in the history of technology and science. It is the archetype of all those strains that exist between the technologist and the scientist, in which professional secrets play such varied roles.

Organization of the Arts in the Past

The arts organize themselves into crafts and guilds of artisans and technicians. They pass on the skills to their apprentices, they improve the arts, and they tend to have trade secrets. They also tend to generate and maintain theories that add understanding to skill. The so-called Hippocratic writings can be read with greater insight and penetration if they are understood as either exoteric or esoteric attempts to come to terms with two aspects of medicine—the arts of diagnosis, prognosis, and therapy on the one hand, and the sciences of anatomy and physiology on the other. Sometimes the writer is persuading prospective patients of the competence of the physician. Sometimes he is trying to persuade his professional colleagues of the truth of his theories. Galen later struggles through with the help of Plato and Aristotle to the establishment of theoretical medicine and the self-recognition of the profession. Primitive as Galenic medicine may seem at present, and corrupt as it may have been at certain stages, it was the only medicine practiced for fifteen hundred years. It set the standards for its own profession and for the other professions of law, theology, and teaching up to our time, when it seems that any sense of profession except its economic aspects of monopoly is evanescent.

The Pythagorean society, as far as we know it, seems to be the preprofessional prototype of what professions came to be. Its chief theoretical holdings seem to have been mathematical at a time when mathematics almost alone led the pursuit of theoretical knowledge. The word mathematics originally meant learning and things learned. The society was a novel kind of cult which generated and transmitted learning. This function was chiefly performed by the esoteric part of the brotherhood. It taught a somewhat vulgarized version of its knowledge to the exoteric members of the society which seemed to have made up the citizenry of a city, Crotona. The vulgarized learning was ethical and political in nature. The society was responsible for what it decided to investigate and what it decided to teach. It added to its membership and transmitted its learning from one generation to another. It is quite clear that the Hippocratic oath for medicine formalizes an imitation and adaptation of the Pythagorean learning to a more specialized body of learning. In both cases there was a tendency to unify a body of knowledge and to respect theory as magisterial and architectonic for a self-governing community of scholars, teachers, and practitioners of an art or body of arts. The secrecy of the cult is a sign at once of the lack of and the need for some formal institutional structure which the social environment of the time did not supply or recognize.

Connected with the development of certain arts into professions by the development of appropriate bodies of theoretical knowledge is a medieval distinction between the kinds of arts that was not emphasized explicitly by the Greeks. Certain arts which had reached a professional status were practiced on human beings, who also had artistic capacities. In these human subjects of the arts there were primary natural and secondary natural processes which if left to themselves might accomplish their ends, but if aided by the professional would accomplish their ends more easily and more fully. Medicine and teaching were the frequently discussed examples of such arts. They were called cooperative arts because they were understood to be cooperating with rational natures.

But the term cooperative suggests that there might be another kind of distinction among the arts, those that can be practiced by the individual without needing aid and those that cannot be practiced without the cooperation of other artists. Building or architecture is a borderline case since presumably single men have often built their own houses and barns. But there are building operations that are unlikely to be undertaken and completed without the collaboration of many artists, such as the building of temples like the Parthenon or the cathedrals, which in many cases occupied several generations of many men. The brotherhood of free masonry was a society with secrets and standards of skill that recognized this kind of organized or collective art as its responsibility. Similarly, the building of harbors, aqueducts, and roads calls for the orderly organization of many men and many arts.

There have been many kinds of organization that have met the need for order in the divisions of labor. Typically the Romans allocated this organizing function to the army for their greater public works, as the Greeks and Egyptians had trained and managed slaves. Self-governing guilds have often passed through their technological periods on their way from religious cults to polities. We tend to forget this ubiquitousness of organization in the arts since Adam Smith established the science of economics on the thesis that the market both dictates the division of labor and gives a semblance of organization to it. We forget that he also was giving the first description of the organized factory. Although he warned against the organization of the arts by the corporation, he actually laid the ground for its modern predominance. The market now appears to be a rather long, lively, and fruitful interim of disorganization of the much longer continuum of highly organized technology.

Present Organization of the Arts

I know no competent discussion of the question that this series of organizations of the arts seems to pose. There is the often quoted maxim of Marx and Engels that the means of production determine social and even cultural forms.

Neither they nor anybody else has worked out the dialectical or other details of such determination. The question I am posing is somewhat simpler. What are the forms of organization that the arts demand for themselves? The Egyptians seem to say a slave system; the Greeks at one stage say a polity; the Romans say an army; the Middle Ages and many other ages say guilds; the eighteenth and nineteenth centuries seem to say the market; we seem to be saying now that the corporation is the answer. These seem to be little more than historical correlations. They do not answer the question that is now inescapable. We are very far and rapidly moving farther away from the individual craftsman who needs only to pay attention to his own art. The present organization of the arts, what we call our technological system, has passed beyond the powers of any of the preceding forms of organization. How shall we bring it into order and effective service to the human community?

Bright-eyed observers tell us that the answer is already given and that we are more than half-way accepting it in practice. The human arts will be completely built into automatic machines. Human beings are in principle in the position of the Pharaohs, freed from labor. Technology and art which can no longer be turned over to slaves can be given to machines. Peter Drucker tells us that we unknowingly have trained ourselves to behave so much like machines that the substitution of machines for men is easy. During the final stages of automation we shall have to increase our engineering skills and scientific knowledge in order to finish the job in style and keep the apparatus in condition, but our main business will be those activities that free men have traditionally called leisure. Whether this be good prediction or prescription or not, we need to know a little more about how and why the possibility of automation confronts us.

The so-called Industrial Revolution has had many dimensions and causes. On the technological level the substitution of machines for tools seems to have been the essential change. The revolution has been the progressive passage from "manufacture" to "machino-facture." This process can be understood as the extension and penetration of the principle of the division of labor into technics. As labor was broken down into elementary crafts and integrated by exchange in the market, so complex operations in a given craft were broken down and reassembled—sometimes with mechanical operations substituted for manual operations, sometimes with the multiplication and integration of many more operations than the original craftsman could manage. Usually this organization of operations was connected with the new prime movers—steam and gas engines, motors and generators—to supply the greater demand for energy. The factory was the progressive institution that could push the analysis of the jobs and organize both men and machines into larger patterns of organized operations.

Two parallel transformations can be seen in this factory system. The concatenation of machines and the use of the new prime movers progressively removed the human agent-artist from the linkages between the machines. The products of the separate arts or crafts were no longer the ends of the operations but rather materials for stages in the overall production process. To be sure there was an integrated end-product, but the many intermediate ends became mere means.

The consequent or parallel development would be what Europeans call rationalization, the formulation of the rules that control the productive process. In the crafts a great deal could be left to unverbalized and unmeasured skills. After fine analysis of jobs their reintegration needs both elaborate verbal and mathematical expression, not merely to pass on the crafts from one generation to another as in the past, but now for the coordination and management of the going technical process. This amounts to an emphatic return to the original ancient meaning of *technologia,* the giving of rules to the arts. It also connects with the development of the new techniques in analytical mathematics or algebra.

The invention and establishment of algebra as the art of mathematical analysis blots out many of the distinctions that older mathematics had maintained. For instance, arithmetic and geometry are no longer distinct. So when the factory is rationalized, it is hard to maintain any distinction between men and machines. They both are values in equations of efficiency. This suggests two alternative possibilities which have worried us for a long time. Shall machines be taught to think, as in computers; or shall men become robots? Automation offers its solution of the dilemma. It will teach machines to think and also to do the work, so that emancipated man may occupy himself with the liberal arts, politics, the fine arts, and the divine arts . . . until such time as men wish to automate these arts as well and free themselves for still higher existence.

What of the Future?

We are here on the edge of something eerie and not a little puzzling to the pragmatic mind. Let's see if we can establish our position, as the navigator says, and maybe plot a course or two. The kind of abstraction that makes us see automation as an all-encompassing net is like the thinking that the Pythagoreans did when they first happened upon the tricks of measurement. The story goes that they did this with the monochord, a string that vibrated in a musical tone according to its length—the shorter the length, the higher the tone, and the lengths could be expressed in whole numbers or integers. The Pythagorean mind quickly generalized, thinking of many things that could be reduced to numbers, and quickly concluded in a kind of arithmetical astrology that all things are numbers or shadows of numbers. It is hard for even a modern pragmatic mind to deny what is after all the basis for the great successes of mathematical physics. The Pythagoreans identified this process of measurement with music, and in celebration of their discovery made poetry about the music of the spheres, not failing to continue their development of the science of mathematics even to our day.

Plato took the Pythagorean theme as commonplace, and went on to develop a theory of ideas which had the same trick of abstraction and hypostasis in it, but included along with numbers many other kinds of ideas. It is a childish kind of Platonist that turns the trick into metaphysics, as Plato says in the great confrontation of the young Socrates with the venerable Parmenides. But Plato recognized the moment in the long life of dialectic as important and even essential to the intellectual enterprise. The theory of ideas, reformulated as the Logos doctrine, has had as long a life as the Pythagorean doctrine. Its development has been somewhat more versatile, serving law as natural law, serving theology as a person of the Trinity, supplying common sense with its root in reason.

As I suggested earlier, Plato might with his skill in abstraction and hypostasis have "found" a system of the arts, a set of rules for the arts, that would have reflected as in a mirror the firmament of his ideas, the ideas of his theory of ideas. There is a suspicion in his allegory of the cave that there was a level of artificial objects which made a kind of logistic pattern midway between the shadows on the wall and the ideal stars in the heavenly firmament. If he had been faced with the panoply of artificial technical operations, processes, and products among which we live, he surely would have been led to construct something like the technical phenomenon that we find in Jacques Ellul's *La Technique.*

There is a great difference between Anglo-American and continental European thought about technology. The German and French philosophical traditions are more hospitable to many-storeyed imagination and speculation. Whatever we may think of Hegel and Marx as artificers of ideologies, their speculative skill and boldness, and their recognition of levels of being and action are not to be ignored or condemned. So French thought has always had a clear strong line of analytical thought established and controlled by the Cartesian discovery or invention of algebra or analytic geometry. Descartes generalized his discovery and hypostatized it in his own style. He called it the Great Art, *Ars Magna,* and thought of it as the universal method for the intellectual enterprise. He saw algebra, or analytic mathematics, as the master architectonic art of all the arts.

The essential power of all these abstractions and hypostases resides in their objective universality. They give clarity and logical mobility to speculation or theory. This kind of thinking gave us the great hypothetical worlds of modern physics: the systems of the worlds of gravitation, heat, and electricity within which our more empirical investigations still explore. So the idea of technology or of the technical phenomenon of Ellul offers us the schema for a logistic of the arts, a genuine theory of technology.

Suppose we accept some such abstraction, hypostatization, and universalizing of technology. What shall we make of it intellectually and practically? There seem to be three possibilities with some variation in each of the three. We may accept the technical logistic as a system of determinism, a kind of artificial fate which we have brought upon ourselves or in which we have been trapped. A superficial reading of Ellul tempts one to this conclusion. But Ellul himself warns us that this is not his intention, unrelenting as his method may seem.

We may see it as a system that has developed piecemeal without our intention except with regard to its slowly developing parts. A key to this view seems to be hidden in Ellul's many comments that the system has reduced all things, including our ends and purposes, to means. In other language this means that we have developed an unlimited, autonomous, universal system of exploitation. Both the things in nature and in human nature come under the sweep of unrelenting exploitation. We have been pious in the later stages about exempting men from the exploitation; they must be conceived as ends in themselves. But the only assurance that this is still respected is that we have kept free contracts and the democratic processes. The irony of this may be that we have supposed that men themselves will not knowingly submit to exploitation. There is much evidence to the contrary. We may have at least temporarily been enchanted to submit our bodies and souls to the contingent processes of self-reduction to means. This comes close to my understanding of the alarm that Ellul is sounding. He is trying to wake the prisoners in Plato's cave and incite them to throw off the chains of their empirical piecemeal thinking about themselves.

If this is a good guess, the wide-awake conclusion might be that we should accept the logistic clarities and necessities in the technical phenomenon and do a thorough job of searching out the articulations in it, of looking to the shape of the integrated system, with a view to remaking by disassembly and reassembly the most familiar and least understood phenomena of our time.

It may be recalled that this abstraction of technics from the arts, the rules from the actual makings, leaves the agent-artist and his ends out. For the sake of organization they are assimilated as merely means. Thus we are left with an apparent automaton, an almost integrated and automated technical system. But it is merely a stage in development and dialectical understanding. If we are to deal with it competently, we must find some way of bringing back human artists and human ends into the powerful order that the technological system presents to our amazed and puzzled view. My suggestion is that we trace the human role that the system involves and detect where we have surrendered our judgments and our wills. If we can find these points of default, we may be able to recover our truly scientific understandings, our objective knowledge of our ends and the ends of nature, and our individual and common wills. This might give us back our reverence and love of nature as well as our shrewd ingenuities in exploiting it.

The Preparation of Leather and Parchment by the Dead Sea Scrolls Community

J. B. Poole and R. Reed

The Dead Sea Scrolls Community

The first discovery of manuscripts in the Dead Sea area may have occurred as early as the third century of our era, when Hebrew and Greek books were said to have been found in a jar[1]; but it was in 1947, or possibly in 1945, that the documents now almost universally called the "Dead Sea Scrolls" were found near the Wady Qumran, at the eastern edge of the Judaean plateau, about 1¼ miles from the northwest corner of the Dead Sea. The story of the discovery of the Cave 1 documents and of the intrigues surrounding their purchase will scarcely bear a further repetition. Since the first find, eleven caves near the Wady Qumran

This article attempts to reconstruct the leather and parchment technology of the people who owned the Dead Sea Scrolls and to relate this to their beliefs and practices. It is an outgrowth of research performed at the Department of Leather Industries (now the Proctor Department of Food and Leather Science) of the University of Leeds, the aim of which was to clarify the nature, origin, techniques of manufacture, and the age of a number of blank scroll fragments provided by the Jordanian Department of Antiquities.

Dr. Poole received his doctorate from Leeds for his research on the Dead Sea Scrolls and is now with a London firm of industrial consultants; Dr. Reed, of the University of Leeds, has published numerous papers dealing with the structure and chemistry of skin and other forms of connective tissue and has twice served as secretary of the British Electron Microscopy Group.

This article was published in the Winter 1962 issue of Technology and Culture *(Vol. 3, No. 1): 1–26. It has been slightly condensed for publication in this anthology.*

143

have yielded important manuscript collections, and less significant finds have come from about thirty other caves. The physical closeness of all these caves strongly indicates that the deposition of their contents cannot be ascribed to coincidence; thus the term "Dead Sea Scrolls" is now applied to all these documents.

Up to 1958, over five hundred manuscripts had been recognized among the scrolls, nearly all in an incomplete state.[2] They comprise all the Old Testament books (with the exception of Esther), apocryphal and pseudepigraphic material, and a miscellany of other works, such as disciplinary codes, commentaries on the prophets, and psalteries.

Near to Cave 1 are the ruins of Khirbet Qumran; these had attracted little attention before the discovery of the scrolls, but after that event it was suspected that a connection might exist between this site and the scrolls. Hence excavations were undertaken,[3] the results of which clearly show that Qumran was a self-contained community center before its final destruction in 68 A.D. The water conservation system, drawing its supply from an aqueduct stretching to the Judaean hills, was elaborate in the extreme; in addition to a pottery, a bakery, a laundry, there were storerooms and stables, together with industrial facilities whose functions have not yet been elucidated. The spiritual life of the community is believed to have centered around a long assembly room, orientated toward Jerusalem. Few of the rooms within the buildings appear to have been monastic cells; the community is thought to have lived in tents in a region extending two miles north, and the same distance south, of Qumran.[4]

The connection between the scrolls and Qumran is not based solely on physical proximity. Thus twenty-five caves yielded pottery of similar paste, form, and decoration, which had almost certainly a common origin in the Qumran pottery[5]; inkwells found at Qumran contained ink similar to that used for writing some of the scrolls, whilst inscribed potsherds from the ruins (where no scrolls were found) possess the same Hebrew script as that on some of the scrolls. Perhaps the most striking link is that there is considerable agreement between the archaeological dating of the site and the palaeographic dating of the scrolls. The earliest community level[6] has been dated in the second half of the second century B.C. and the last at 68 A.D.,[7] whilst the scrolls have been assigned to the period 300 B.C. to 70 A.D.[8] Moreover, it seems significant that the earliest sectarian manuscripts cannot be dated *before* the second half of the second century B.C.,[9] i.e., the date of the oldest community level at Khirbet Qumran.

Such a close association between the scrolls and Khirbet Qumran is now generally assumed to be well grounded, but the contrary argument has been forcefully stated by Del Medico,[10] who claims that the caves are *genizoth,* or storerooms for documents and writings which were either no longer fit for use or were canonically unacceptable. Such documents could not be destroyed but had to be left to decay.[11] Del Medico goes further and denies that any community could have lived at Qumran because of the air pressure and the aridity of the terrain, a statement contrary to the experience of Masterman:[12]

> The whole of these ruins [Khirbet Qumran] stand on a commanding position, surrounded on all sides, and especially to the south, by steep declivities; at one point at the north-west corner, however, a narrow neck connects it with the plateau to the west. From this site every part of the 'Ain Feshkhah oasis and all its approaches can be overlooked; it is, also, a fresher, healthier situation than any spot in the plain below. I found a fresh breeze there when on all the lower ground it was hot and still. The site is just such a one as would have been chosen in, say, Roman times to protect the springs and the road passing through the district to the south. . . .

One hypothesis regarding the relationship between the Qumran community and the scroll caves, and which has much to commend it, insofar as it would in some measure reconcile the current majority view with that of Del Medico, does not appear to have been fully considered.[13] This is to regard the scroll caves, not as *genizoth* in the normal sense, but as a long-term plan of storing the written treasures of the Jewish people in a safe situation, remote from the more active centers of political upheaval. Such a plan may have been conceived well before the fall of the Temple of Jerusalem in 70 A.D., whilst Qumran might have been chosen because its community had already demonstrated that it was quite practicable both to exist and to practice its religion, relatively undisturbed by the outside world. Thus with the growing uncertainty in Jerusalem, Qumran may have assumed importance as the repository of the treasures of the Jewish faith and hence the means of safeguarding for posterity the cultural heritage of Judaism. This brings us directly to the question of the place occupied by the Qumran community in contemporary Judaism. That the community was Jewish in character is evident from the content of the Dead Sea Scrolls and from the large number of biblical manuscripts found amongst them. Again, such indications of practical Judaism as *tefillin*[14] were found, whilst a further clue is provided by the alternative name for Khirbet Qumran—Khirbet el Yehud;[15] this latter point seems most significant since no evidence of prolonged occupation has been found datable after the residence of the community had been brought to an end and which is clearly Jewish.

Present-day Judaism is the spiritual heir of the Pharisaic element cotemporaneous with the Qumran community. Two years after the destruction of the buildings of Qumran and following the fall of Jerusalem in 70 A.D., the whole pattern of Judaism was violently and permanently changed, leaving only the Pharisees as the survivors. Thus the survival of Judaism as a distinct cultural entity in the modern world owes much to the Pharisees. The view most commonly advanced is that the Qumran community was composed of Essenes, one of the Jewish sects (the others being the Pharisees and the Sadducees) said by Josephus to have flourished about the middle of the second century B.C.[16] Josephus, Philo, and Pliny are our primary sources of information regarding the Essenes; a number of secondary and later sources are somewhat repetitious of their accounts.[17] It cannot be denied that with the publication of the scrolls and the excavation of Qumran, the identification of the community with the Essenes has much to commend it. The similarities in beliefs and practices are many and cover, for instance, such matters as the admission of new members, organization, and the ownership of property. Moreover, Qumran is geographically situated in the region where the Essenes are said to have dwelt. There are, however, apparent differences between the two societies, for example, with regard to pacifism and to sacrifices,[18] but these may well be due to evolutionary trends within the community, as is evidenced by comparing the Zadokite Document with the Manual of Discipline.[19] The latter shows a greatly enhanced organizational rigidity when compared with the former.[20]

Since the Essenes are referred to at some length by Josephus, Philo, and Pliny, it is curious that they are almost absent from Jewish and Christian writings, whereas there is no shortage of references to the Pharisees, notwithstanding the fact that the two societies appear to have been of comparable size.[21] Some part of the explanation of this silence concerning the Essenes may arise from the possibility that the Essenes and the Pharisees had a common origin in the Hasidim, the pietistic group which arose in the first half of the second century B.C. to combat the Hellenization of Judaism under the Seleucids. This tendency initiated the Maccabaean revolt. After this successful revolt, the Hasmoneans took over both the monarchic and high-priestly functions, which eventually led to a severance of the already strained relations between the Hasmoneans and the Hasidim. From this time, the second half of the second century B.C., is probably to be dated the schism within the Hasidim which gave rise to the Essenes and the Pharisees.[22] It will be recalled that this period coincides with the dating assigned to the oldest community levels at Qumran.

The differences between the Essenes and the Pharisees were not ones of principle, but rather concerned the methods by which their aims should be realized; however, the Essenes appear as the more extreme party, many of them withdrawing from society in pursuit of their ideals.[23] Schürer concluded: "If Essenism in general can be regarded as a purely Jewish formation, it is certainly most simple to view it as the climax of the Pharisaic tendency."[24] Nevertheless, in outward practices, the resemblances between the Essenes and the Pharisees are quite striking, e.g., in the rules for the novitiate, communal organization, assembly, and ritualistic lustrations.[25] These similarities have been offered as explanation for the paucity of references to the Essenes, the cleavage between the two groups only appearing when regarded from some distance in time and from the standpoint of an alien culture.[26] Such an explanation appears feasible when it is borne in mind that within Pharisaism itself there was a considerable diversity of opinion (the distorted view presented in the New Testament arose from the fact that the formalistic followers of Shammai were in the ascendancy at the time of Christ); R. Nathan (*floruit* second century A.D.) said that there were eight kinds of Pharisees, the Essenes being the celibate ones.[27]

Some writers have considered the Qumran community as an example of the Pharisaic *haburah,* a society in which there was considerable sympathy between members regarding the observance of the ritual laws of everyday life.[28] Lieberman believes[29] that the differences between the *haburah* and the Qumran community are no greater than those which existed within the body of the Pharisees, al-

though the latter were usually less strict in their practices—
a view in accordance with that of Rabin.[30] Rabin's view,
however, is difficult to reconcile with the archaeological
evidence, as Baumgarten[31] points out; nevertheless, this
author accepts Rabin's major tenet that the Dead Sea
Scrolls reflect a rigorous Pharisaism and he shows that to
consider the Essenes as hyper-Pharisees, as, for instance,
Schürer did, is to lend support to the identification of the
Qumran community as Essenes, since the deviations from
orthodox Pharisaism in the scrolls are consistently in a
hyper-Pharisaic direction.[32]

Though the above account is of necessity simplified, it
seems certain that the Qumran community consisted of the
more strict elements of the Pharisaic sect.

The Nature and Growth of the Rabbinic Literature

In the preceding section we have discussed the Qumran
community and its probable position in contemporary
Judaism. We have emphasized that it is owing to the Pharisees that Judaism survived the destruction in 70 A.D., for
they were able to transmit to the people a living faith, distinct from the "establishment" ideology of the Sadducees,
and structured with a comprehensive set of teachings and
regulations whose authority resided ultimately in the Mosaic Pentateuch itself.

Here we are not concerned with the moral and ethical
teachings of which the Pharisees were the guardians, but
only with those other teachings whose scope extended to
cover every facet of everyday life, including the preparation of leather and of parchment. The exposition of the
Written Law or *Torah* (i.e., the Pentateuch) had been carried on from the earliest times; however, it was Ezra, in
the fifth century B.C., who elevated the exposition itself to
an importance comparable with the *Torah* in which it was
founded.[33] The growth of this material is analogous to that
of English common law: the primal laws upon which the
later rulings are based are not rendered nugatory in consequence of the enactment of the latter, but rather the whole
corpus of the law should be seen as a growing entity.

Two expositional methods were used, which gave rise to
the Oral Law, as distinct from the Written Law. The first
method was to teach the Oral Law as an exposition of a
Biblical text, which gave rise either to a *Midrash Halachah*
(legal teaching) or a *Midrash Aggadah* (ethical or devo-
tional teaching). The second method was to teach the Oral

Law independently of a Biblical text; this was the *Mishnah* method. Although the *Midrash* method was the older of the two, the *Mishnah* method gradually became more important as a large number of traditional practices became established without possessing a prior Mosaic basis and because the Sadducees rejected the Pharisaic interpretations of the Written Law. Up to the destruction of the Temple in 70 A.D., the Oral Law had been truly oral in order that it should not be confused with the Written Law, but by that time its volume had grown too great for a single memory to assimilate it, and varying interpretations were appearing.[34] This body of teachings was first effectively codified by Rabbi Judah the Prince, at the end of the second century A.D., to give the work known as the "Mishnah." Rabbinic schools, themselves tracing their origins to the Pharisees, developed in Babylonia as well as in Palestine and for these, R. Judah's "Mishnah" itself became a subject for exegetical study together with other, similar works containing teachings which R. Judah had excluded. The results of these labors produced the "Gemara" (literally, "completion"): R. Judah's Mishnah and the Gemara—the second being a commentary upon the first—constitute the "Talmud." Of the two redactions of the Talmud, the Jerusalem version dates from about the middle of the fourth century A.D. and the Babylonian, some three times the size of the former, from the fifth century.[35]

During medieval times much effort was expended by Jewish scholars in compiling compendia of the Oral Law, the most well-known of which is the "Mishneh Torah" (second Torah) of Moses ben Maimon (Maimonides, 1135–1204). This work was envisaged as a complete legal system giving the final decision in the case of each law, together with customs and explanations from the time of Moses until the editing of the "Talmudim," even though some of this material was no longer relevant to his own times, e.g., the laws relating to the Jerusalem Temple.[36]

It has been necessary to give this outline of the growth of the Rabbinic literature to explain, in part, the assumptions we shall make in the final section of our attempt to reconstruct the technology used by the Qumran community in making leather and parchment. We have seen that there is good reason for believing that this community was hyper-Pharisaic or Essenic in character and that they would probably believe most strongly in the validity of the Oral Law. It follows that the older material, where we can detect this, preserved in the "Mishnah," "Talmudim," and "Mishneh

Torah" serves to supplement our knowledge of the practices of the Qumran community derived from the Dead Sea Scrolls. In what follows we have drawn particularly upon Maimonides as a source of information. From the brief survey we have made during our work of the rabbinic literature in translation, it appears that Maimonides put much of the Oral Law into a written form for the first time; as far as this relates to skin technology, it is established that the regulations can be dated at least to Tannaitic times (i.e., to the first and second centuries A.D.) and probably to Biblical days.[37]

A Reconstructed Technology

Much Jewish culture and technology was, no doubt, adopted from neighboring countries such as Egypt and Mesopotamia, but here we attempt to reconstruct the crafts of leather and parchment making solely from Jewish sources. The Old Testament is wholly lacking in details of preparative methods for leather and parchment, although we infer from the references to skin and leather articles that its authors were not destitute of such knowledge. For instance, skins were used for clothing (Lev. xiii, 48; Num. xxxi, 20; 2 Kings i, 8; Mark i, 6), tent coverings (Ex. xxvi, 14), bottles (Judith x, 5), sandals (Ezek. xvi, 10; seal or porpoise skin being used, not badger as in the King James version), and shields (Is. xxi, 5).

In the Jewish attitude toward leather making and associated trades reflected in the rabbinic writings, an apparent contradiction is obvious. The preparation of animal skins for parchment is an honorable calling, presumably because the material was frequently used for the holy books, but the work of the tanner was despised. Farrar has pointed out that the New Testament statement of Peter's living with a tanner (Acts ix, 43) is an indication of how far, by aligning himself with the Christians, he had moved away from Jewish orthodoxy.[38] In the rabbinic writings explicit condemnation of the tanner's work may be found:

> It was taught: Rabbi said: No craft can disappear from the world—happy is he who sees his parents in a superior craft and woe unto him who sees his parents in a mean craft. The world cannot exist without a perfume-maker or a tanner—happy is he whose craft is that of a perfume-maker and woe unto him who is a tanner by trade.[39]

This condemnatory attitude clearly arises because of the materials used by the tanner. The handling of dead animals would confer on the laborer considerable ritual uncleanness[40]; similarly, the use of dung in the treatment of skins would be viewed with antipathy by the orthodox. The attitude of the Essenes in this respect, as described by Josephus,[41] was not exceptional; new members were given spades with which to bury excrement, as required by the Mosaic law (Deut. xxiii, 12–14).[42] In the Dead Sea Scrolls[43] we find a comparable outlook.

> When wood, stone or dust is contaminated by human uncleanness, the degree of contamination is to be determined by the rules governing that particular form of uncleanness; and it is by this standard that all contact with them is to be gauged.

> When a dead body lies in a house, every utensil—even a nail or a peg in the wall—is to be regarded as defiled, just as much as implements of work.

The smells associated with tanneries were undoubtedly responsible for the regulations controlling their siting.[44]

> Carcases, graves and tanneries may not remain within a space of 50 cubits from the town. A tannery may be set up only on the east side of the town. R. Akiba says: It may be set up on any side but the west, but it may not be within a distance of 50 cubits.

The prevailing wind in Palestine is from the northwest; 50 cubits are about 22 yards. It should be noted also that a synagogue could not be sold for use as a tannery, bathhouse, immersion pool, or urinal.[45]

At this point the relevance of the foregoing to the Qumran community may be examined. A number of lines of evidence suggest that the treatment of skins to produce either leather or parchment was not carried out in the main community buildings: in neither of the two "industrial" quarters has a tannery been recognized, although many of the constituent rooms have pits, vats, or cobbled floors (suggesting that wet work was carried out there); no deposits of organic matter have been detected; the elaborate conservation system indicates that water was a valuable commodity and it seems unlikely that much of it could have been spared for tanning purposes; in any case, the character of the community would appear to have been too strict to permit the practice. However, there is clear evidence that the community possessed a number of subsidiary establishments in the vicinity of Qumran. These served functional rather than residential ends, and probably helped in preserving the self-reliant nature of the community. The most important site is near the Feshkha spring, about two miles south of Qumran; here there were stalls for animals, barns or sheds, and an intricate, interconnected system of pits or tanks, which de Vaux[46] has tentatively identified as a tannery. This author has described and illustrated the findings with his customary thoroughness, and it is unnecessary to repeat his account in detail. Zeuner has suggested that the pits may have been used for fish breeding, in order to augment the food supply of the community.[47]

In our own examination of a number of samples from the 'Ain Feshkha installation, we determined the chemical elements present and compared them with those of some scroll fragments from Qumran Cave 4. We were, however, unable to establish any link between the two sites. Moreover, no traces of vegetable tanning materials or of animal debris indicative of skin processing were found. Thus it would appear that the installation at 'Ain Feshkha was not used for the preparation of animal skins. Even so, the overall appearance of the site as revealed by the archaeologists is strongly suggestive of a tannery, especially since close to the elaborately linked pits, and unmistakably associated with them, are two stone beams which may have been used in the manual unhairing of skins.

Turning to the preparative methods which were anciently used in Palestine, we find that primitive unhairing of the flayed hides and skins was carried out by treading them, either in the public thoroughfare or by using them as doormats[48]; this inefficient process later gave way to one in which the wet skins were struck by sticks, as in ancient Greek practice.[49] Alternatively, unhairing was achieved by soaking the skins in water and allowing some enzymatic action to take place[50]; Krauss has suggested that the New Testament statement that Simon's tannery was by the sea may be explained by the need for water to carry out this soaking.[51] Some skins were also unhaired with excrement, which would encourage the development of hair-loosening enzymes, as is apparent from the following passage: "It is forbidden to read the *Shema*[52] where one is facing ordures of human beings, dogs or swine, even if hides are immersed in them (for tanning) or any other filth that emits an odour as foul as these."[53] It should be noted also that the more affluent tanners were distinguished from their poorer brethren by the fact that the latter themselves collected the dung necessary for their work, whereas the former did not.[54]

Vegetable materials also appear to have been used as unhairing agents, e.g., mulberry leaves or bryony.[55] When steeped in water, such materials give rise to enzyme systems which are particularly effective in loosening the hair from the skin and also in preparing the latter for tannage or other subsequent processes. Perhaps this Jewish practice is based on earlier Egyptian methods, for it is known that skins, after the removal of adhering flesh, were treated on the flesh (under) side with an aqueous infusion of the pounded stalks of *Periploca secamone*.[56] (This particular plant, however, is not native to Palestine.[57])

Krauss assumes that the salt with which skins are often described as being treated, would probably be mixed with alum, and would thus give rise to what is known as alum-tawed leather.[58] If alum tawing was indeed carried out, the presence of salt would be necessary, but it seems more likely that the salt was employed as a curing agent, the function of which would be to halt the degenerative changes which normally occur in flayed skin.

After salting, a treatment with meal or flour followed as a "leavening" process. The fermentation of such materials not only brought about unhairing, but also swelled or "raised" the skins and prepared them for subsequent processing by bringing their constituent fibers into a clean and open condition.[59] It may be noted that fermenting cereals and vegetable systems, such as mulberry and other leaves, are akin in that they give rise to acid solutions containing enzymes, known as carbohydrases or mucopolysaccharidases. Such enzymes have a degradative action on the mucous cement-like materials, which envelop the protein fibers of skin. Krauss considered that the use of liquors based on fermenting vegetable matter would, in addition to unhairing them, also *invariably* bring about some degree of simultaneous tannage.[60] Such is not the case, as unhairing with fermenting cereals is accompanied by no tanning action. Where, however, materials such as mulberry leaves and twigs are used, which contain vegetable tannin, it is likely that, besides unhairing, some degree of tannage will occur.

Farmer[61] has suggested that wheat may have been grown at another of the subsidiary establishments of Qumran, namely Khirbet Abu Tabbak, in the Buqei'a, about five miles west of Khirbet Qumran. (There are traces of a track connecting these two places.) Thus flour, derived from this source, may have been available to the Qumran community. In this connection it should be noted that flour paste was used by the Egyptians to stick together the long strips of papyrus when making a roll or sheet. The latter was then dried in the sun and polished by rubbing with cedarwood oil. This oil was used by the Palestinian Jews in the preparation of parchments (see below); hence it is probable that the application of flour to animal skins was also based on this earlier Egyptian practice.

It is most significant that there are no references in the rabbinic literature to the use of lime liquors for unhairing animal skins, whereas nowadays this is the most common method. Since in those passages of the Mishneh Torah dealing with the preparation of skin, copious details are given of the materials to be employed, it would seem that the use of lime for unhairing purposes was unknown to the Jews, even as late as medieval times. In our examination of the deposits from the 'Ain Feshkha installation, one pit, *locus* 24, contained a white material attached to the plaster with which the whole system was originally daubed; this material was chalk (calcium carbonate). This could be taken as evidence for the use of lime in the preparation of skins, but this conclusion is unsupported by any other findings.

Later, we found evidence which indicated that liming was not used by the Qumran community in preparing their scroll material. Among the skin fragments from Cave 4 in our possession were both leathers and parchments (see below); we ashed samples of both and also samples of a modern sole leather and modern parchments, and analyzed the residues by arc-emission spectroscopy. From an analysis of the spectra it was clear that the modern parchments contained calcium as the predominant metal and that they contained more calcium than the modern sole leather. These results are to be expected from a knowledge of the processes used in making modern parchments and sole leathers. After treatment with the lime liquor to bring about unhairing, a skin destined for parchment is merely washed superficially before being dried under tension on a wooden frame. In sole-leather manufacture, the limed hides are given slightly more de-liming than in parchment making, before they are taken into tannage. On the other hand, the Cave 4 parchments did not contain calcium as the predominant metal and showed approximately the same amount of this metal as did the Cave 4 leathers.

We assume, therefore, that skins were first cured with salt and afterward treated with flour or other vegetable material in order to unhair them, clean them, and loosen their fiber structure. This seems most likely, as three types of skin were distinguished: *mazza* (literally, "unleavened") was treated neither with salt nor with flour, *hippa* was treated with salt only, whilst *diftera* was treated both with salt and with flour.[62] When making leather, tannage would follow the stages already outlined. Tannage is considered to have been carried out in the same pit, or vessel, used for the flour or other cereal treatment, the fermentation being initiated by the acid-impregnated vessel.[63] This is at variance with de Vaux's suggested reconstruction of the scheme of operations at the 'Ain Feshkha "tannery," where the two processes of unhairing and of tannage were thought to have been carried out in separate pits.

According to Krauss, an infusion of gall-apple powder was used as the (vegetable) tanning agent,[64] but it is unlikely that this was used exclusively by the Jewish people at that time. It is even more unlikely that it was used by the Qumran community, which, as we have seen, led a secluded existence in a somewhat infertile environment. All the following plants, which survive in a dry situation and which have been used as tanning agents,[65] could however have been available to this community: tamarisk, palm, willow, acacia, henna, sumac, grape, sea lavender, plumbago, and aleppo pine.

From other parts of Palestine the following materials would also be available: poplar, myrtle, pomegranate, walnut, rock rose, elm, pistacia, juniper, geum, agrimony, strawberry tree, dock, polygonum, white water lily, and oak.

Some selective reduction of these lists is possible. Thus the property of astringency, possessed to some degree by all vegetable tannins, is noted in Babylonian and Assyrian texts[66] for oak galls, sumac, acacia, and pomegranate. These four tannins may, therefore, have been used in Mesopotamia, though a Carchemish text suggests that oak and sumac were principally used.[67] By cultural interchange, the use of these materials would probably also be known in Palestine.

The vegetable tannins are a class of substances found in many plants which have the common property of converting skin into leather. They are derived either from the dihydric phenol *catechol* or from the trihydric phenol *pyrogallol,* and consequently tannins are divided into two classes which take their names from these phenols. The two classes show marked differences in properties. Thus members of the pyrogallol class are often called "hydrolyzable" tannins because on heating with acids, they give rise to gallic acid or its derivatives and sugars. On the other hand, the catechol tannins resist such hydrolysis and usually the action of acid produces more complex and insoluble substances; hence this class is alternatively known as "condensed." During tannage, some of the pyrogallol tannins—the ellagi-tannins—give rise to deposits of highly insoluble ellagic acid in the leather, whilst the catechol class deposit dark-brown substances known as phlobaphenes. As far as one can judge, most of the tannins used by the earliest makers of leather were of the hydrolyzable class. We have hydrolyzed a number of leather fragments from Cave 4 and separated the products by paper chromatography.[68] In each case, gallic acid was readily identifiable, thus proving that pyrogallol (hydrolyzable) tannins had been used. Of the plants listed above, several are known to liberate gallic acid on hydrolysis.[69] Hence it is likely that this shortened list includes the vegetable tannins used in the making of the Dead Sea Scrolls. It is interesting to note that this list also includes the tannin sources mentioned in the Mesopotamian texts and further, that if Mesopotamian practice were followed by the Qumran community, the principal tannins used would most probably have been oak and sumac.

The Nature of Skin Writing Materials among the Jews

"Books" were in existence before the establishment of the monarchy[70] and, since codices were unknown at that time (*ca.* 1000 B.C.), it must be assumed that they were scrolls. Later we find explicit reference to book rolls.[71] Driver considers that these were made of papyrus rather than of skin, since in references to the cutting and burning of such rolls, he believes that these acts would be more easily performed upon papyrus.[72] However, by the end of the third century B.C., there is evidence that skins were used for the Holy Scrolls in the pseudepigraphic "Letter of Aristeas," which describes the reception of Jewish scrolls by Ptolemy Philadelphus:[73]

When they entered with the gifts which had been sent with them, and the valuable parchments, on which the Law was inscribed in gold in Jewish characters, for the parchment was wonderfully prepared and the connexion between the sheets had been effected as to be invisible, the king, as soon as he saw them began to ask them about the books. And when they had taken the rolls out of their coverings and unfolded the pages [sic], the king stood still for a long time and then making an obeisance seven times, he said, "I thank you my friends, and I thank him that sent you still more, and most of all God, whose oracles these are."

Diftera, prepared as described, with salt and flour, probably was a rough, cheap parchment suitable for everyday use,[74] and, as such, would be unacceptable for ritual use, since doubt would exist concerning the manner of its preparation and possibly even of the animal species involved. In time, more of the stages of preparing skins as writing materials became the care of the scribe himself (and has remained so, to some extent at least), possibly to prevent such doubts arising. It was essential to establish that a skin used to make parchment for a ritual purpose came from a clean animal:

Scrolls, *Tefillin* and *Mezuzahs*[75] are not written on the hide of a domestic or wild beast that is unclean or on the skin of an unclean bird. They are written on the hides of domestic or wild beasts that are clean or on the skins of clean birds, even when the flesh of these animals may not be eaten owing to their not having been slaughtered according to the ritual or being found to have suffered from a lesion of mortal character. Nor are Scrolls, *Tefillin* or *Mezuzahs* written on the skin of a fish, even if it be of a clean species, because of its foulness; for this is not removed by tanning.[76]

The laws of clean and unclean animals appear in the Biblical books of Leviticus (xi, 1–23) and Deuteronomy (xiv, 3–20) in virtually identical lists. The creatures proscribed in Leviticus may be classed as follows: large land animals (vv. 3–8), water creatures (vv. 9–12), birds (vv. 13–19), and winged creeping creatures (vv. 20–23).

With the fourth class we are not concerned for obvious reasons, and, similarly, with the second, for the reason given above. The identification of the creatures in the first and third classes (especially the latter) is conjectural[77]; however, we have been unable to show the presence of bird skin fragments in our samples from Cave 4 (see later) and therefore we shall limit the discussion to the first class only.

Those land animals are clean which both chew the cud and divide the hoof; they are unclean unless they satisfy both these requirements; thus, in effect, all beasts of prey are eliminated. The Revised Standard Version of Deuteronomy gives the following modern identifications:

> Clean: ox, sheep, goat, hart, gazelle, roebuck, wild
> goat, ibex, antelope, mountain sheep.
> Unclean: camel, hare, rock badger, swine.

A cornerstone in Del Medico's argument that the scroll caves are *genizoth* (see above) is that the Qumran region is uninhabitable. Birds cannot bear the air pressure in the Ghôr, and so insects flourish, making life burdensome;[78] furthermore, the region is sterile, animals could not have been bred locally, and the Feshkha spring could not have supported a farm community.[79] It is interesting to compare these hypotheses with Masterman's description of the region, following two of the many visits he made in the early years of this century:

> Early in the year, in January and February, Bedawin descend into this part of the plain [near 'Ain el-Feshkhah] and flocks of goats and sheep and also camels may be seen on all hands. The Bedawin at this time inhabit caves in the hills around. The 'Ain Feshkhah oasis itself has been tenanted for some eight months now by two men . . . who are in charge of a large herd of cattle, belonging to the Sultan, which thrive in the reeds.[80]

> The most interesting and characteristic animal of the district is the coney. Both in the *Wady Kumrân* and in the rocks between *'Ain Feshkhah* and *Râs Feshkhah* there are abundant evidences of their habitations, and on some occasions we have seen them . . . Then the gazelle is commonly and the ibex occasionally encountered. The names *'Ain Ghuzâl* and *'Ain Ghuzlân*, given to springs in the district, and also the Bedouin name of the mountain just above the *Hajar el-Asbah*—the 'Mountains of the Ibex'—mark the haunts of these

beautiful animals. At sunset the ubiquitous jackal may be encountered on every side. The Bedouin state that wild boar exist in the marshes, and I have seen their footprints.[81]

There is some evidence that at the time of the Qumran community, the Palestinian climate was moister than it has been in recent years.[82] It may be then fairly assumed that Masterman's accounts give a good idea of the potentiality of the Qumran region at the time of the community's existence to support both wild and domesticated animals. The stalls for animals found at 'Ain Feshkha have already been mentioned; these suggest the rearing rather than the hunting of animals, although, of course, they do not indicate the particular species involved. This information is most likely provided by the many deposits of animal bones found at Qumran.[83] The theological significance of these deposits, and whether they were preserved in accordance with a local sacrificial cultus, is not the concern here. The bones date mainly from the first occupational level (from the middle of the second century B.C. to 31 B.C.) and are of lambs, kids, and adult sheep and goats, with a few calves and cows and oxen, that is, all ritually clean animals. In our own examination of the grain surface patterns of forty scroll and leather fragments from Cave 4, we showed that twenty-seven were either hairy sheep or goat skins, whilst one was possibly pig skin and another deer skin. Specimens of eighteen scroll fragments were given to Dr. M. L. Ryder, late of the Wool Industries Research Association in Leeds, to aid his work on the evolution of the domestic sheep.[84] He was able to confirm our findings and increased the total number of identifications. Thus in all, thirty-two sheep or goat skin fragments and one calf skin fragment were clearly recognized, besides the two doubtful identifications already mentioned. Only five fragments were completely unidentifiable. Hence these findings provide striking evidence for the use of "clean" animals in the life of the Qumran community.

The oldest comprehensive source of information, concerning the preparation of the Holy Scrolls is the extra-canonical, minor Talmudic tractate "Soferim," which is basically a manual for scribes. In its present form it dates from the eighth century A.D., although much of its material is considerably older, as is shown by the absence of sources for a number of the regulations.[85] Thus we learn from Soferim that skins for the Scrolls of the Law, *tefillin,* and *mezuzoth* could be purchased anywhere before they were prepared, but if already prepared, they could be purchased only from reliable persons who could vouch that the preparation had been performed correctly.[86] Soferim gives no preparative methods for probably the reason that it itself was redacted at such a late date: the information was so widely known that there was no need to commit it to writing. However, Maimonides here effectively supplements Soferim:

> There are three kinds of parchment, *Gewil, Kelaf* and *Duxustus*. How are these made? A hide of domestic cattle or wild beast is taken. First its hair is removed. It is then pickled in salt, afterwards prepared with flour and subsequently tanned with gall-wood or similar materials which contract the pores of the hide and make it durable. And this it is that is called *Gewil* (parchment of whole hide leather).
>
> If, after removing the hair, the hide had been split through its thickness into two parts, so as to make of it two skins, one thin, namely that which had been next to the hair; the other thick, namely that which had been next to the flesh, and if these were prepared with salt, then with flour and afterwards with gall-wood or similar substance, the skin which had been next to the hair is called *Kelaf* (outer skin parchment) and that which had been next to the flesh is called *Duxustus* (inner skin parchment).
>
> It is an article dating back to Moses . . . that the Scroll of the Law should be written on *Gewil* . . . and the writing should be on the side which had been next to the hair. The *Tefillin* should be written on *Kelaf* . . . and the writing should be on the side which had been nearer to the flesh; and the *Mezuzah* should be written on *Duxustus* . . . on the side which had been nearer the hair. If on a *Kelaf* one writes on the side that had been next to the hair, or, on a *Gewil* or *Duxustus,* one writes on the side that had been next to the flesh, the Scroll, *Tefillin* or *Mezuzah* so written is unfit for use.
>
> Though this the rule dating back to Moses . . . yet if one writes a Scroll of the Law on parchment made from the *exterior* half of a split hide it is fit for use. The reason why a complete hide is mentioned is in order to exclude that made from the inner half of a split hide. If one wrote on it a Scroll of the Law it is

unfit for use. So also, if one wrote a *Mezuzah* on the external half of a split hide or on an undivided hide, the *Mezuzah* is fit for use. The inner half of a split hide is recommended to be used only as a *Mitzvuh* (specially approved).

The undivided skin for a Scroll of the Law and the external half of the hide to be used for *Tefillin* or for a Scroll of the Law must be tanned for their specific purposes. If they were not expressly so tanned, they are not fit for use. Accordingly, if a non-Israelite or a Samaritan tanned them, they are unfit for use. . . .

The skin for a *Mezuzah* does not require to be tanned for that express purpose.[87]

It is clear from this passage that the initial processes in the preparation of parchment are the same as those for leather, namely the treatment with salt and flour; liming is not mentioned (see above). The final gall-wood treatment will be considered later. The scrupulous care exercised in the correct and valid performance of the preparation and in the allocation of the several qualities of parchment to their proper use are also evident. The names given by Maimonides to the different types of parchment are variously interpreted elsewhere. *Gewil* is generally taken to be synonymous with skin and leather, unsplit and with the hair either rubbed or scraped away. *Kelaf* is sometimes considered synonymous with parchment and usually a superior material to *duxustus,* although occasionally this reading is reversed and *duxustus* is considered the finer material, resembling vellum.[88]

Although none of the Cave 4 fragments in our possession bore any recognizable words or letters, sixteen of them were ruled for writing purposes. These ruled fragments were examined microscopically and all were judged to have been ruled on their grain (hair) surfaces. If the Qumran community adhered to the rabbinic regulations recorded by Maimonides, it follows that these sixteen fragments were all of *gewil,* rather than *kelaf* or *duxustus.* It would also seem to follow that these fragments are all of sacred manuscripts, if not the Pentateuchal books themselves. Moreover, the absence of *kelaf* fragments may possibly be significant, suggesting that the development of this type of parchment may postdate the existence of the Qumran community.

It seems clear that Maimonides' references to the final "tanning" of *gewil,* etc., signify no more than a superficial treatment of the parchment and not a thorough penetration of it by a tanning agent. This interpretation was first suggested to us by the following passage:[89]

> In case a Scroll of the Law had one of its sheets torn, the rent is sewn up, if it extends to two lines; if it extends to three lines, it is not sewn up. This rule applies to a scroll *the gall dressing of which is no longer perceptible* (our italics). But if the parchment still shows that it has been dressed, then, even when the rent extends to three lines, it is sewn up.

Here "gall dressing" and elsewhere "tan" (see above) are translations of the same Hebrew word *'aphas* from *'aphsā—* gall nut.[90] It appears that the objects of this dressing, besides giving a more durable writing surface, were to produce a more attractive one[91] and to ensure that the writing could not be erased.[92] (In this last connection, Partington's statement[93] may be noted, *viz.*, it is "almost impossible to write upon parchment with a carbon ink.") Our interpretation was confirmed by a Jewish scribe.[94] The gall dressing was applied to both surfaces of *gewil* (similarly with *kelaf* and *duxustus,* these terms being used in Maimonides' senses), and involved fresh galls, although it was impossible to discover of what tree. The gall juices also have a cleansing action on the parchment which proves very useful.

We were able to examine a modern *gewil* scroll[95] and found that it closely resembled many of our Cave 4 fragments, an observation made originally by Sukenik after examining the Cave 1 manuscripts[96]; the *gewil* was a darkish, yellow-brown color and had not the stiff resilience of modern parchment, possessing a soft handle and a much fuller substance. *Gewil* is indeed a parchment, but its likeness to an old glove or boot leather is undeniable, and this no doubt explains why there is much confusion in the literature regarding the nature of the Dead Sea Scrolls.

By hydrolyzing a number of scroll fragments and examining the resulting solutions by paper chromatographic methods, we have shown that in almost every sample, gallic and ellagic acid were present. Two modern parchments were similarly examined and were found free of these two acids. This evidence suggests that the tannin with which the scrolls were originally dressed was an ellagi-tannin, a member of a subclass of the "pyrogallol" or "hydrolyzable" tannins (see above), which can deposit traces of ellagic acid in the skin. That such a vegetable tannin was applied only superficially is shown in transverse sections of Cave 4 fragments as seen in the microscope, compared with a section of an English parchment of 1692 A.D. The dark, uniform bands at the grain and flesh surfaces of the scroll sections have no counterpart in the more recent parchment. Moreover, the localization of vegetable tannin in the surface layers has been confirmed by chemical and by electron microscope studies. These findings therefore suggest most strongly that the scroll parchments were processed in direct accordance with the regulations set down by Maimonides, whereby tannin is applied superficially in order to obtain a better writing surface and to render difficult any alteration in the written word. The indication that ellagi-tannins were mostly used for this purpose also enables us further to limit the various plants and trees which could have been used as sources of tannin.

There is a basic similarity in structure between the scroll sections and those of the modern parchment. All have the same, flattened fibrous weave, indicative of stretching, rolling, or pressing during processing. In modern practice, the unhaired and superficially washed skin is dried under tension on a wooden frame. Although in the rabbinic accounts there is no mention of such a practice, there is abundant evidence that it was, in fact, also used in the past, particularly in medieval times.

The practice of applying a tannin dressing to parchment seems to have been peculiar to the Jews, and we have found no indication that it has ever been used in Western Europe.[97] Even the method now used in Israel[98] is virtually identical with current European practice, which appears to stem directly from the monastic, guild, and craft traditions developed in the Middle Ages. Medieval accounts of parchment making pay much attention to the production of a smooth writing surface, but the latter was achieved mainly by mechanical processes. However, though chemical treatments were also involved, there is no record that vegetable tannins were ever used for this purpose. Why this uniquely Jewish method of parchment making was lost to history and was not acquired later in medieval times, when parchment technology was again thoroughly developed (though in an empirical fashion), is by no means clear. Perhaps it is another example of the degree to which the Jewish nation was dispersed following the destruction of the Temple, with a consequent loss of their technological practices.

After the making of the scroll material, the preparation of its writing surface, and the completion of the writing, it was furnished with rollers or pillars (*umbilici*), presumably both for ease of handling and to prevent the hands coming into contact with the parchment.[99] Soferim gives no indication of how the parchment sheets were joined or of how the rollers were affixed, but Maimonides states that for both these purposes sinews were taken from the heels of domestic or wild, clean animals; they were softened by pounding, between stones, for example, until they became "like flax," after which they were spun or twined and used for sewing.[100] Some of the fragments in our possession bore remnants of stitching material, but examination showed that these were all of vegetable origin and most probably derived from flax. This finding is most interesting as it is the only apparent departure from the rabbinic regulations which we have encountered. It might be added also that the stitching was very fine and regular, displaying great care in workmanship.

Because of their sacred character, worn-out scrolls could not be destroyed. They were stored away in jars to decay (in *genizoth*), preferably near the remains of dead scholars.[101] However, when it was required to preserve the scrolls, they were sometimes treated with cedar wood oil before being placed into earthenware jars.[102] As we have mentioned, this use of cedar wood oil may have been acquired from Egypt, since in that country it had long been used to preserve *papyri* and to obtain a smooth surface for writing.

We began this survey of the leather and parchment technology available to the Qumran community by assuming that the validity of the rabbinic, Pharisaic regulations would prove a useful guide. With the sole exception mentioned, we have found this assumption to be justified, and hence we believe this to be evidence of the antiquity of these regulations. Moreover, the methods of producing the various kinds of parchment prescribed in the rabbinic literature appear to be unique to the Jews. As stated, even today, only a few scribes appear to follow strictly these rabbinic regulations, the majority making their material according to modern Western European methods, which were developed later and independently. Finally, the finding that the Dead Sea Scrolls were apparently made with strict adherence to the rabbinic regulations lends strong support to the view that they were made, owned, and maintained by a highly orthodox Jewish sect, most probably of a hyper-Pharisaic or Essenic nature.

REFERENCES

[1] P. E. Kahle, *The Cairo Genizah* (London, 1947), pp. 161–162.

[2] F. M. Cross, *The Ancient Library of Qumram and Modern Biblical Studies* (Garden City, N.Y., 1958), p. 2.

[3] R. de Vaux, "Preliminary reports of the excavations at Khirbet Qumran," *Revue Biblique,* LX (1953), pp. 83–106; LXI (1954), pp. 206–236; LXIII (1956), pp. 532–577.

[4] Cross, *op. cit.,* p. 41.

[5] *Ibid.,* p. 15 and note.

[6] An earlier, Israelite level has been detected and dated in the eighth or seventh century B.C.; it has been suggested that this is the City of Salt mentioned in Joshua xv, 62.

[7] R. de Vaux, *Revue Biblique,* 63 (1956), p. 569.

[8] M. Burrows, *The Dead Sea Scrolls* (London, 1956), pp. 83–101; Cross, *op. cit.,* pp. 88–90; D. Burton, J. B. Poole, & R. Reed, "A New Approach to the Dating of the Dead Sea Scrolls," *Nature,* 184 (1959), pp. 533–534.

[9] Cross, *op. cit.,* pp. 88–91.

[10] H. E. Del Medico, *The Riddle of the Scrolls,* translated by H. Garner (London, 1958).

[11] Cf. Jeremiah xxxii, 14.

[12] E. W. G. Masterman, *Palestine Exploration Fund Quarterly Statement,* 1902, pp. 161–162.

[13] T. H. Gaster, personal communication.

[14] Small leather cases containing parchment strips bearing portions of the Pentateuch and worn on the head and arm.

[15] Masterman, *loc. cit.,* p. 160.

[16] *Antiquitatum Judiacarum,* XIII, v, 9.

[17] See, most conveniently, C. D. Ginsburg, *The Essenes* (London, reprinted 1955).

[18] Burrows, *op. cit.,* pp. 279 ff., 284 ff.

[19] T. H. Gaster, *The Dead Sea Scriptures* (Garden City, N.Y., 1957).

[20] R. P. C. Hanson, *A Guide to the Scrolls* (London, 1958), pp. 62–64.

[21] Josephus says there were 4,000 Essenes (*A.J.,* XVIII, i, 5); the number of Pharisees has been put at 6,000 in a Palestinian Jewish population of half a million (W. S. Lasor, *Amazing Dead Sea Scrolls and the Christian Faith* [Chicago, 1956], p. 191).

[22] F. F. Bruce, *Second Thoughts on the Dead Sea Scrolls* (London, 1956), pp. 99–101.

[23] J. W. Lightley, *Jewish Sects and Parties in the Time of Jesus* (London, 1925), pp. 279–280.

[24] E. Schürer, *A History of the Jewish People in the Time of Jesus Christ* (Edinburgh, 1893), Second Division, Vol. II, p. 209.

[25] Ginsburg, *op. cit.,* pp. 21–22.

[26] *Ibid.,* p. 73.

[27] *Ibid.,* p. 22.

[28] For instance, C. Rabin, *Scripta Judaica II: Qumran Studies* (Oxford, 1957).

[29] S. Lieberman, "The Discipline in the So-called Dead Sea Manual of Discipline," *J. Biblical Literature,* 71 (1952), pp. 199–206.

[30] *Op. cit.,* pp. 69–70.

[31] J. M. Baumgarten, Review of Chaim Rabin, *Qumran Studies, J. Biblical Literature* (1958), 77 (1958), pp. 249–257.

[32] Although at least one author has detected criticism of Pharisees in the Dead Sea Scrolls: Y. Yadin, *The Message of the Scrolls* (London, 1957), p. 169.

[33] I. Epstein, *Judaism: A Historical Presentation* (London, 1959), pp. 84–85.

[34] *Ibid.,* pp. 114–115.

[35] *Ibid.,* pp. 121–128.

[36] Article "Moses ben Maimon" in *The Jewish Encyclopaedia* (New York and London, 1901–1907, 12 vols.).

[37] Rabbi C. B. Chavel; personal communication.

[38] F. W. Farrar, *The Life and Work of St. Paul* (London, n.d.), Vol. I, p. 264.

[39] *The Babylonian Talmud,* ed. I. Epstein (London, 1935–1953; 36 vols.), Tractate Kiddushin, pp. 82a–82b; see also *The Mishnah,* ed. H. Danby (Oxford, 1933), Ketuboth 7:10.

[40] See Lev. xi.

[41] *De Bello Judaico,* II, viii, 7.

[42] Ginsburg, *op. cit.,* p. 45, n. 36.

[43] Gaster, *op. cit.,* Zadokite Document, xii, 15–18.

[44] Mishnah Baba Bathra, 2:9.

[45] Mishnah, Megillah, 3:2.

[46] *Revue Biblique* 66 (1959), pp. 230–237.

[47] *Palestine Exploration Quarterly,* Jan.–June, 1960, p. 27.

[48] S. Krauss, *Talmudische Archäologie* (Leipzig, 1911), Vol. II, pp. 260–261; Mishnah, Betzah 1:5.

[49] Krauss, *op. cit.,* p. 260; R. J. Forbes, "Leather in Antiquity" in *Studies in Ancient Technology,* Vol. V (Leiden, 1957), p. 47.

[50] Krauss, *loc. cit.*

[51] *Ibid.;* Acts x, 6.

[52] Deut., vi, 4, 5.

[53] Maimonides, *The Mishneh Torah,* Book II, trans. M. Hyamson (New York, 1949), "Laws concerning reading of the *Shema,"* III, 6.

[54] Babylonian Talmud, Tractate Ketuboth, p. 77a and notes.

[55] Krauss, *op. cit.,* p. 261; cf. Forbes, *loc. cit.*

[56] J. Hastings *et al., A Dictionary of the Bible* (Edinburgh, 1902), Vol. IV, p. 677.

[57] G. E. Post, *Flora of Syria, Palestine and Sinai,* 2 vols. (Beirut, 1932–1933).

[58] Krauss, *loc. cit.*

[59] D. Burton, R. Reed and F. O. Flint, *J. Society of Leather Trades Chemists,* 37 (1953), pp. 82–87.

[60] *Loc. cit.*

[61] W. R. Farmer, "Λ Postscript to the Economic Basis of the Qumran Community," *Theologische Zeitschrift*, 12 (1956), pp. 56–58.

[62] Babylonian Talmud, Tractate Shabbath, p. 79a and notes.

[63] Krauss, *op. cit.*, p. 262.

[64] *Op. cit.*, p. 261. See also refs. 69, 71, 79, and 81.

[65] The main sources were: Post, *op. cit.*; Forbes, *op. cit.*; F. N. Howes, *Vegetable Tanning Materials* (London, 1953); J. R. Partington, *The Origins and Development of Applied Chemistry* (London, 1935); M. Nierenstein, *The Natural Organic Tannins* (London, 1934); H. B. Tristram, *The Natural History of the Bible* (London, 9th edn., 1898); *Idem, The Survey of Western Palestine. The Fauna and Flora of Palestine* (London, 1884).

[66] Partington, *op. cit.*, p. 314.

[67] M. Levey, "'Tanning Technology in Ancient Mesopotamia," *Ambix,* 6 (1957), p. 44.

[68] See R. J. Block, R. LeStrange and G. Zweig, *Paper Chromatography* (New York, 1952) for a general introduction to the technique.

[69] Nierenstein, *op. cit.*, pp. 108–110, n. 32; 172–174.

[70] G. R. Driver, *Semitic Writing* (London, 1948), p. 88; Num. xxi, 14; 2 Sam. i, 18; Josh. viii, 31; xxiii, 6.

[71] Jer. xxxvi, 2, 4; Ezek. ii, 9; Is. xxxiv, 4.

[72] *Op. cit.*, p. 82. See Jer. xxxvi, 23.

[73] R. H. Charles, *The Apocrypha and Pseudepigrapha of the Old Testament* (Oxford, 1913), Vol. II, "Pseudepigrapha," pp. 110–111.

[74] Krauss, *op. cit.*, p. 263.

[75] Pieces of parchment bearing the passages Deut. vi, 4–9, xi, 13–21 and generally obligatory for the doorposts of Jewish dwellings.

[76] Maimonides, *op. cit.*, "Laws concerning Phylacteries, the *Mezuzah* and the Scroll of the Law," I, 10.

[77] J. R. Dummelow (ed.), *A Commentary on the Holy Bible* (London, 1910).

[78] *Op. cit.*, p. 20.

[79] *Ibid.*, pp. 77–80.

[80] *Palestine Exploration Fund Quarterly Statement,* 1902, p. 166.

[81] *Ibid.*, 1904, p. 92.

[82] See for instance D. Baly, *The Geography of the Bible* (London, 1957), pp. 71 ff.

[83] R. de Vaux, *loc. cit., Revue Biblique,* 63 (1956), pp. 549–550.

[84] M. L. Ryder, "Follicle Arrangement in Skin from Wild Sheep, Domestic Sheep and in Parchment," *Nature,* 182 (1958), pp. 781–783.

[85] Article: "Soferim" in *The Jewish Encyclopaedia.*

[86] *Masseketh Soferim,* ed. I. W. Slotki, I, 3 and n. 9; we acknowledge the kindness of Mr. S. M. Bloch, of the Soncino Press Ltd., London, who loaned to us a proof copy of this work.

[87] *Op. cit.,* "Laws concerning Phylacteries, the *Mezuzah* and the Scroll of the Law," I, 6–9, 11.

[88] See articles: "Vellum," "Scroll of the Law" and "Manuscripts" in *The Jewish Encyclopaedia.*

[89] Maimonides, *op. cit.,* IX, 15.

[90] J. M. Allegro; personal communication.

[91] Rabbi C. B. Chavel; personal communication.

[92] Babylonian Talmud, Tractate Gittin, p. 11a.

[93] *Origins and Development of Applied Chemistry,* pp. 206–208. Cf. Maimonides, *op. cit.,* I, 4.

[94] Rabbi A. M. Perlman; personal communication.

[95] We acknowledge the kindness of Mohel G. Heilpern in permitting us to do this.

[96] E. L. Sukenik, *The Dead Sea Scrolls of the Hebrew University* (Jerusalem, 1955), p. 25.

[97] Cf. other work from this laboratory: H. Saxl, M.Sc. Thesis, The University of Leeds, 1954.

[98] Rabbi Zev W. Gotthold; personal communication.

[99] *Masseketh Soferim,* II, 5.

[100] *Op. cit.,* III, 9; IX, 14.

[101] Jer. xxxii, 14; *Masseketh Soferim,* V, 14, 15; Maimonides, *op. cit.,* X, 3.

[102] *The Assumption of Moses,* ed. R. H. Charles (London, 1897), pp. 7–8.

Sullivan's Skyscrapers As the Expression of 19th-Century Technology

Carl W. Condit

It is now a matter of common consent that Louis Sullivan (1856–1924) was the first great modern architect, the first to create a new and powerful vocabulary of forms derived from the major cultural determinants of his age. He was the most imaginative and the most articulate figure among a small group of creative men in Europe and America who, suddenly around 1890, struck out in a new direction with the deliberate intention of breaking once and for all with the traditional architectural forms of the classical and medieval heritage. In Europe the movement called itself *Art Nouveau,* its initiator being the Belgian architect Baron Victor Horta (1861–1947). In the United States it was at first confined largely to Chicago, where the fire of 1871 prepared the way for one of the most exuberant outbursts of creative activity in nineteenth-century architecture. The leadership of this movement, now known as the Chicago school, was initially in the hands of William Le Baron Jenney (1832–1906), but by 1890 it had passed to Sullivan and his engineering partner, Dankmar Adler (1844–1900). Within a single decade Adler and Sullivan moved rapidly,

Dr. Condit, of Northwestern University, is the author of The Rise of the Skyscraper, American Building, The Chicago School of Architecture, *and a two-volume prize-winning work on* American Building Art. *He is one of the founders of the Society for the History of Technology and one of the editors of* Technology and Culture.

This article is based on a paper read at the meeting of the Midwest Junto of the History of Science at the University of Illinois in April 1959. It appeared in the first issue of Technology and Culture: *78–93, and has been widely reprinted.*

if irregularly, from close dependence on past architectural styles to an organic form which derived its character from the industrial and scientific culture which had swept everything before it in the Western world.

By the last decade of the century Sullivan had developed in preliminary terms his organic theory of building art, a system which was later to be presented at length in his major writings, *Kindergarten Chats* (1901–02) and *The Autobiography of an Idea* (1922). The philosophy of architecture offered in these works contains extensive ethical and social elements as well as formal and aesthetic ones. Since the doctrine has been discussed, analyzed, and interpreted in detail by historians and critics, we need not here inquire into it at length. Our purpose is to find, if we can, the broad symbolic meaning of Sullivan's major works, for which it may be useful to summarize some of the fundamental ideas in his system of thought.

Sullivan's interest in structural engineering—in part, of course, the product of professional necessity—early developed into a wide-ranging enthusiasm for science as a whole. It centered mainly in biology, from which his organic theory in part stemmed, but it included the new physical theories as well. He read Darwin, Huxley, Spencer, and Tyndall at length and was well acquainted with the writers who were then developing the seminal theories of building art in the past century, chiefly Ruskin, Morris, and Viollet-le-Duc. What distinguishes Sullivan's thought is his profound grasp of the social basis, the responsibility, and the problem of the arts in a technical and industrial society. He felt that he had discovered the rule with no exceptions (to use his own phrase) in the concept "form follows function," but the key to his philosophy lies in the proper understanding of the word *function*. An organic architecture, he believed, is one which grows naturally or organically out of the social and technical factors among which the architect lives and with which he must work. These factors embrace not only the technical and utilitarian problems of building but also the aspirations, ideals, and needs, both material and psychological, of mankind. Thus *functionalism* involved for him something much wider and deeper than utilitarian and structural considerations, as important as these are.

To Sullivan the creation of a genuine architectural style was not a matter of historical styles or of dipping into a vocabulary of contemporary forms and details in order to secure a style which the architect might feel to be consonant with the life of his time. The architect must first recognize the importance of true aesthetic expression for the symbolic recreation, the harmonization, and the emotional enrichment of the many practical and intellectual elements of contemporary civilization. In European and American society at the end of the nineteenth century such an art would begin, by necessity, with the fundamentals: industry, technology, and science. It is the task of the architect, as Sullivan conceived it, to take the products of techniques, on the one hand, and the logic and order of a scientific technology, on the other, and mold them into a form uniting both in a single aesthetic expression. An architecture so developed means the humanization through aesthetic statement of the often cold and nonhuman facts of industrial techniques.

The early application of this complex philosophy of the organic to a specific building problem appeared in a document which has become a classic of modern theory, "The Tall Office Building Artistically Considered," first published in *Lippincott's Magazine* in 1896. Scattered throughout *The Autobiography of an Idea,* which is Sullivan's final testament, are many sentences of an epigrammatic character that summarize his thought; for example, "As the people are within, so the buildings are without," or again, "It is the task of the architect to build, to express the life of his own people."

The realization of this program in actual commissions reached its mature form in the four largest and most impressive buildings which Adler and Sullivan designed. The first is the Auditorium Building, now Roosevelt University in Chicago, designed in 1886–87 and opened in 1889. The design of this great building, with its huge masonry bearing walls, was much influenced by Richardson (1838–86) and the Romanesque-like forms which he handled so brilliantly. But it marks a transition toward the open and dynamic wall forms that were soon to become the distinguishing feature of Sullivan's work. In the year following the completion of the Auditorium he struck out in a new direction to produce one of the most remarkable exhibitions of sheer architectural originality in his own or any age, the steel-framed Wainwright Building in St. Louis (1890-91). A few years later the formal character of this structure was refined and enriched in the Guaranty, later Prudential, Building in Buffalo (1894–95). In the last of his large commissions, before the poverty and neglect of his later years, he turned in still another direction and produced a radically different kind of expression in the Carson Pirie Scott Store in Chicago, built in three parts over the years from 1899 to 1906, although designed as early as 1896.

Behind these steel-framed buildings lay a long preparation in the history of iron construction. The use of iron as a structural material goes back to classical antiquity, but it did not appear wholly emancipated from masonry until the construction of Darby and Pritchard's cast-iron arch over the River Severn at Coalbrookdale, England, in 1775–79. The first building with interior columns and beams of iron was William Strutt's Calico Mill at Derby (1793). It was thirty-five years before iron members appeared in American buildings and 1850 before complete iron construction was established, largely through the work of the New York inventors and builders Daniel Badger and James Bogardus. The remainder of the century saw steady progress in the techniques of cast and wrought iron and later steel framing, reaching its culmination in the skyscrapers of New York and Chicago, in which all the essential features of the modern commercial building were given a practical demonstration.

Thus, by 1890, the technical means of a new building art were available to Sullivan. It remained for him to transmute the structural solutions to these unprecedented functional requirements into a symbolic art. His organic philosophy had already come to exist at least in an inchoate form, but so broad an approach to architectural design could not lead directly to a specific kind of formal expression. The key to the process of transmutation may be found, I think, in those passages of Sullivan's writings in which he gives voice to his feelings about particular architectural and structural achievements of his age. We have already mentioned his debt to Richardson. There is a chapter in *Kindergarten Chats*—Number VI, "The Oasis"—in which he acknowledges this debt, and it is the first of various passages that lead us to an understanding of Sullivan's inner purpose. He describes for us, in ironic and impressionistic metaphors, his strong emotional reaction to Richardson's Marshall Field Wholesale Store in Chicago (1885–87) and to what it stands for.

Let us pause, my son, at this oasis in our desert. Let us rest awhile beneath its cool and satisfying calm, and drink a little at this wayside spring. . . .
You mean, I suppose, that here is a good piece of

architecture for me to look at—and I quite agree with you.

No; I mean here is a *man* for you to look at. A man that walks on two legs instead of four, has active muscles, heart, lungs, and other viscera; a man that lives and breathes, that has red blood; a real man, a manly man; a virile force—broad, vigorous and with a whelm of energy—an entire male.

I mean that stone and mortar, here, spring into life, and are no more material and sordid things, but, as it were, become the very diapason of a mind rich-stored with harmony. . . .

Four square and brown, it stands, in physical fact, a monument to trade, to the organized commercial spirit, to the power and progress of the age, to the strength and resource of individuality and force of character; spiritually, it stands as the index of a mind, large enough, courageous enough to cope with these things, master them, absorb them and give them forth again, impressed with the stamp of large and forceful personality; artistically, it stands as the creation of one who knows well how to choose his words, who has somewhat to say and says it—and says it as the outpouring of a copious, direct, large and simple mind.[1]

It is clear that Sullivan was profoundly moved by Richardson's building, but even a fine work of architecture did not arouse in him the powerful emotions that were evoked by the great achievements of the bridge engineers in the nineteenth century. There are several illuminating passages in *The Autobiography,* among the most remarkable in the book, in which he tries to analyze his emotional and philosophical response to these monuments of pure structural form. The first records a childhood experience in which he saw a chain suspension bridge over the Merrimack River (possibly Finley's Bridge of 1810, near Newburyport, Massachusetts). The description is loaded with the most extreme expressions of feeling.

Mechanically he ascended a hill . . . musing, as he went, upon the great river Merrimac. . . . Meanwhile something large, something dark was approaching unperceived; something ominous, something sinister that silently aroused him to a sense of its presence. . . . The dark thing came ever nearer, nearer in the stillness, became broader, looming, and then it changed itself into full view—an enormous terrifying mass that overhung the broad river from bank to bank. . . .

He saw great iron chains hanging in the air. How could iron chains hang in the air? He thought of Julia's fairy tales and what giants did. . . . And then he saw a long flat thing under the chains; and this thing too seemed to float in the air; and then he saw two great stone towers taller than the trees. Could these be giants? . . . [A page follows in which Sullivan records how he ran frightened to his father to tell him that the giants would eat him.]

So [his father] explained that the roadway of the bridge was just like any other road, only it was held up over the river by the big iron chains; that the big iron chains did not float in the air but were held up by the stone towers over the top of which they passed and were anchored firmly into the ground at each end beyond the towers; that the road-bed was hung to the chains so it would not fall into the river. . . . On their way to rejoin Mama, the child turned backward to gaze in awe and love upon the great suspension bridge. There, again, it hung in the air—beautiful in power. The sweep of the chains so lovely, the roadway barely touching the banks. And to think it was made by men! How great must men be, how wonderful; how powerful, that they could make such a bridge; and again he worshipped the worker.[2]

In later years, on his way to becoming an established architect in partnership with one of the great building engineers of his time, Sullivan came to understand how these miracles were accomplished. Then he was prepared to pay his fullest tribute to the bridge engineers and to record it again in his *Autobiography*.

About this time two great engineering works were under way. One, the triple arch bridge to cross the Mississippi at St. Louis, Capt. Eades [*sic*], chief engineer; the other, the great cantilever bridge which was to cross the chasm of the Kentucky River, C. Shaler Smith, chief engineer, destined for the use of the Cincinnati Southern Railroad. In these two growing structures Louis's soul became immersed. In them he lived. Were they not his bridges? Surely they were his bridges. In the pages of the *Railway Gazette* he saw them born, he watched them grow. Week by week he grew with them. Here was Romance, here again was man, the great adventurer, daring to think, daring to have faith, daring to do. Here again was to be set forth to view man in his power to create beneficently. Here were two ideas differing in kind. Each was emerging from a brain, each was to find realization. One bridge was to cross a great river, to form the portal of a great city, to be sensational and architectonic. The other was to take form in the wilderness, and abide there; a work of science without concession. Louis followed every detail of design, every measurement; every operation as the two works progressed from the sinking of the caissons in the bed of the Mississippi, and the start in the wild of the initial cantilevers from the face of the cliff. He followed each, with the intensity of personal identification, to the finale of each. Every difficulty he encountered he felt to be his own; every expedient, every device, he shared in. The chief engineers became his heroes; they loomed above other men. The positive quality of their minds agreed with the aggressive quality of his own. In childhood his idols had been the big strong men who *did* things. Later on he had begun to feel the greater power of men who could *think* things; later the expansive power of men who could *imagine* things; and at last he began to recognize as dominant the will of the Creative Dreamer: he who possessed the power of vision needed to harness Imagination, to harness the intellect, to make science do his will, to make the emotions serve him—for without emotion—nothing.[3]

There is a distinct strain of romanticism in this passionate devotion to the builder, perhaps even a Nietzschean quality in the worship of creative power. For Sullivan came to see in science and technology the triumphant assertion of man's will expressing itself in a wholly new way. As he himself put it, "Louis saw power everywhere; and as he grew on through his boyhood, and through the passage to manhood, and to manhood itself, he began to see the powers of nature and the powers of man coalesce in his vision into an IDEA *of power*. Then and only then he became aware that this idea was a *new idea*—a complete reversal and inversion of the commonly accepted intellectual and theological concept of the nature of man."[4]

Thus Sullivan conceived of a bridge as the personal testament of a man, a testament expressing a unification of the highest energies and skills of the age. What distinguishes these achievements is not only the technical virtuosity that men like Eads and Smith commanded, but the integration of many streams of technical and scientific progress in the nineteenth century. For it was the age that saw the transformation of building from an empirical and pragmatic technique into an exact science. Since Sullivan chose the St. Louis and Dixville bridges as his examples, we may use them as representatives of the transformation that made possible their design and construction. Eads Bridge (1868–1874) is the earlier of the two, and so we may begin with an analysis which reveals how such structures brought to focus the various scientific and technical currents.

The general staff of Eads Bridge consisted of James B. Eads as chief engineer, Charles Pfeiffer and Henry Flad as principal assistants, and William Chauvenet, Chancellor of Washington University, as mathematical consultant. The choice of Eads, who had never built a bridge before, as head of the St. Louis project rested in large part on his intimate knowledge of the river. For the builder who proposed to found his piers on the rock far below its bed, it was a formidable obstacle indeed. The pilots could read its surface with remarkable skill for the hidden snags and bars that once menaced them, but only Eads knew at first hand its fluid, shifting, treacherous bed. By means of the diving bell which he had invented, he was able to investigate the bottom directly, and he had seen its depth change from 20 to 100 feet at obstacles in the bed as the result of the scouring action of currents. For the first time the topographical and geological surveys of the bridge site could be carried on to a certain extent under water.

With the design of his bridge substantially completed and sufficient capital available, Eads began clearing the site and constructing caissons in the summer of 1867, but difficulties with his iron and steel contractor soon required a suspension of operations. Eads had already decided to substitute steel for the traditional cast iron in the arches and thus became the first to introduce the stronger metal into American building techniques. The Carnegie-Kloman Company at first found it impossible to roll pieces with the physical properties that Eads demanded. The earlier cast steel samples had already failed in the testing machines. At this point Eads insisted on the costly and hitherto unused chrome-steel, an innovation which was to have wide implications for structural and mechanical engineering. Equally important was the application of the methods of experimental science to the investigation of the physical properties of the metal. Eads Bridge is the first major structure in the United States in which testing machines, which had been developed over the previous thirty years in Europe, played a vital role in the successful completion of the bridge.

The initial problem solved, construction was resumed in 1868. Eads began with the east, or Illinois, pier, where the maximum depth of bedrock offered the most serious challenge. The pneumatic caisson was a necessity, and thus Eads became the first to introduce its use in the United States, anticipating Roebling by a year. It had been used in Europe since 1849, when Lewis Cubitt and John Wright developed it for the construction of the piers of a span at Rochester, England. Eads built a cylindrical iron-shod caisson of massive timbers heavily reinforced with iron bands. Its diameter was 75 feet, the working chamber 8 feet deep. Within this huge enclosure the masonry pier was built up, the weight of the masonry forcing the cutting edge into the river bed. The Eads caisson extended continuously up to the water surface, successive rings being added as it sank lower. The caisson for the Illinois pier reached bedrock at 123 feet below water level at the time of construction. Five months of excavation and pumping were required to uncover the foundation rock. Since the top stratum of the bedrock rises steadily from the east to the west bank, the caissons for the center and west piers had to be sunk to a progressively smaller depth, reaching a minimum of 86 feet at the St. Louis pier. The river piers and the masonry arches of the west approach were completed in 1873.

The construction of the steel arches and the wrought-iron superstructure was a relatively simple matter after the dangerous work on the piers and required only about one-fifth of the time. In this part of the project Eads introduced another of his important innovations. The tubular arches were erected without falsework by the method of cantilevering them out from the piers to the center of the span. All arches were built out simultaneously from their piers so that the weights of the various cantilevers would balance each other and on completion the horizontal thrust of adjacent arches would cancel each other. With the arches in place, the spandrel posts and the two decks were erected upon them. The bridge was completed and opened to traffic in 1874. The finished structure between abutments is divided into three spans, the one at the center 520 feet long, the two at the sides 502 feet each, the rise for all of them

45 feet. Eight tubular arches, four for each deck, constitute the primary structure of each span. Wrought-iron spandrel posts and transverse bracing transmit the load of the two decks to the arches. The upper one carries a roadway, the lower a double-track railroad line.

The arches of Eads Bridge are the hingeless or fixed-end type and hence are statically indeterminate structures. It is possible that Chauvenet was familiar with the recent work of French theorists in the solution of problems arising from arches of this kind, and certainly he knew of the many carefully designed wrought-iron arches which had been built by French engineers before 1865. The successful attack on problems of indeterminacy was the product of an international effort in which a great many mathematicians had a hand, among them the great English physicist James Clerk Maxwell. The chief figure in the development of methods of stress analysis for fixed and two-hinged arches was Jacques Antoine Bresse, the first edition of whose *Applied Mechanics* was published at Paris in 1859. But for all the mathematical computations of Eads and Chauvenet, they relied to a great extent—as the engineers always did until the last decade of the century—on empirical approximations and gross overbuilding. Eads calculated that his bridge would be capable of sustaining a total load of 28,972 tons uniformly distributed—about four times the maximum that can be placed upon it—and of withstanding the force of any flood, ice jam, or tornado that the Mississippi Valley could level against it. Now in its ninth decade of active service, the bridge carries a heavy traffic of trucks, buses, automobiles, and the freight trains of the Terminal Railroad of St. Louis.

Sullivan could hardly have chosen a better example to represent the new power of his age. As a work of structural art Eads Bridge remains a classic. In its method of construction and its material, in the testing of full-sized samples of all structural members and connections, in the thoroughness and precision of its technical and formal design, and in the close association of manufacturer and builder, it stands as a superb monument to the building art. It is, moreover, an architectural as well as an engineering achievement. Eads was careful to reduce his masonry elements to the simplest possible form, depending on the rich texture of the granite facing to provide the dignity and sense of restrained power that he was consciously striving for. Nowhere does the masonry extend above the line of the parapet to distract attention from the overall profile, the major parts, and their relation to each other. The tight curve of the arches is the primary visual as well as structural element, and Eads knew that the best he could do was to give full expression to the combination of stability and energy implicit in the form.

The Dixville Bridge posed an entirely different and somewhat less formidable problem. The solution, moreover, belongs strictly to the nineteenth century, the particular truss form employed having been abandoned before the beginning of the twentieth century. It was the decision to use the cantilever principle for a large railway bridge, for which there was only the slightest of precedents, that excited Sullivan's interest. The occasion in this case was the necessity of bridging the Kentucky River at Dixville, Kentucky, for the Cincinnati Southern Railway. The engineer in charge of the project was L. F. G. Bouscaren (1840–1904), chief engineer of the railroad company and designer of its Ohio River Bridge at Cincinnati, a structure which was built simultaneously with the Dixville Bridge and which contained the longest simple truss span in the world at the time of its erection. There is no question that Bouscaren deserves as much credit as Charles Shaler Smith for the Dixville project, but Sullivan and the rest of posterity have always honored Smith and forgotten the other half of the team.

The chief problem at Dixville was that of erecting the trusses. The Kentucky River gorge at this point is 1,200 feet wide and 275 feet deep. The river has always been subject to flash floods of disastrous proportions, a maximum rise of 40 feet in one day having already been recorded when the two engineers made their preliminary survey. The use of falsework under such conditions was out of the question. Smith originally planned to build a continuous Whipple-Murphy deck truss, 1,125 feet long, extending over three spans of 375 feet each.[5] At this point Bouscaren made the proposal that hinges be introduced into the truss at two points, one at each end of the bridge between the shore and the nearest pier. The use of hinges in a continuous beam or truss was first proposed by the German engineer and theorist Karl Culmann in his *Graphical Statics* (1866). The introduction of hinges and the resulting transformation of the parts of the beam on either side of the support into cantilevers has the consequence that the action of the member more nearly conforms to the theoretical curve of stress distribution, or stress trajectory, as it is sometimes called. The first large cantilever bridge whose design seems clearly to have been influenced by Culmann's theory was Heinrich Gerber's bridge over the Main River at Hassfurt, Germany (1867). The structure excited wide interest and was undoubtedly known to Smith and Bouscaren.

Several other factors, however, led to Bouscaren's decision. In addition to improvement in the efficiency of the truss action, the engineers were concerned to prevent the excessively high stresses which would have occurred in the continuous truss as a consequence of pier settlement. Further, the successful construction of Eads Bridge by the method of cantilevering the arches out from the abutments and piers suggested not only a similar mode of construction for continuous trusses but also the possibility of using the cantilever as a permanent structural form. By adopting Bouscaren's suggestion the designers turned the bridge into a combination of types which were, in succession from shore to shore, a semi-floating span fixed at the abutment and hinged at the free end, a 75-foot cantilever, a simple truss acting as anchor span to the cantilevers, and so on in reverse order to the opposite shore. The material of the structure throughout was wrought iron. There was, as we noted, little precedent for a bridge of this kind, and Smith and Bouscaren staked their reputations on it. They saw it through to successful completion, but the increasing weight of traffic required that it be replaced in 1911 by a steel bridge built on the same masonry.

Sullivan's intuitive grasp of the meaning of these bridges was perfectly sound. The union of science and technology which made them possible was the creation of men who possessed a rare combination of faculties: they were men who could imagine and think things and who, when they translated the products of imagination into physical fact, did so on a heroic scale. It was difficult not to be impressed, however little one understood the methods of their achievement. Sullivan was profoundly moved, and he knew that he would have to create a building art which could give voice to these powerful feelings and thus evoke them in others.

The Auditorium Building marks the initial step, although what is visible both inside and out at first seems to have little to do with the great achievements of the Age of Iron. Perhaps it is the fact of the building itself, rather than what it says in detail, that heralds a new epoch in architectural form. The exterior walls, as magnificent as they are in their architectonic power, are masonry bearing elements disposed in the long-familiar system of stout piers and arcades.

Inside this uniform block, however, is the most extraordinary diversity of internal volumes that one can house in a single structure. The huge theater, seating 4,000 people, is surrounded by a block of offices on the west and south and by a hotel on the east. Offices, hotel rooms, lobbies, small dining rooms, and other utilitarian facilities are carried on a complete system of framing composed of cast-iron columns and wrought-iron beams. The great vault of the theater is hung from a series of parallel elliptical trusses which are suspended in turn from horizontal trusses immediately under the roof. The same construction, on a smaller scale, supports the vault of the main dining room. Above the theater at one end still another type of truss is used to support the ceiling of the rehearsal room. The vaulted enclosures are in no way like the traditional barrel vaults of Roman and medieval building. They are great wide-span cylinders of elliptical or segmental section which were derived from the huge trainsheds of the nineteenth-century railway station and thus constitute a metamorphosis of a quasi-monumental utilitarian form into one element of an aesthetic complex. The system of truss framing in the Auditorium grew out of the inventions of the bridge engineers. Under the once-brilliant colors and intricate interweaving patterns of Sullivan's ornament, none of this construction is visible, yet the light screens with their plastic detail, and the multiplicity of shapes and volumes could not have been created without the structural means that Adler employed. As a matter of fact, the Auditorium embraces every basic structural technique available to the nineteenth-century builder.

The three purely commercial buildings—the Wainwright, the Prudential, and the Carson store—rest on the more advanced structural technic of complete steel framing without masonry-bearing members of any kind other than the concrete column footings, but because of the uniformity and relative simplicity of their interior spaces they are much less complex in their construction than the Auditorium. Yet it was precisely here that Sullivan saw his opportunity: now he could take full advantage of the steel frame in the treatment of the elevations, the obvious parts of the building that everyone had to see. What he was trying to articulate in the three buildings was not simply structure and utility, which the bridge engineers had done in their wholly empirical forms, but rather his complex psychological response to the structural technics that the engineers employed so boldly. The idea underlying the Wainwright and

Prudential buildings is clearly summed up in the celebrated passage of the *Autobiography* on the skyscraper. "The lofty steel frame makes a powerful appeal to the architectural imagination where there is any. . . . The appeal and the inspiration lie, of course, in the element of loftiness, in the suggestion of slenderness and aspiration, the soaring quality of a thing rising from the earth as a unitary utterance, Dionysian in beauty."[6]

The formal character of the Prudential Building (which is a larger and more refined counterpart of the Wainwright and thus may be taken to represent the essential quality of both buildings) is an organic outgrowth of its utilitarian functionalism, but it is in no way confined by it. Above an open base, designed chiefly for purposes of display, rises a uniform succession of office floors, identical in function and hence appearance, topped by an attic floor which carries heating returns and elevator machinery and whose external treatment provides a transition to the flat slab that terminates the upward motion of the whole block. Elevator shafts, plumbing, and mechanical utilities are concentrated in an inner core. Sustaining all roof, floor, wall, and wind loads is an interior steel frame covered with fireproof tile sheathing. All this constitutes the empirical answer to utilitarian necessity.

Beyond the empirical form, however, are the wholly aesthetic elements that transform structure into symbolic art. The great bay-wide windows of the base are carefully designed not only to reveal structure but to separate it clearly from all subsidiary elements and thus to give forceful utterance to its potentially dramatic quality. The columns which stand out so clearly are the forerunners of Le Corbusier's *pilotis,* now so common in contemporary building that they have become a cliché. Above the open base appears the single most striking feature, the pronounced upward vertical movement achieved by the closely ranked pier-like bands of which every other one clothes the true structural column, the alternate piers being purely formal additions without bearing function.

The basic theme of this light screen is movement, the dynamic transcendence of space and gravitational thrust, qualities Sullivan long before felt in the "floating" chains and roadway of the Merrimack suspension bridge. In a broader sense the theme suggests the underlying energy of a world of process, of evolutionary growth in living things, or the dynamic of the electric field in physics. The bridge, like the building, is not seen by Sullivan as a static thing but as something which leaps over its natural obstacle and thus becomes a living assertion of man's skill operating through his simultaneous dependence upon and command over nature. Again Sullivan's intuition led him into the right path, for this is exactly how the bridge behaves. We can sense this directly in the suspension bridge with its wire cables and suspenders: it seems alive, constantly quivering under its changing load. Although we can neither see it nor feel it, exactly the same thing is occurring in the dense and massive members of the big railroad truss as the internal stress continuously adjusts itself to the moving weight that it sustains. It took a century and a half of painstaking scientific inquiry to discover this hidden and vital activity.

The rich and intricate ornament that covers the two office buildings, an ornament created by Sullivan which died with him, offers a much more difficult problem of interpretation. It is so subjective that it is scarcely possible to find objective experiences that might have led to the feelings out of which it grew. In its complex, somewhat abstract naturalism, it appears to symbolize the biologically organic. While Sullivan sometimes allowed his ornament to flow in uncontrolled and undifferentiated profusion over much of the surface, he was generally careful to observe the limits of architectural ornamentation. By spreading it in low relief over whole elevations, and by confining a particular pattern to the surface of a certain kind of structural member— column, or spandrel beam—he was able to distinguish in a striking way the separate structural surfaces. Thus his ornament enhances the major elements of the structure and further heightens the vivid sense of movement. It also seems to suggest the diversity underlying the unitary organic statement.

In the Carson Pirie Scott Store Sullivan turned to an entirely different kind of expression, one derived from the dominant mode of the work of the Chicago school. Where he used the close vertical pattern in the older buildings, in the department store he opened the main elevations into great cellular screens which exactly express the neutral steel cage behind them. The form was dictated initially by the requirement for maximum natural light in the store, but again, in many subtle ways, he translated the practical functionalism into art. If the theme of the Wainwright and the Prudential is movement, that of the Carson store is power. Here the elaborate interplay of tension and compression, of thrust and counterthrust in the bridge truss is given a heightened and dramatic statement by means so delicate as almost to escape notice—the careful calculation of the depth of the window reveals and the breadth of the terra-cotta envelope on the columns and spandrels, the narrow band of ornament that enframes the window, the even narrower band that extends continuously along each sill and lintel line to give the whole façade a tense, subdued horizontality. The base in its sheath of ornament is an exact reversal of that of the office buildings. In the Carson store it is a weightless screen, glass and opaque covering (cast iron) forming one unbroken plane, making the cellular wall above literally seem to float free of the earth below it.

In the last analysis Sullivan's civil architecture is a celebration of technique, as is most of the contemporary architecture of which he was the foremost pioneer. But Sullivan had carried the expression far beyond the rather sterile geometry that characterizes most building today. If his work seems limited beside the vastly richer symbolism of medieval and baroque architecture, we may at least say that he was responding to the one coherent order that was discernible in the contradictory currents of nineteenth- and twentieth-century culture. In the absence of a cosmos in which man was conceived to be the central figure, the scientific technology on which building increasingly depended became the one sure basis of architectural and civic art. It is Sullivan's achievement to have understood how this basis could be transmuted into an effective and valid artistic statement.

REFERENCES

[1] Louis Sullivan, *Kindergarten Chats* (New York, 1947), pp. 28–30. By permission of Wittenborn, Schultz, Inc.

[2] *The Autobiography of an Idea* (New York, 1926), pp. 82–85. By permission of the American Institute of Architects.

[3] *Ibid.*, pp. 246–248.

[4] *Ibid.*, p. 248.

[5] The Whipple-Murphy truss was a variation on the form patented by Squire Whipple in 1847. Its distinguishing characteristic is the fact that the diagonal members of the web slope in one direction in any one half of the truss and cross two panels formed by the posts, instead of the usual one. A continuous truss is one that is carried on intermediate supports as well as those at the ends, that is, it extends continuously over more than two supports. In a deck-truss bridge the deck, or track level, is located at the level of the top chord of the truss.

[6] *The Autobiography of an Idea,* pp. 313–314.

Part III
Man and Machines

Introduction

Man is fascinated by his machines. On the one hand, he admires them as objects of his own creation; they are extensions of man's hereditary organic equipment and products of his own creative imagination and skills, allowing him to amplify his muscle power and multiply his mental potentialities. On the other hand, he sometimes fears his own machines; they seem to take on inhuman qualities and are regarded as divorced from the essence of humanity; indeed, some of man's infernal machines—for example, nuclear bombs—threaten to destroy all mankind, and his use—or misuse—of technology appears to threaten the quality of human life, as well as the life of all creatures, by environmental changes which disturb the ecological balance. What is more, man fears that some of the monster automatons which he has recently created will make him expendable as a worker—and even deprive him of his humanity.

The curiously ambivalent attitude of man toward the machines of his own creation has given rise to much discussion on the man-machine interface. Are machines something apart and distinct from man, or do they form a part of man? How does he make or use them to perform his will, or, once made, do machines have a will of their own?

Peter Drucker leaves no question about the importance of technology in enabling man to evolve and survive. "Man, alone of all animals," he says, "is capable of purposeful, nonorganic evolution; he makes tools." But, he points out, the approach from human biological evolution shows that technology is much more than tools, processes, and products; technology is about work: "the specifically human activity by means of which man pushes back the limitations of the iron biological law which condemns all other animals to devote all their time and energy to keeping themselves alive for the next day, if not for the next hour." Tools and techniques indicate what and how work is done, but the structure and organization of work in turn influence the tools and techniques. The subject matter of the study of technology, he concludes, is human work; we can only understand the development of a tool or technic by knowing and understanding its relationship to work. Technology thus deals with more than machines; it is the study of man's development of machines and man's use of them in terms of work.

Lewis Mumford views the matter quite differently. To him, technology has not been the main operative agent in man's development; tools and machines have been overrated as to their impact on the emergence of our species. Instead, he emphasizes the elaboration of symbolic culture —language, abstract thought, ritual, song-and-dance forms —as answering more imperious needs than those involving control of the external environment through work and tools. Indeed, he despairs over the fact that technology has made human life work-centered or power-centered. Modern industrial technology, copying the collective "megamachines" which enabled pharaohs to build the pyramids, has systematically disassociated work from the rest of life. Though the labor machine, both in antiquity and the present, lends itself to vast constructive enterprises and an outpouring of goods, it has also resulted in military machines which bring on colossal destruction and human extermination. He thus calls for a "life-centered technology" and demands the reestablishment of human autonomy rather than a continuing development of those human functions that merely serve the machine.

Contradicting Mumford's view of the dichotomy between man and machine is Bruce Mazlish's attempt to eliminate the "discontinuity" between man and machines by pointing out that machines have now evolved to the point where man's nature is one with his tools and machines. If man feels threatened by his machines and is "alienated" (that is, out of harmony with himself), it is because he is out of harmony with the machines that are part of himself. The new computers which man has produced have a feedback mechanism and a generalizing reason which demonstrates their continuity with man's own physical and mental capabilities. The contemporary revolution in computers and automation puts man on the threshold of breaking past the discontinuity between himself and machines.

The remaining selections in Part Three provide examples

of man–machine relationships which illustrate human ingenuity, imagination, and skill employed with the purpose of achieving various human goals and aspirations. In the process, some of our authors destroy old clichés regarding the evolution of man's techniques and his use of machines.

C. St. C. Davison, for example, reaches far back into history to contradict the postulate of earlier historians that man's ability to move heavy objects followed an evolution from sledge, through rollers, to wheels. In their eagerness to assume that technology followed this logical path, scholars have ignored the evidence and assumed that rollers *must* have preceded the wheel, and they endeavored to find evidence of this in bas-reliefs and pictures depicting the movement of heavy blocks of stone in early Assyria and Egypt. Using the same evidence, Davison is able to demolish the notion that loose rollers formed a stage in the evolution between the sledge and wheeled wagons; instead, he proves that lubricated boards were both more practical and efficient than rollers would have been, and that they did indeed provide the means for transporting sixty-ton statues. In other words, the evolution from sledge to wheeled wagon was direct and immediate, and not through an intermediate stage of rollers.

While Davison destroys the technological myth of rollers, Lynn White, Jr., demonstrates that the legend of Eilmer of Malmesbury might indeed be fact. Pursuing man's age-old aspiration to fly, Eilmer, an Anglo-Saxon Benedictine monk, actually succeeded in flying some 600 feet at the beginning of the eleventh century. Unfortunately, poor Eilmer had neglected to put a tail on himself, and his failure in aerodynamics caused his downfall—and two broken legs! In the process of fitting this episode into the history of heavier-than-air flight, White also succeeds in showing the originality and ingenuity of technology in the medieval period.

Finally, Wayne D. Rasmussen, in his case study of a modern advance in agricultural technology, the mechanical tomato harvester, shows how contributions from many different technologies—a systems approach, or a "package" of technical applications—help to increase efficiency. But perhaps most interesting is the fact that not only were techniques and machines adjusted to the peculiar requirements of the tomato, but the tomato itself was re-designed by the breeding of new varieties which lent themselves to handling by machine. In other words, technological ingenuity resulted in the redesigning of nature in order to fit the requirements of the machine!

Work and Tools

Peter F. Drucker

I

Man, alone of all animals, is capable of purposeful, non-organic evolution; he makes tools. This observation by Alfred Russell Wallace, co-discoverer with Darwin of the theory of evolution, may seem obvious if not trite. But it is a profound insight. And though made some seventy or eighty years ago, its implications have yet to be thought through by biologists and technologists.

One such implication is that from a biologist's (or a historian's) point of view, the technologist's identification of tool with material artifact is quite arbitrary. Language, too, is a tool, and so are all abstract concepts. This does not mean that the technologist's definition should be discarded. All human disciplines rest after all on similarly arbitrary distinctions. But it does mean that technologists ought to be conscious of the artificiality of their definition and careful lest it become a barrier rather than a help to knowledge and understanding.

This is particularly relevant for the history of technology, I believe. According to the technologist's definition of "tool," the abacus and the geometer's compass are normally considered technology, but the multiplication table or a table of logarithms are not. Yet this arbitrary division makes all but impossible the understanding of so important a subject as the development of the technology of mathematics. Similarly the technologist's elimination of the fine arts from his field of vision blinds the historian of technology to

This article originally appeared in the first issue of Technology and Culture: *28–37, and has been reprinted in Peter F. Drucker,* Technology, Management and Society *(New York, 1970). For a note on Dr. Drucker, see p. 41.*

an understanding of the relationship between scientific knowledge and technology. (See, for instance, volumes III and IV of Singer's monumental *History of Technology*.) For scientific thought and knowledge were married to the fine arts, at least in the West, long before they even got on speaking terms with the mechanical crafts: in the mathematical number theories of the designers of the Gothic cathedral,[1] in the geometric optics of Renaissance painting, or in the acoustics of the great Baroque organs. And Lynn White, Jr., has shown in several recent articles that to understand the history and development of the mechanical devices of the Middle Ages we must understand something so nonmechanical and nonmaterial as the new concept of the dignity and sanctity of labor which St. Benedict first introduced.

Even within the technologist's definition of technology as dealing with mechanical artifacts alone, Wallace's insight has major relevance. The subject matter of technology according to the preface to *A History of Technology* is "how things are done or made"; and most students of technology to my knowledge agree with this. But the Wallace insight leads to a different definition: the subject matter of technology would be "how man does or makes." As to the meaning and end of technology, the same source, again presenting the general view, defines them as "mastery of his (man's) natural environment." Oh no, the Wallace insight would say (and in rather shocked tones): the purpose is to overcome man's own natural, i.e., animal, limitations. Technology enables man, a land-bound biped, without gills, fins or wings, to be at home in the water or in the air. It enables an animal with very poor body insulation, that is, a subtropical animal, to live in all climate zones. It enables one of the weakest and slowest of the primates to add to his own strength that of elephant or ox, and to his own speed that of the horse. It enables him to push his life span from his "natural" twenty years or so to threescore years and ten; it even enables him to forget that natural death is death from predators, disease, starvation, or accident, and to call death from natural causes that which has never been observed in wild animals: death from organic decay in old age.[2]

These developments of man have, of course, had impact on his natural environment—though I suspect that until recent days the impact has been very slight indeed. But this impact on nature outside of man is incidental. What really matters is that all these developments alter man's biological capacity—and not through the random genetic mutation of biological evolution but through the purposeful nonorganic development we call "technology."

What I have called here the "Wallace insight," that is, the approach from human biology, thus leads to the conclusion that technology is not about things: tools, processes, and products. It is about work: the specifically human activity by means of which man pushes back the limitations of the iron biological law which condemns all other animals to devote all their time and energy to keeping themselves alive for the next day, if not for the next hour. The same conclusion would be reached, by the way, from any approach, for instance, from that of the anthropologist's "culture," that does not mistake technology for a phenomenon of the physical universe. We might define technology as human action on physical objects or as a set of physical objects characterized by serving human purposes. Either way the realm and subject matter of the study of technology would be human work.

II

For the historian of technology this line of thought might be more than a quibble over definitions. For it leads to the conclusion that the study of the development and history of technology, even in its very narrowest definition as the study of one particular mechanical artifact (either tool or product) or a particular process, would be productive only within an understanding of work and in the context of the history and development of work.

Not only must the available tools and techniques strongly influence what work can and will be done, but how it will be done. Work, its structure, organization, and concepts must in turn powerfully affect tools and techniques and their development. The influence, one would deduce, should be so great as to make it difficult to understand the development of the tool or of the technique unless its relationship to work was known and understood. Whatever evidence we have strongly supports this deduction.

Systematic attempts to study and to improve work only began some seventy-five years ago with Frederick W. Taylor. Until then work had always been taken for granted by everyone—as it is still, apparently, taken for granted by most students of technology. "Scientific management," as Taylor's efforts were called misleadingly ("scientific work study" would have been a better term and would have avoided a great deal of confusion), was not concerned with technology. Indeed, it took tools and techniques largely as given and tried to enable the individual worker to manipulate them more economically, more systematically, and more effectively. And yet this approach resulted almost immediately in major changes and development in tools, processes, and products. The assembly line with its conveyors was an important tool change. An even greater change was the change in process that underlay the switch from building to assembling a product. Today we are beginning to see yet another powerful consequence of Taylor's work on individual operations: the change from organizing production around the doing of things to things, to organizing production around the flow of things and information, the change we call "automation."

A similar, direct impact on tools and techniques is likely to result from another and even more recent approach to the study and improvement of work: the approach called variously "human engineering," "industrial psychology," or "industrial physiology." Scientific management and its descendants study work as operation; human engineering and its allied disciplines are concerned with the relationship between technology and human anatomy, human perception, human nervous system, and human emotion. Fatigue studies were the earliest and most widely known examples; studies of sensory perception and reaction, for instance of airplane pilots, are among the presently most active areas of investigation, as are studies of learning. We have barely scratched the surface here; yet we know already that these studies are leading us to major changes in the theory and design of instruments of measurement and control, and into the re-design of traditional skills, traditional tools, and traditional processes.

But of course we worked on work, if only through trial and error, long before we systematized the job. The best example of scientific management is after all not to be found in our century: it is the alphabet. The assembly line as a concept of work was understood by those unknown geniuses who, at the very beginning of historical time, replaced the aristocratic artist of warfare (portrayed in his last moments of glory by Homer) by the army soldier with his uniform equipment, his few repetitive operations, and his regimented drill. The best example of human engineering is still the long handle that changed the sickle into the scythe, thus belatedly adjusting reaping to the evolutionary change that had much earlier changed man from crouching quadruped into upright biped. Everyone of these developments in work had immediate and powerful impact on tools, process, and product, that is, on the artifacts of technology.

The aspect of work that has probably had the greatest impact on technology is the one we know the least about: the organization of work.

Work, as far back as we have any record of man, has always been both individual and social. The most thoroughly collectivist society history knows, that of Inca Peru, did not succeed in completely collectivizing work; technology, in particular, the making of tools, pottery, textiles, cult objects, remained the work of individuals. It was personally specialized rather than biologically or socially specialized, as is work in a beehive or in an ant heap. The most thoroughgoing individualist society, the perfect market model of classical economics, presupposed a tremendous amount of collective organization in respect to law, money and credit, transportation, and so on. But precisely because individual effort and collective effort must always be calibrated with one another, the organization of work is not determined. To a very considerable extent there are genuine alternatives here, genuine choices. The organization of work, in other words, is in itself one of the major means of that purposeful and nonorganic evolution which is specifically human; it is in itself an important tool of man.

Only within the very last decades have we begun to look at the organization of work.[3] But we have already learned that the task, the tools, and the social organization of work are not totally independent but mutually influence and affect one another. We know, for instance, that the almost pre-industrial technology of the New York women's dress industry is the result not of technological, economic, or market conditions but of the social organization of work which is traditional in that industry. The opposite has been proven, too: When we introduce certain tools into locomotive shops, for instance, the traditional organization of work, the organization of the crafts, becomes untenable; and the very skills that made men productive under the old technology now become a major obstacle to their being able to produce at all. A good case can be made for the hypothesis that modern farm implements have made the Russian collective farm socially obsolete as an organization of work, have made it yesterday's socialist solution of farm organization rather than today's, let alone tomorrow's.

This interrelationship between organization of work, tasks, and tools must always have existed. One might even speculate that the explanation for the mysterious time gap between the early introduction of the potter's wheel and the so very late introduction of the spinning wheel lies in the social organization of spinning work as a group task performed, as the Homeric epics describe it, by the mistress working with her daughters and maids. The spinning wheel with its demand for individual concentration on the machinery and its speed is hardly conducive to free social intercourse; even on a narrowly economic basis, the governmental, disciplinary, and educational yields of the spinning bee may well have appeared more valuable than faster and cleaner yarn.

If we know far too little about work and its organization scientifically, we know nothing about it historically. It is not lack of records that explains this, at least not for historical times. Great writers—Hesiod, Aristophanes, Virgil, for instance—have left detailed descriptions. For the early empires and then again for the last seven centuries, beginning with the high Middle Ages, we have an abundance of pictorial material: pottery and relief paintings, woodcuts, etchings, prints. What is lacking is attention and objective study.

III

The political historian or the art historian, still dominated by the prejudices of Hellenism, usually dismisses work as beneath his notice; the historian of technology is "thing-focused." As a result we not only still repeat as fact traditions regarding the organization of work in the past which both our available sources and our knowledge of the organization of work would stamp as old wives' tales, but we also deny ourselves a fuller understanding of the already existing and already collected information regarding the history and use of tools.

One example of this is the lack of attention given to materials-moving and materials-handling equipment. We know that moving things—rather than fabricating things—is the central effort in production. But we have paid little attention to the development of materials-moving and materials-handling equipment.

The Gothic cathedral is another example. H. G. Thomson in *A History of Technology* (II, 384) states, for instance, flatly, "there was no exact medieval equivalent of the specialized architect" in the Middle Ages; there was only "a master mason." But we have overwhelming evidence to the contrary (summarized, for instance, in Simson[4]); the specialized, scientifically trained architect actually dominated. He was sharply distinguished from the master mason by training and social position. Far from being anonymous, as we still commonly assert, he was a famous man, sometimes with an international practice ranging from Scotland to Poland to Sicily. Indeed, he took great pains to make sure that he would be known and remembered, not only in written records but above all by having himself portrayed in the churches he designed in his full regalia as a scientific geometer and designer—something even the best known of today's architects would hesitate to do. Similarly we still repeat early German Romanticism in the belief that the Gothic cathedral was the work of individual craftsmen. But the structural fabric of the cathedral was based on strict uniformity of parts. The men worked to moulds which were collectively held and administered as the property of the guild. Only roofing, ornaments, doors, statuary, windows, and so on, were individual artists' work. Considering both the extreme scarcity of skilled people and the heavy dependence on local, unskilled labor from the countryside, to which all our sources attest, there must also have been a sharp division between the skilled men who made parts and the unskilled who assembled them under the direction of a foreman or a gang boss. There must thus have been a fairly advanced materials-handling technology which is indeed depicted in our sources but neglected by the historians of technology with their uncritical Romanticist bias. And while the moulds to which the craftsman worked are generally mentioned, no one, to my knowledge, has yet investigated so remarkable a tool, and one that so completely contradicts all we otherwise believe we know about medieval work and technology.

I do not mean to suggest that we drop the historical study of tools, processes, and products. We quite obviously need to know much more. I am saying first that the history of work is in itself a big, rewarding, and challenging area which students of technology should be particularly well equipped to tackle. I am saying also that we need work on work if the history of technology is truly to be history and not just the engineer's antiquarianism.

IV

One final question must be asked: Without study and understanding of work, how can we hope to arrive at an understanding of technology?

Singer's great *History of Technology* abandons the attempt to give a comprehensive treatment of its subject with 1850; at that time, the editors tell us, technology became so complex as to defy description, let alone understanding. But it is precisely then that technology began to be a central force and to have major impact both on man's culture and on man's natural environment. To say that we cannot encompass modern technology is very much like saying that medicine stops when the embryo issues from the womb. We need a theory that enables us to organize the variety and complexity of modern tools around some basic, unifying concept.

To a layman who is neither professional historian nor professional technologist, it would moreover appear that even the old technology, the technology before the great explosion of the last hundred years, makes no real sense and cannot be understood, can hardly even be described, without such basic concepts. Every writer on technology acknowledges the extraordinary number, variety, and complexity of factors that play a part in technology and are in turn influenced by it: economy and legal system, political institutions and social values, philosophical abstractions, religious beliefs, and scientific knowledge. No one can know all these, let alone handle them all in their constantly shifting relationship. Yet all of them are part of technology in one way or another, at one time or another.

The typical reaction to such a situation has of course always been to proclaim one of these factors as *the* determinant—the economy, for instance, or the religious beliefs. We know that this can only lead to complete failure to understand. These factors profoundly influence but do not determine each other; at most they may set limits to each other or create a range of opportunities for each other. Nor can we understand technology in terms of the anthropologist's concept of culture as a stable, complete, and finite balance of these factors. Such a culture may exist among small, primitive, decaying tribes, living in isolation. But this is precisely the reason why they are small, primitive, and decaying. Any viable culture is characterized by capacity for internal self-generated change in the energy level and direction of any one of these factors and in their interrelationships.

Technology, in other words, must be considered as a system,[5] that is, a collection of interrelated and intercommunicating units and activities.

We know that we can study and understand such a system only if we have a unifying focus where the interaction of *all* the forces and factors within the system registers some discernible effect, and where in turn the complexities of the system can be resolved in one theoretical model. Tools, processes, products, are clearly incapable of providing such focus for the understanding of the complex system we call "technology." It is just possible, however, that work might provide the focus, might provide the integration of all these interdependent, yet autonomous variables, might provide one unifying concept which will enable us to understand technology both in itself and in its role, its impact on and relationships with values and institutions, knowledge and beliefs, individual and society.

Such understanding would be of vital importance today. The great, perhaps the central, event of our times is the disappearance of all non-Western societies and cultures under the inundation of Western technology. Yet we have no way of analyzing this process, of predicting what it will do to man, his institutions and values, let alone of controlling it, that is, of specifying with any degree of assurance what needs to be done to make this momentous change productive or at least bearable. We desperately need a real understanding, and a real theory, a real model of technology.

History has never been satisfied to be a mere inventory of what is dead and gone—that indeed is antiquarianism. True history always aims at helping us understand ourselves, at helping us make what shall be. Just as we look to the historian of government for a better understanding of government, and to the historian of art for a better understanding of art, so we are entitled to look to the historian of technology for a better understanding of technology. But how can he give us such an understanding unless he himself has some concept of technology and not merely a collection of individual tools and artifacts? And can he develop such a concept unless work rather than things becomes the focus of his study of technology and of its history?

REFERENCES

[1] S. B. Hamilton only expresses the prevailing view of technologists when he says (in Singer's *A History of Technology,* IV, 469) in respect to the architects of the Gothic cathedral and their patrons that there is "nothing to suggest that either party was driven or pursued by any theory as to what would be beautiful." Yet we have overwhelming and easily accessible

evidence to the contrary; both architect and patron were not just "driven," they were actually obsessed by rigorously mathematical theories of structure and beauty. See, for instance, Sedlmayer, *Die Entstehung der Kathedrale* (Zurich, 1950); Von Simson, *The Gothic Cathedral* (New York, 1956); and especially the direct testimony of one of the greatest of the cathedral designers, Abbot Suger of St. Denis, in *Abbot Suger and the Abbey Church of St. Denis,* ed. Erwin Panofsky (Princeton, 1946).

[2] See on this P. B. Medawar, the British biologist, in "Old Age and Natural Death" in his *The Uniqueness of the Individual* (New York, 1957).

[3] Among the studies ought to be mentioned the work of the late Elton Mayo, first in Australia and then at Harvard, especially his two slim books: *The Human Problems of an Industrial Civilization* (Boston, 1933) and *The Social Problems of an Industrial Civilization* (Boston, 1945); the studies of the French sociologist Georges Friedmann, especially his *Industrial Society* (Glencoe, Illinois, 1955); the work carried on at Yale by Charles Walker and his group, especially the book by him and Robert H. Guest: *The Man on the Assembly Line* (Cambridge, Mass., 1952). I understand that studies of the organization of work are also being carried out at the Polish Academy of Science but I have not been able to obtain any of the results.

[4] Von Simson, *op. cit.,* pp. 30 ff.

[5] The word is here used as in Kenneth Boulding's "General Systems Theory—The Skeleton of Science," *Management Science,* II, no. 3 (April 1956), 197, and in the publications of the Society for General Systems Research.

Technics and the Nature of Man

Lewis Mumford

The last century, we all realize, has witnessed a radical transformation in the entire human environment, largely as a result of the impact of the mathematical and physical sciences upon technology. This shift from an empirical, tradition-bound technics to an experimental scientific mode has opened up such new realms as those of nuclear energy, supersonic transportation, cybernetic intelligence, and instantaneous planetary communication.

In terms of the currently accepted picture of the relation of man to technics, our age is passing from the primeval state of man, marked by his invention of tools and weapons for the purpose of achieving mastery over the forces of nature, to a radically different condition, in which he will not only have conquered nature but detached himself completely from the organic habitat. With this new megatechnics, man will create a uniform, all-enveloping structure, designed for automatic operation. Instead of functioning actively as a tool-using animal, man will become a passive, machine-serving animal whose proper functions, if this process continues unchanged, will either be fed into a machine, or strictly limited and controlled for the benefit of depersonalized collective organizations. The ultimate tendency of this development was correctly anticipated by Samuel Butler more than a century ago; but it is only now

This paper is Mr. Mumford's address at the Smithson Bicentennial celebration, held in Washington, D.C., in September 1965, to commemorate the birth of James Smithson. The paper was published in Technology and Culture *(Vol. 7, No. 3): 303–17, by permission of the Smithsonian Institution. For a note on Mr. Mumford, see p. 50.*

LEWIS MUMFORD

that his playful fantasy shows many signs of becoming a far-from-playful reality.

My purpose in this paper is to question both the assumptions and the predictions upon which our commitment to the present form of technical and scientific progress, as an end itself, has been based. In particular, I find it necessary to cast doubts upon the generally accepted theories of man's basic nature which have been implicit during the last century in our constant overrating of the role of tools and machines in the human economy. I shall suggest that not only was Karl Marx in error in giving the instruments of production a central place and a directive function in human development, but even the seemingly benign interpretation by Teilhard de Chardin[1] reads back into the whole story of man the narrow technological rationalism of our own age, and projects into the future a final state in which all the further possibilities of human development would come to an end, because nothing would be left of man's original nature that had not been absorbed into, if not suppressed by, the technical organization of intelligence into a universal and omnipotent layer of mind.

Since the conclusions I have reached require, for their background, a large body of evidence I have been marshaling in a still unpublished book, I am aware that the following summary must, by its brevity, seem superficial and unconvincing. At best, I can only hope to show that there are serious reasons for reconsidering the whole picture of both human and technical development upon which the present organization of Western society is based.

Now, we cannot understand the role that technics has played in human development without a deeper insight into the nature of man; yet that insight has itself been blurred during the last century because it has been conditioned by a social environment in which a mass of new mechanical inventions had suddenly proliferated, sweeping away many ancient processes and institutions, and altering our very conception of both human limitations and technical possibilities.

For a century, man has been habitually defined as a tool-using animal. This definition would have seemed strange to Plato, who attributed man's rise from a primitive state as much to Marsyas and Orpheus as to Prometheus and Hephaestos, the blacksmith-god. Yet the description of man as essentially a tool-using and tool-making animal has become so firmly accepted that the mere finding of the fragments of skulls, in association with roughly shaped pebbles, as with S. L. Leakey's Australopithecines, is deemed sufficient to identify the creature as a protohuman, despite marked physical divergences from both earlier apes and later men.[2]

Two substantial errors are imbedded in this general interpretation. The first is the unintentional distortion of evidence, through the fact that the only durable remains of either early man or his hominid ancestors are an extremely scanty supply of bones and stones, presumably tools—though apart from grubbing, pounding, and ripping, one can only guess what purpose they served. The durability of stone artifacts has given this part of man's technical equipment a prominence it could never have claimed if the far richer store of organic materials, which early man shared with many primate ancestors, had been preserved. But it is an illusion to suppose that man's technical development was confined to exploiting stone quarries, to flint-chipping, to the manipulation of tools alone; for the source of man's early technics was the whole natural environment —edible plants and animals, vines, leaves, shells, reeds, twigs, bark, fiber, skins—all of which, save the last, can be utilized without other tools than man's own unaided teeth and hands.

By fastening attention on the surviving stone artifacts, many anthropologists and ethnologists have gratuitously attributed to the shaping and using of tools the enlargement of the human brain and therewith the development of man's higher intelligence, though motor-sensory coordinations involved in this elementary manufacture do not call for or evoke any considerable mental acuteness. Since the subhominids of South Africa had a brain capacity about a third of Homo sapiens, no greater indeed than that of many apes, the capacity to make tools neither called for nor generated early man's rich cerebral equipment, as Ernst Mayr of Harvard has recently pointed out.[3]

The second error in interpreting man's nature is a less pardonable one, since Francis Bacon should long ago have put scientists on guard against it; and that is the current tendency to read back into prehistoric times modern man's overwhelming interest in tools and machines, to the exclusion of equally important items of technical equipment. Tools and weapons are specialized extrapolations of man's own organs for pushing, pounding, crunching, cutting, stabbing—all basic motor activities. No one can doubt that these dynamic processes, which man shares with many other species, formed an essential part of his earliest technical complex.

But just because man's need for tools is so obvious, we must guard against overemphasizing the role of tools hundreds of thousands of years before they became functionally efficient. In treating tool-making as central to the paleolithic economy, ethnologists have underplayed, or neglected, a mass of activities in which many other species were for long far more knowledgeable than man. Despite the contrary evidence put forward by R. U. Sayce,[4] Daryll Forde,[5] and André Leroi-Gourhan,[6] there is still a tendency to identify tools and machines with technology—to substitute a part for the whole. Even in describing only the material components of technics, this practice overlooks the equally vital role of containers: hearths, pits, houses, pots, sacks, clothes, traps, bins, byres, baskets, bags, ditches, reservoirs, canals, cities. These static components play an important part in every technology, not least in our own day, with its high-tension transformers, its giant chemical retorts, its atomic reactors.

In any comprehensive definition of technics, it should be plain that many insects, birds, and mammals had made far more radical innovations in the fabrication of containers, with their intricate nests and bowers, their geometric beehives, their urbanoid anthills and termitaries, than man's ancestors had achieved in the making of tools until the emergence of Homo sapiens. In short, if technical proficiency were alone sufficient to identify potential intelligence, man would for long have rated as a hopeless duffer alongside many other species. The consequences of this perception should be plain, namely, that there was nothing uniquely human in early technology until it was modified by linguistic symbols and aesthetic designs. At that point, the human mind, not just the hand, made a profound difference.

At the beginning, then, I suggest that the human race had achieved no special position by reason of its tool-using or tool-making propensities alone. Or rather, man at the beginning possessed one primary all-purpose tool that was more important than any later assemblage, namely his own mind-activated body, every part of it, not just those motor activities that produced hand axes and wooden spears. To compensate for his extremely primitive working gear, early man had a much more important asset that widened his whole technical horizon: he had far richer biological equipment than any other animals, a body not specialized for any single activity but, precisely because of its extraordinary plasticity, more effective in using a larger portion of both his external environment and his internal psychosomatic resources.

Through man's overdeveloped and incessantly active brain, he had more mental energy to tap than he needed for survival at a purely animal level; and he was accordingly under the necessity of canalizing that energy, not just into food-getting and reproduction, but into modes of living that would convert this energy more directly and constructively into appropriate cultural—that is, symbolic—forms. Cultural work, by necessity, took precedence over manual work; this involved far more than the discipline of hand, muscle, and eye in making and using tools. It likewise demanded a control of all man's biological functions, including his bodily organs, his emotions, his sexual activities, his dreams. Even the hand was no mere horny work tool: it stroked a lover's body, held a baby close to the breast, made significant gestures, or expressed in ordered dance and shared ritual some otherwise inexpressible sentiment about life or death, a remembered past or an anxious future. Tool technics is but a fragment of biotechnics: man's total equipment for life.

On this view, one may well hold it an open question whether the standardized patterns and the repetitive order which came to play such an effective part in the development of tools from an early time on, as Braidwood has pointed out,[7] derive solely from tool-making. Do they not rather derive even more, perhaps, from the forms of ritual, song, and dance forms that exist in a state of perfection among primitive peoples, often in a far more exquisitely finished state than their tools? There is, in fact, widespread evidence, first noted by Hocart,[8] that ritual exactitude in ceremony preceded mechanical exactitude in work, that the first rigorous division of labor came through specialization in ceremonial offices.

These facts help to explain why simple peoples who easily get bored by purely mechanical tasks that might improve their physical well-being will, nevertheless, repeat a meaningful ritual, often to the point of physical exhaustion. The debt of technics to play and to play toys, to myth and fantasy, to magic rite and religious rote, which I called attention to in *Technics and Civilization,* has still to be sufficiently recognized, though J. Huizinga, in *Homo Ludens,* went so far as to treat play itself as the basic formative element in all culture.

Tool-making, in the narrow technical sense, may indeed go back to our hominid African ancestors. But the technical equipment of "Chellean" and Acheulian times remained extremely limited until a more richly endowed creature, with a nervous system nearer to that of Homo sapiens than to any primeval hominid predecessors, had come into existence and brought into operation not alone his hands and legs but his entire body and mind, projecting them not just in tools and utensils but in more purely symbolic nonutilitarian forms.

In this revision of the accepted technological stereotypes, I would go even further: for I submit that at every stage man's technological expansions and transformations were less for the purpose of increasing the food supply or controlling nature than for utilizing his immense actual resources and expressing his latent potentialities in order to fulfill more adequately his own unique superorganic needs. When not threatened by hostile environmental pressures, man's elaboration of symbolic culture answered a more imperious need than that for control over the external environment—and, as one must infer, largely predated it and for long outpaced it. Leslie White deserves credit for giving due weight to this fact by his emphasis on "minding" and "symboling."[9]

On this reading, the invention of language—a culmination of man's more elementary forms of expressing and transmitting meaning—was incomparably more important to further human development than the chipping of a mountain of hand-axes. Besides the relatively simple coordinations required for tool-using, the delicate interplay of the many organs needed for the creation of articulate speech was a far more striking advance and must have occupied a great part of early man's time, energy, and mental concentration, since its collective product, language, was infinitely more complex and sophisticated at the dawn of civilization than the Egyptian or Mesopotamian kit of tools. For only when knowledge and practice could be stored in symbolic forms and passed on by word of mouth from generation to generation was it possible to keep each fresh cultural acquisition from dissolving with the passing moment or the dying generation. Then, and then only, did the domestication of plants and animals become possible. Need I remind you that this decisive technical transformation,

sometimes termed a "revolution," was achieved with no better tools than the digging stick, the ax, the mattock? The
plow, like the cart wheel, came later as a specialized adaptation to the large-scale cultivation of grain.

To consider man as primarily a tool-using animal, then,
is to overlook the main chapters of human prehistory. Even
when one considers the technical milieu alone, this concentration on the dynamic components results in our having
long treated as negligible that vast area of technics in which
man's control of chemical changes through heat and fermentation, leaching and sterilizing—as in brewing, burning,
melting, tanning, cooking—favorably modified the conditions of human existence. Opposed to this stereotype is the
view that man is pre-eminently a mind-using, self-mastering
animal; and the primary locus of all his activities is his
own organism. Until he had made something of himself, he
could make little of the world around him.

In this process of self-discovery and self-transformation,
technics, in the narrow sense, served well as a subsidiary
instrument, but not as the main operative agent in man's
development; for technics was never till our own age dis-
sociated from the larger cultural whole in which man—
as man—has always functioned. Early man's original development was based upon what André Varagnac happily
called "technology of the body": the utilization of man's
highly plastic bodily capacities for the expression of his
still unformed and uninformed mind, before that mind had
yet achieved, through the development of symbols and
images, its own more etherealized technical instruments.[10]
From the beginning, the creation of significant modes of
symbolic expression, rather than more effective tools, was
the basis of Homo sapiens' further development. In approaching this conclusion I happily find myself reaching,
by a quite independent route, Levi-Strauss's conception of
cultural determinants.[11]

Unfortunately, so firmly were the prevailing nineteenth-
century conceptions committed to the notion of man as
primarily *Homo faber,* the tool-maker, rather than *Homo
sapiens,* the mind-maker, that, as you know, the first discovery of the art of the Altamira caves was dismissed as a
hoax because the leading paleo-ethnologists would not admit that the Ice Age hunters whose weapons and tools they
had recently discovered could have had either the leisure or
the mental inclination to produce art—not crude forms, but
images that showed powers of observation and abstraction

of a high order. But when we compare the carvings and paintings of the Aurignacian or Magdalenian finds with their surviving technical equipment who shall say whether it is art or technics that shows the highest development? Even the finely finished Solutré laurel leaf points were plainly a gift of aesthetically sensitive artisans to functional efficiency. The Greek form for "technics" makes no distinction between industrial production and symbolic art; and for the greater part of human history these aspects were inseparable, one side respecting objective conditions and functions, the other responding to subjective needs.

Our age has not yet overcome the peculiar utilitarian bias that regards technical invention as primary and aesthetic expression as secondary or superfluous; and this means that we have still to acknowledge that technics derives from the whole man, in his intercourse with every part of the environment, utilizing every aptitude in himself to make the most of his own biological and ecological potentials.

Even at the earliest stage, trapping and foraging called less for tools than for sharp observation of animal habits and habitats, backed by a wide experimental sampling of plants and a shrewd interpretation of the effects of various foods, medicines, and poisons upon the human organism. And in those horticultural discoveries which, if Oakes Ames was right,[12] must have preceded by many thousands of years the active domestication of plants, taste and formal beauty played a part no less than their food value; so that the earliest domesticates, other than the grains, were often valued for the color and form of their flowers, for their perfume, their texture, their spiciness, rather than merely for nourishment. Edgar Anderson has suggested that the neolithic garden, like the gardens in many simpler cultures today, was probably a mixture of food plants, dye plants, and ornamentals—all treated as equally essential for life.[13]

Similarly, some of early man's most daring technical experiments had nothing whatever to do with the mastery of the external environment: they were concerned with the anatomical modification or the superficial decoration of the human body for sexual emphasis, self-expression, or group identification. The Abbé Breuil found evidence of such practices as early as the Mousterian culture, which served equally in the development of ornament and surgery.[14] Plainly, tools and weapons, so far from dominating man's technical equipment, as the petrified artifacts too glibly suggested, constituted a small part of the biotechnical assemblage; and the struggle for existence, though sometimes severe, did not engross the energy and vitality of early man or divert him from his more central need to bring order and meaning into every part of his life. In that larger effort, ritual, dance, song, painting, carving, and above all discursive language—arts which utilize all the organs of the body—must have for long played a decisive role.

At its points of origin, then, technics was related to the whole nature of man, and that nature played a formative part in the development of every aspect of technology; thus, technics at the beginning was broadly life-centered, not work-centered or power-centered. As in all ecological complexes, other human interests and purposes, other organic needs restrained the overgrowth of any single component. As for the greatest technical feat before our own age, the domestication of plants and animals, this advance owed almost nothing to new tools, though it encouraged the development of clay containers. But it owed much, we now begin to realize—since Eduard Hahn[15]—to an intense subjective concentration upon sexuality in all its manifestations, abundantly visible in cult objects and symbolic art. Plant selection, hybridization, fertilization, manuring, seeding, castration were the products of an imaginative cultivation of sexuality, whose first evidence one finds tens of thousands of years earlier in the emphatically sexual carvings of Paleolithic woman: the so-called Venuses.[16]

But at the point where history, in the form of the written record, becomes visible, that life-centered economy, a true polytechnics, was challenged and in part displaced in a series of radical technical and social innovations. About five thousand years ago a monotechnics devoted to the increase of power and wealth by the systematic organization of work-a-day activities in a rigidly mechanical pattern came into existence. At this moment, a new conception of the nature of man arose and with it a new stress upon the exploitation of physical energies, cosmic and human, apart from the processes of growth and reproduction, came to the fore. In Egypt, Osiris symbolizes the older, life-oriented technics: Atum-Re, the sun-god, who characteristically created the world out of his own semen without female cooperation, stands for the machine-centered one. The expansion of power, through ruthless human coercion and mechanical organization, took precedence over the enhancement of life.

The chief mark of this change was the construction of

the first complex, high-powered machines; and therewith the beginning of a new regimen, accepted by all later civilized societies—though reluctantly by more archaic cultures —in which work at a single specialized task, segregated from other biological and social activities, not only occupied the entire day, but increasingly engrossed the entire lifetime. That was the fundamental departure which, during the last few centuries, has led to the increasing mechanization and automation of all production. With the assemblage of the first collective machines, work, by its systematic dissociation from the rest of life, became a curse, a burden, a sacrifice, a form of punishment; and by reaction this new regimen soon awakened compensatory dreams of effortless affluence, emancipated not only from slavery but from work itself.

The machine I refer to was never discovered in any archaeological diggings for a simple reason: it was composed almost entirely of human parts. These parts were brought together in a hierarchical organization under the rule of an absolute monarch whose commands, supported by a coalition of the priesthood, the armed nobility, and the bureaucracy secured a corpselike obedience from all the components of the machine. Let us call this archetypal machine—the human model for all later specialized machines —the "Megamachine." This new kind of machine was far more complex than the contemporary potter's wheel or bow-drill, and it remained the most advanced type of machine until the invention of the mechanical clock in the fourteenth century.

Only through the deliberate invention of such a high-powered machine could the colossal works of engineering that marked the Pyramid Age in both Egypt and Mesopotamia have been brought into existence, often in a single generation. This new technics came to an early climax in the Great Pyramid at Giza. That structure, as J. H. Breasted pointed out, exhibited a watch-maker's standard of exact measurement.[17] By operating as a mechanical unit, the 100,000 men who worked on that pyramid generated 10,000 horsepower; this human mechanism alone made it possible to raise that colossal structure with the use of only the simplest stone and copper tools—without the aid of such otherwise indispensable machines as the wheel, the wagon, the pulley, the derrick, or the winch.

Two things must be noted about this new mechanism, because they identify it through its historical course down to the present. The first is that the organizers of this machine derived their power and authority from a cosmic source. The exactitude in measurement, the abstract mechanical order, the compulsive regularity of this Megamachine sprang directly from astronomical observations and abstract scientific calculations; this inflexible, predictable order, incorporated in the calendar, was then transferred to the regimentation of the human components. By a combination of divine command and ruthless military coercion, a large population was made to endure grinding poverty and forced labor at dull repetitive tasks in order to ensure "life, prosperity, and health" for the divine or semidivine ruler and his entourage.

The second point is that the grave social defects of the human machine were partly offset by its superb achievements in flood control and grain production, which laid the ground for an enlargement in every area of human culture: in monumental art, in codified law, and in systematically pursued and permanently recorded thought. Such order, such collective security and abundance as was achieved in Mesopotamia and Egypt, later in India, China, in the Andean and Mayan cultures, was never surpassed until the Megamachine was re-established in a new form in our own time. But conceptually the machine was already detached from other human functions and purposes than the increase of mechanical power and order: with mordant symbolism, its ultimate products in Egypt were tombs and mummies, while later in Assyria the chief testimonial to its efficiency was typically a waste of destroyed cities and poisoned soils.

In a word, what modern economists lately termed the Machine Age had its origin, not in the eighteenth century, but at the very outset of civilization. All its salient characteristics were present from the beginning in both the means and the ends of the collective Megamachine. So Keynes's acute prescription of pyramid building as an essential means of coping with the insensate productivity of a highly mechanized technology applies both to the earliest manifestations and the present ones; for what is a space rocket but the precise dynamic equivalent, in terms of our present-day theology and cosmology, of the static Egyptian pyramid? Both are devices for securing, at an extravagant cost, a passage to heaven for the favored few.

Unfortunately, though the labor machine lent itself to vast constructive enterprises, which no small-scale community could even contemplate, much less execute, the most conspicuous result has been achieved through military machines, in colossal acts of destruction and human extermination, acts that monotonously soil the pages of history, from the rape of Sumer to the blasting of Rotterdam and Hiroshima. Sooner or later, I suggest, we must have the courage to ask ourselves: Is this association of inordinate power and productivity with equally inordinate violence and destruction a purely accidental one?

Now the misuse of Megamachines would have proved intolerable had they not also brought genuine benefits to the whole community by raising the ceiling of collective human effort and aspiration. The least of these advantages was the gain in efficiency derived from concentration upon rigorously repetitive motions in work, already indeed introduced in the grinding and polishing processes of neolithic toolmaking. This inured civilized man to long spans of regular work, with higher productive efficiency per unit. But the social by-product of this new discipline was, perhaps, even more significant; for some of the psychological benefits, hitherto confined to religious ritual, were transferred to work. The sterile repetitive tasks imposed by the Megamachine, which in a pathological form we associate with a compulsion neurosis, nevertheless served, like all ritual and restrictive order, to lessen anxiety and to defend the worker himself from the often demonic promptings of the unconscious, no longer held in check by the traditions and customs of the neolithic village.

In short, mechanization and regimentation, through labor armies, military armies, and ultimately through the derivative modes of industrial and bureaucratic organization, supplemented and increasingly replaced religious ritual as a means of coping with anxiety and promoting psychic stability in mass populations. Orderly, repetitive work provided a daily means of self-control, more pervasive, more effective, more universal than either ritual or law. This hitherto unnoticed psychological contribution was a possibility more important than those gains in productive efficiency which were too often offset by absolute losses in war and conquest. Unfortunately, the ruling classes, which claimed immunity from manual labor, were not subject to this discipline; hence, as the historical record testifies, their disordered fantasies too often found an outlet in reality through destruction and extermination.

Having indicated the beginnings of this process, I must regrettably pass over the actual institutional forces that have been at work during the last five thousand years, and leap, all too suddenly, into the present age, in which the ancient forms of biotechnics are being either suppressed or supplanted, and in which the continued enlargement of the Megamachine itself has become, with increasing compulsiveness, the condition of scientific and technical advance, if not the main purpose of human existence. But if the clues I have been attempting to expose prove helpful, many aspects of the scientific and technical transformation of the last three centuries will call for reinterpretation and judicious reconsideration. For at the very least, we are now bound to explain why the whole process of technical development has become increasingly coercive, totalitarian, and—subjectively speaking—compulsive and irrational.

Before accepting the ultimate translation of all organic processes, biological functions, and human aptitudes into an externally controllable mechanical system, increasingly automatic and self-expanding, it might be well to re-examine the ideological foundations of this whole system, with its concentration upon centralized power and external control. We must, in fact, ask ourselves if the probable destination of this system is compatible with the further development of specifically human potentialities.

Consider the alternatives now before us. If man were actually, as current theory still supposes, a creature whose use of tools alone played the largest formative part in his development, on what valid grounds do we now propose to strip mankind of the wide variety of autonomous activities historically associated with agriculture and manufacture, leaving the residual mass of workers with the trivial task of watching buttons and dials and responding to one-way communication and remote control? If man actually owes his intelligence mainly to his tool-using propensities, by what logic do we now take his tools away, so that he will become a functionless, workless being, conditioned to accept only what the Megamachine offers him—an automaton within a larger system of automation, condemned to compulsory consumption, as he was once condemned to compulsory production? What in fact will be left of human life if one function after another is either taken over by the machine or else genetically suppressed, if not surgically removed?

But if the analysis of human development in relation to technics sketched out in this paper proves sound, there is an even more fundamental criticism to be made; for we must then go on to question the present effort to shift the locus of human activity from the organic environment and the human group to the Megamachine, and eventually reduce all forms of life and culture to those that can be translated into the current system of scientific abstractions, and transferred on a mass basis to machines and electronic apparatus. We are now in a position to question the dubious assumptions that have too long been treated as axioms, for the system of thought upon which they are still based antedated by three centuries anything like the present comprehension—scientific, humanistic, and historical—of man's nature and special gifts. From the Pyramid Age to the so-called Nuclear Age, we note, the inventors and controllers of the Megamachine have been haunted by delusions of omniscience and omnipotence; and these delusions are not less irrational now that they have at their disposal all the formidable resources of exact science and a high-energy technology.

The Nuclear Age conception of "absolute power" and infallible intelligence, exercised by a military-scientific elite, corresponds to the Bronze Age conception of Divine Kingship; and both belong to the same infantile magico-religious scheme as ritual human sacrifice. Living organisms can use only limited amounts of energy, as living personalities can utilize only limited quantities of knowledge and experience. "Too much" or "too little" is equally fatal to organic existence. Even too much abstract knowledge, insulated from feeling, from moral evaluation, from historical experience, from responsible purposeful action can produce a serious unbalance in both the personality and the community. Organisms, societies, human persons are nothing less than delicate devices for regulating energy and putting it at the service of life. To the extent that our megatechnics ignores these fundamental insights into the nature of organisms and human personalities, it is prescientific in its attitude toward the human personality, even when not actively irrational. When the implications of this weakness are taken in, a deliberate large-scale dismantling of the Megamachine, in all its institutional forms, must surely take place, with a redistribution of power and authority to smaller units, under direct human control.

If technics is to be brought back again into the service of human culture, the path of advance will lead, not to the further expansion of the Megamachine, but to the development of all those parts of the organic environment and the human personality that have been suppressed in order to magnify the offices of the pure intelligence alone and therewith to maximize its coercive collective exercise and quantitative productivity. The deliberate expression and fulfilment of human potentialities requires a quite different approach from that bent solely on the control of natural forces and the modification of human nature in order to facilitate and expand the system of control. We know now that play and sport and ritual and dream fantasy, no less than organized work, have exercised a formative influence upon human culture and even upon technics. But make-believe cannot for long be a sufficient substitute for productive work; only when play and work form part of a larger cultural whole, as in Tolstoi's picture of the mowers in *Anna Karenina,* can the many-sided requirements for full human growth be satisfied.

Instead of liberation *from* work being the chief contribution of mechanization and automation, I would suggest that liberation *for* work—for educative, mind-forming work, self-rewarding even on the lowest physiological level—may become the most salutary contribution of a life-centered technology. This may prove an indispensable counterbalance to universal automation, partly by protecting the displaced worker from boredom and suicidal desperation, only temporarily relievable by anesthetics and sedatives, partly by giving play to constructive impulses, autonomous functions, meaningful activities.

Relieved from abject dependence upon the Megamachine, the whole world of biotechnics will at once become open to man; and those parts of his personality that have been crippled or paralyzed by insufficient use should again come into play. Automation is indeed the proper end of a purely mechanical system; and in its place, subordinate to other human purposes, automation will serve the human community no less effectively than the reflexes, the hormones, and autonomic nervous system—nature's earliest experiment in automation—serve the human body. But autonomy is the proper end of organisms; and further technical development must aim at reestablishing autonomy at every stage of human growth by giving play to every part of the human personality, not merely to those functions that serve the machine.

I realize that in opening up these difficult questions I am not in a position to provide ready-made answers, nor do I suggest that such answers will be easy to fabricate. But it is time that our present wholesale commitment to the machine, which arises largely out of our one-sided interpretation of man's early technical development, should be replaced by a fuller picture of both human nature and the technical milieu, as both have evolved together.

REFERENCES

[1] Pierre Teilhard de Chardin, *The Phenomenon of Man* (New York, 1959).

[2] Kenneth P. Oakley, *Man the Tool-Maker* (5th ed.; London, 1963).

[3] Ernst Mayr, *Animal Species and Evolution* (Cambridge, Mass., 1963).

[4] R. U. Sayce, *Primitive Arts and Crafts: An Introduction to the Study of Material Culture* (Cambridge, 1933).

[5] C. Daryll Forde, *Habitat, Economy, and Society: A Geographical Introduction to Ethnology* (London, 1945).

[6] André Leroi-Gourhan, *Milieu et Techniques* (Paris, 1945).

[7] Robert J. Braidwood, *Prehistoric Men* (5th ed.; Chicago, 1961).

[8] A. N. Hocart, *The Progress of Man* (London, 1933).

[9] Leslie A. White, *The Science of Culture* (New York, 1949). Also, more briefly in "Four Stages in the Evolution of Minding," in Sol Tax (ed.), *The Evolution of Man* (Chicago, 1960).

[10] André Varagnac, *De la Préhistoire au Monde Moderne: Essai d'une Anthropodynamique* (Paris, 1954).

[11] Claude Levi-Strauss, *Structural Anthropology* (New York, 1963).

[12] Oakes Ames, *Economic Annuals and Human Cultures* (Cambridge, 1939).

[13] Edgar Anderson, *Plants, Man and Life* (Boston, 1952).

[14] Henri Breuil and Raymond Lantier, *Les Hommes de la Pierre ancienne* (Paris, 1951).

[15] Eduard Hahn, *Das Alter der Wirtschaftlichen Kultur: Ein Ruckblick und ein Ausblick* (Heidelberg, 1905).

[16] Gertrude R. Levy, *The Gate of Horn: A Study of the Religious Conceptions of the Stone Age* (London, 1948).

[17] James Henry Breasted, *The Conquest of Civilization* (New York, 1926).

The Fourth Discontinuity

Bruce Mazlish

A famous cartoon in *The New Yorker* magazine shows a large computer with two scientists standing excitedly beside it. One of them holds in his hand the tape just produced by the machine, while the other gapes at the message printed on it. In clear letters, it says, "Cogito, ergo sum," the famous Cartesian phrase, "I think, therefore I am."

My next cartoon has not yet been drawn. It is a fantasy on my part. In it, a patient wild of eye and hair on end, is lying on a couch in a psychiatrist's office talking to an analyst, who is obviously a machine. The analyst-machine is saying, "Of course I'm human—aren't you?"[1]

These two cartoons are a way of suggesting the threat which the increasingly perceived continuity between man and the machine poses to us today. It is with this topic that I wish to deal now, approaching it in terms of what I shall call the "fourth discontinuity." In order, however, to explain what I mean by the "fourth discontinuity," I must first place the term in a historical context.

In the eighteenth lecture of his *General Introduction to Psychoanalysis,* originally delivered at the University of Vienna between 1915 and 1917, Freud suggested his own place among the great thinkers of the past who had outraged man's naïve self-love. First in the line was Copernicus, who taught that our earth "was not the center of the

Professor of history at the Massachusetts Institute of Technology, Dr. Mazlish is the co-author, with J. Bronowski, of The Western Intellectual Tradition. *In addition, he has written many articles and books dealing with historiography, the relations of psychology to history, and a broad range of cultural and technological topics. This article was published in* Technology and Culture *(Vol. 8, No. 1): 1–15.*

universe, but only a tiny speck in a world-system of a magnitude hardly conceivable." Second was Darwin, who "robbed man of his peculiar privilege of having been specially created, and relegated him to a descent from the animal world." Third, now, was Freud himself. On his own account, Freud admitted, or claimed, that psychoanalysis was "endeavoring to prove to the 'ego' of each one of us that he is not even master in his own house, but that he must remain content with the veriest scraps of information about what is going on unconsciously in his own mind."

A little later in 1917, Freud repeated his sketch concerning the three great shocks to man's ego. In his short essay, "A Difficulty in the Path of Psychoanalysis," he again discussed the cosmological, biological, and now psychological blows to human pride and, when challenged by his friend Karl Abraham, admitted, "You are right in saying that the enumeration of my last paper may give the impression of claiming a place beside Copernicus and Darwin."[2]

There is some reason to believe that Freud may have derived his conviction from Ernst Haeckel, the German exponent of Darwinism, who in his book *Natürliche Schöpfungsgeschichte* (1889) compared Darwin's achievement with that of Copernicus and concluded that together they had helped remove the last traces of anthropomorphism from science.[3] Whatever the origin of Freud's vision of himself as the last in the line of ego-shatterers, his assertion has been generally accepted by those, like Ernest Jones, who refer to him as the "Darwin of the Mind."[4]

The most interesting extension of Freud's self-view, however, has come from the American psychologist, Jerome Bruner. Bruner's version of what Freud called his "transvaluation" is in terms of the elimination of discontinuities, where discontinuity means an emphasis on breaks or gaps in the phenomena of nature—for example, a stress on the sharp differences between physical bodies in the heavens or on earth or between one form of animal matter and another —instead of an emphasis on its continuity. Put the other way, the elimination of discontinuity, that is, the establishment of a belief in a continuum of nature, can be seen as the creation of continuities, and this is the way Bruner phrases it. According to Bruner, the first continuity was established by the Greek physicist-philosophers of the sixth century, rather than by Copernicus. Thus, thinkers like Anaximander conceived of the phenomena of the physical world as "continuous and monistic, as governed by the common laws of matter."[5] The creating of the second continuity, that between man and the animal kingdom was, of course, Darwin's contribution, a necessary condition for Freud's work. With Freud, according to Bruner, the following continuities were established: the continuity of organic lawfulness, so that "accident in human affairs was no more to be brooked as 'explanation' than accident in nature"; the continuity of the primitive, infantile, and archaic as coexisting with the civilized and evolved; and the continuity between mental illness and mental health.

In this version of the three historic ego-smashings, man is placed on a continuous spectrum in relation to the universe, to the rest of the animal kingdom, and to himself. He is no longer discontinuous with the world around him. In an important sense, it can be contended, once man is able to accept this situation, he is in harmony with the rest of existence. Indeed, the longing of the early nineteenth-century romantics, and of all "alienated" beings since, for a sense of "connection" is fulfilled in an unexpected manner.

Yet, to use Bruner's phraseology, though not his idea, a fourth and major discontinuity, or dichotomy, still exists in our time. It is the discontinuity beween man and machine. In fact, my thesis is that this fourth discontinuity must now be eliminated—indeed, we have started on the task—and that in the process man's ego will have to undergo another rude shock, similar to those administered by Copernicus (or Galileo), Darwin, and Freud. To put it bluntly, we are now coming to realize that man and the machines he creates are continuous and that the same conceptual schemes, for example, that help explain the workings of his brain also explain the workings of a "thinking machine." Man's pride, and his refusal to acknowledge this continuity, is the substratum upon which the distrust of technology and an industrialized society has been reared. Ultimately, I believe, this last rests on man's refusal to understand and accept his own nature—as a being continuous with the tools and machines he constructs. Let me now try to explain what is involved in this fourth discontinuity.

The evidence seems strong today that man evolved from the other animals into humanity through a continuous interaction of tool, physical, and mental-emotional changes. The old view that early man arrived on the evolutionary scene, fully formed, and then proceeded to discover tools and the new ways of life which they made possible is no longer acceptable. As Sherwood L. Washburn, professor of anthropology at the University of California, puts it, "From the rapidly accumulating evidence it is now possible to speculate with some confidence on the manner in which the way of life made possible by tools changed the pressures of natural selection and so changed the structure of man." The details of Washburn's argument are fascinating, with its linking of tools with such physical traits as pelvic structure, bipedalism, brain structure, and so on, as well as with the organization of men in cooperative societies and the substitution of morality for hormonal control of sexual and other "social" activities. Washburn's conclusion is that "it was the success of the simplest tools that started the whole trend of human evolution and led to the civilizations of today."[6]

Darwin, of course, had had a glimpse of the role of tools in man's evolution.[7] It was Karl Marx, however, who first placed the subject in a new light. Accepting Benjamin Franklin's definition of man as a "tool-making animal," Marx suggested in *Das Kapital* that "the relics of the instruments of labor are of no less importance in the study of vanished socio-economic forms, than fossil bones are in the study of the organization of extinct species." As we know, Marx wished to dedicate his great work to Darwin —a dedication rejected by the cautious biologist—and we can see part of Marx's reason for this desire in the following revealing passage:

> Darwin has aroused our interest in the history of *natural technology,* that is to say in the origin of the organs of plants and animals as productive instruments utilised for the life purposes of those creatures. Does not the history of the origin of the productive organs of men in society, the organs which form the material basis of every kind of social organisation, deserve equal attention? Since, as Vico [in the *New Science* (1725)] says, the essence of the distinction between human history and natural history is that the former is the work of man and the latter is not, would not the history of *human technology* be easier to write than the history of natural technology? Technology reveals man's dealings with nature, discloses the direct productive activities of his life, thus throwing light upon social relations and the resultant mental conceptions.[8]

Only a dogmatic anti-Marxist could deny that Marx's brilliant imagination had led him to perceive a part of the continuity between man and his tools. Drawn off the track, perhaps, by Vico's distinction between human and natural history as man-made and God-made, Marx might almost be given a place in the pantheon of Copernicus, Darwin, and Freud as a destroyer of man's discontinuities with the world about him. Before our present-day anthropologists, Marx had sensed the unbreakable connection between man's evolution as a social being and his development of tools. He did not sense, however, the second part of our subject, that man and his tools, especially in the form of modern, complicated machines, are part of a theoretical continuum.

The *locus classicus* of the modern insistence on the fourth discontinuity is, as is well known, the work of Descartes. In his *Discourse on Method*, for example, he sets up God and the soul on one side, as without spatial location or extension, and the material-mechanical world in all its aspects, on the other side. Insofar as man's mind or soul participates in reason—which means God's reason—man knows this division or dualism of mind and matter, for, as Descartes points out, man could not know this fact from his mere understanding, which is based solely on his senses, "a location where it is clearly evident that the ideas of God and the soul have never been."[9]

Once having established his God, and man's participation through reason in God, Descartes could advance daringly to the very precipice of a world without God. He conjures up a world in imaginary space and shows that it must run according to known natural laws. Similarly, he imagines that "God formed the body of a man just like our own, both in the external configuration of its members and in the internal configuration of its organs, without using in its composition any matter but that which I had described [i.e., physical matter]. I also assumed that God did not put into this body any rational soul [defined by Descartes as "that part of us distinct from the body whose essence . . . is only to think"]."

Analyzing this purely mechanical man, Descartes boasts of how he has shown "what changes must take place in the brain to cause wakefulness, sleep, and dreams; how light, sounds, odors, taste, heat, and all the other qualities of external objects can implant various ideas through the medium of the senses . . . I explained what must be understood by that animal sense which receives these ideas, by memory which retains them, and by imagination which can change them in various ways and build new ones from them." In what way, then, does such a figure differ from real man? Descartes confronts his own created "man" forth-rightly; it is worth quoting the whole of his statement:

> Here I paused to show that if there were any ma-chines which had the organs and appearance of a monkey or of some other unreasoning animal, we would have no way of telling that it was not of the same nature as these animals. But if there were a ma-chine which had such a resemblance to our bodies, and imitated our actions as far as possible, there would always be two absolutely certain methods of recogniz-ing that it was still not truly a man. The first is that it could never use words or other signs for the purpose of communicating its thoughts to others, as we do. It indeed is conceivable that a machine could be so made that it would utter words, and even words appropriate to physical acts which cause some change in its organs; as, for example, if it was touched in some spot that it would ask what you wanted to say to it; if in another, that it would cry that it was hurt, and so on for similar things. But it could never modify its phrases to reply to the sense of whatever was said in its presence, as even the most stupid men can do. The second method of recognition is that although such machines could do many things as well as, or perhaps even better than, men, they would infallibly fail in certain others, by which we would discover that they did not act by understanding, but only by the disposition of their or-gans. For while reason is a universal instrument which can be used in all sorts of situations, the organs have to be arranged in a particular way for each particular action. From this it follows that it is morally impossi-ble that there should be enough different devices in a machine to make it behave in all the occurrences of life as our reason makes us behave.

Put in its simplest terms, Descartes' two criteria for dis-criminating between man and the machine are that the lat-ter has (1) no feedback mechanism ("it could never modify its phrases") and (2) no generalizing reason ("reason is a universal instrument which can be used in all sorts of situa-tions"). But it is exactly in these points that, today, we are no longer able so surely to sustain the dichotomy. The work of Norbert Wiener and his followers, in cybernetics, indi-cates what can be done on the problem of feedback. Inves-tigations into the way the brain itself forms concepts are basic to the attempt to build computers that can do the same, and the two efforts are going forward simultaneously, as in the work of Dr. W. K. Taylor of University College, London, and of others. As G. Rattray Taylor sums up the matter: "One can't therefore be quite as confident that computers will one day equal or surpass man in concept-forming ability as one can about memory, since the trick hasn't yet been done; but the possibilities point that way."[10] In short, the gap between man's thinking and that of his thinking machines has been greatly narrowed by recent research.

Descartes, of course, would not have been happy to see such a development realized. To eliminate the dichotomy or discontinuity between man and machines would be, in effect, to banish God from the universe. The rational soul, Descartes insisted, "could not possibly be derived from the powers of matter . . . but must have been specially created." Special creation requires God, for Descartes' reasoning is circular. The shock to man's ego, of learning the Darwinian lesson that he was not "specially created," is, in this light, only an outlying tremor of the great earthquake that threatened man's view of God as well as of himself. The obstacles to removing not only the first three but also the fourth discontinuity are, clearly, deeply imbedded in man's pride of place.

How threatening these developments were can be seen in the case of Descartes' younger contemporary, Blaise Pascal. Aware that man is "a thinking reed," Pascal also realized that he was "engulfed in the infinite immensity of spaces whereof I know nothing, and which knows nothing of me." "I am terrified," he confessed. To escape his feeling of terror, Pascal fled from reason to faith, convinced that reason could not bring him to God. Was he haunted by his own construction, at age nineteen, of a calculating machine which, in principle, anticipated the modern digital computer? By his own remark that "the arithmetical machine produces effects which approach nearer to thought than all the actions of animals"? Ultimately, to escape the anxiety that filled his soul, Pascal commanded, "On thy knees, powerless reason."[11]

Others, of course, walked where angels feared to tread. Thus, sensationalist psychologists and epistemologists, like Locke, Hume, or Condillac, without confronting the problem head on, treated the contents of man's reason as being formed by his sense impressions. Daring thinkers, like La Mettrie in his *L'Homme machine* (1747) and Holbach, went all the way to a pure materialism. As La Mettrie put it in an anticipatory transcendence of the fourth discontinuity, "I believe thought to be so little incompatible with organized matter that it seems to be a property of it, like Electricity, Motive Force, Impenetrability, Extension, etc."[12]

On the practical front, largely leaving aside the metaphysical aspects of the problem, Pascal's work on calculating machines was taken up by those like the eccentric nineteenth-century mathematician Charles Babbage, whose brilliant designs outran the technology available to him.[13] Thus it remained for another century, the twentieth, to bring the matter to a head and to provide the combination of mathematics, experimental physics, and modern technology that created the machines that now confront us and that reawaken the metaphysical question.

The implications of the metaphysical question are clear. Man feels threatened by the machine, that is, by his tools writ large, and feels out of harmony with himself because he is out of harmony—what I have called discontinuous— with the machines that are part of himself. Today, it is fashionable to describe such a state by the term "alienation." In the Marxist phraseology, we are alienated from ourselves when we place false gods or economies over us and then behave as if they had a life of their own, eternal and independent of ourselves, and, indeed, in control of our lives. My point, while contact can be established between it and the notion of alienation, is a different one. It is in the tradition of Darwin and Freud, rather than of Marx, and is concerned more with man's ego than with his sense of alienation.

A brief glimpse at two "myths" concerning the machine may illuminate what I have in mind. The first is Samuel Butler's negative utopia, *Erewhon,* and the second is Mary Shelley's story of Frankenstein. In Butler's novel, published in 1872, we are presented with Luddism carried to its final point. The story of the Erewhonian revolution against the machines is told in terms of a purported translation from a manuscript, "The Book of the Machines," urging men on to the revolt and supposedly written just before the long civil war between the machinists and the antimachinists, in which half the population was destroyed. The prescient flavor of the revolutionary author's fears can be caught in such passages as follows:[14]

"There is no security"—to quote his own words— "against the ultimate development of mechanical consciousness, in the fact of machines possessing little consciousness now. A mollusc has not much consciousness. Reflect upon the extraordinary advance which machines have made during the last few hundred years, and note how slowly the animal and vegetable kingdoms are advancing. The more highly organized machines are creatures not so much of yesterday as of the last five minutes, so to speak, in comparison with past time. Assume for the sake of argument that conscious beings have existed for some twenty million years: see what strides machines have made in the last thousand! May not the world last twenty million years longer? If so, what will they not in the end become? Is it not safer to nip the mischief in the bud and to forbid them further progress?

"But who can say that the vapour-engine has not a kind of consciousness? Where does consciousness begin, and where end? Who can draw the line? Who can draw any line? Is not everything interwoven with everything? Is not machinery linked with animal life in an infinite variety of ways? The shell of a hen's egg is made of a delicate white ware and is a machine as much as an egg-cup is; the shell is a device for holding the egg as much as the egg-cup for holding the shell: both are phases of the same function; the hen makes the shell in her inside, but it is pure pottery. She makes her nest outside of herself for convenience' sake, but the nest is not more of a machine than the egg-shell is. A 'machine' is only a 'device.' "

Then he continues:

"Do not let me be misunderstood as living in fear of any actually existing machine; there is probably no known machine which is more than a prototype of future mechanical life. The present machines are to the future as the early Saurians to man. The largest of them will probably greatly diminish in size. Some of the lowest vertebrata attained a much greater bulk than has descended to their more highly organized living representatives, and in like manner a diminution in the size of machines has often attended their development and progress."

Answering the argument that the machine, even when

more fully developed, is merely man's servant, the writer contends:

> "But the servant glides by imperceptible approaches into the master; and we have come to such a pass that, even now, man must suffer terribly on ceasing to benefit the machines. . . . Man's very soul is due to the machines; it is a machine-made thing; he thinks as he thinks, and feels as he feels, through the work that machines have wrought upon him, and their existence is quite as much a *sine qua non* for his, as his for theirs. This fact precludes us from proposing the complete annihilation of machinery, but surely it indicates that we should destroy as many of them as we can possibly dispense with, lest they should tyrannize over us even more completely."

And, finally, the latent sexual threat is dealt with:

> "It is said by some with whom I have conversed upon this subject, that the machines can never be developed into animate or quasi-animate existences, inasmuch as they have no reproductive systems, nor seem ever likely to possess one. If this be taken to mean that they cannot marry, and that we are never likely to see a fertile union between two vapour-engines with the young ones playing about the door of the shed, however greatly we might desire to do so, I will readily grant it. But the objection is not a very profound one. No one expects that all the features of the now existing organizations will be absolutely repeated in an entirely new class of life. The reproductive system of animals differs widely from that of plants, but both are reproductive systems. Has nature exhausted her phases of this power?"

Inspired by fears such as these, which sound like our present realities, the Erewhonians rise up and destroy almost all their machines. It is only years after this supposed event that they are sufficiently at ease so as to collect the fragmentary remains, the "fossils," of the now defunct machines and place them in a museum. At this point, the reader is never sure whether Butler's satire is against Darwin or the anti-Darwinists, probably both, but there is no question of the satire when he tells us how machines were divided into "their genera, subgenera, species, varieties, subvarieties, and so forth" and how the Erewhonians "proved the existence of connecting links between machines that seemed to have very little in common, and showed that many more such links had existed, but had now perished." It is as if Butler had taken Marx's point about *human technology* and stood it on its head!

Going even further, Butler foresaw the threatened ending of the fourth discontinuity, just as he saw Darwin's work menacing the third of the discontinuities we have discussed. Thus, we find Butler declaring, in the guise of his Erewhonian author,

> I shrink with as much horror from believing that my race ever be superseded or surpassed, as I should do from believing that even at the remotest period my ancestors were other than human beings. Could I believe that ten hundred thousand years ago a single one of my ancestors was another kind of being to myself, I should lose all self-respect, and take no further pleasure or interest in life. I have the same feeling with regard to my descendants and believe it to be one that will be felt so generally that the country will resolve upon putting an immediate stop to all further mechanical progress, and upon destroying all improvements that have been made for the last three hundred years.

The counter-argument, that "machines were to be regarded as a part of man's own physical nature, being really nothing but extra-corporeal limbs. Man [is] a machinate mammal," is dismissed out of hand.

Many of these same themes—the servant-machine rising against its master, the fear of the machine reproducing itself (fundamentally, a sexual fear, as Caliban illustrates, and as our next example will show), the terror, finally, of man realizing that he is at one with the machine—can be found attached to an earlier myth, that of Frankenstein. Now passed into our folklore, people frequently give little attention to the actual details of the novel. First, the name Frankenstein is often given to the monster created, rather than to its creator; yet, in the book, Frankenstein is the name of the scientist, and his abortion *has no name*. Second, the monster is *not* a machine but a "flesh and blood" product; even so informed a student as Oscar Handlin makes the typical quick shift, in an echo of Butler's fears, when he says, "The monster, however, quickly proves himself the superior. In the confrontation, the machine gives the orders."[15] Third, and last, it is usually forgotten or over-looked that the monster turns to murder *because* his creator, horrified at his production, refuses him human love and kindness. Let us look at a few of the details.

In writing her "Gothic" novel in 1816–17, Mary Shelley gave it the subtitle, "The Modern Prometheus."[16] We can see why if we remember that Prometheus defied the gods and gave fire to man. Writing in the typical early nineteenth-century romantic vein, Mary Shelley offers Frankenstein as an example of "how dangerous is the acquirement of knowl-edge"; in this case, specifically, the capability of "bestow-ing animation upon lifeless matter." In the novel we are told of how, having collected his materials from "the dis-secting room and the slaughterhouse" (as Wordsworth has said of modern science, "We murder to dissect"), Franken-stein eventually completes his loathsome task when he in-fuses "a spark of being into the lifeless thing that lay at my feet." Then, as he tells us, "now that I had finished, the beauty of the dream vanished, and breathless horror and disgust filled my heart." Rushing from the room, Franken-stein goes to his bedchamber, where he has a most odd dream concerning the corpse of his dead mother—the whole book as well as this passage cries out for psychoanalytic interpretation—from which he is awakened by "the wretch —the miserable monster whom I had created." Aghast at the countenance of what he has created, Frankenstein es-capes from the room and out into the open. Upon finally returning to his room with a friend, he is relieved to find the monster gone.

To understand the myth, we need to recite a few further details in this weird, and rather badly written, story. Frank-enstein's monster eventually finds his way to a hovel at-tached to a cottage, occupied by a blind father and his son and daughter. Unperceived by them, he learns the elements of social life (the fortuitous ways in which this is made to occur may strain the demanding reader's credulity), even to the point of reading *Paradise Lost*. Resolved to end his un-bearable solitude, the monster, convinced that his virtues of the heart will win over the cottagers, makes his presence known. The result is predictable: horrified by his appear-ance, they duplicate the behavior of his creator, and flee. In wrath, the monster turns against the heartless world. He kills, and his first victim, by accident, is Frankenstein's young brother.

Pursued by Frankenstein, a confrontation between creator and created takes place, and the monster explains his road to murder. He appeals to Frankenstein in a torrential address:

"I entreat you to hear me, before you give vent to your hatred on my devoted head. Have I not suffered enough that you seek to increase my misery? Life, although it may only be an accumulation of anguish, is dear to me, and I will defend it. Remember, thou hast made me more powerful than thyself; my height is superior to thine; my joints more supple. But I will not be tempted to set myself in opposition to thee. I am thy creature, and I will be even mild and docile to my natural lord and king, if thou wilt also perform thy part, the which thou owest me. Oh, Frankenstein, be not equitable to every other, and trample upon me alone, to whom thy justice, and even thy clemency and affection is most due. Remember, that I am thy creature; I ought to be thy Adam; but I am rather the fallen angel, whom thou drives from joy for no misdeed. Everywhere I see bliss, from which I alone am irrevocably excluded. I was benevolent and good; misery made me a fiend. Make me happy, and I shall again be virtuous."

Eventually, the monster extracts from Frankenstein a promise to create a partner for him "of another sex," with whom he will then retire into the vast wilds of South America, away from the world of men. But Frankenstein's "compassion" does not last long. In his laboratory again, Frankenstein indulges in a long soliloquy:

"I was now about to form another being, of whose dispositions I was alike ignorant; she might become ten thousand times more malignant than her mate; and delight, for its own sake, in murder and wretchedness. He had sworn to quit the neighborhood of man, and hide himself in deserts; but she had not; and she, who in all probability was to become a thinking and reasoning animal, might refuse to comply with a compact made before her creation. They might even hate each other; the creature who already lived loathed his own deformity, and might he not conceive a greater abhorrence for it when it came before his eyes in the female form? She also might turn with disgust from him to the superior beauty of man; she might quit him, and he be again alone, exasperated by the fresh provocation of being deserted by one of his own species.

"Even if they were to leave Europe, and inhabit the deserts of the new world, yet one of the first results of those sympathies for which the demon thirsted would be children, and a race of devils would be propagated upon the earth who might make the very existence of the species of man a condition precarious and full of terror. Had I right, for my own benefit, to inflict this curse upon everlasting generations?"

With the monster observing him through the window, Frankenstein destroys the female companion on whom he had been working. With this, the novel relentlessly winds its way to its end. In despair and out of revenge, the monster kills Frankenstein's best friend, Clerval, then Frankenstein's new bride, Elizabeth. Fleeing to the frozen north, the monster is tracked down by Frankenstein (shades of Moby Dick?), who dies, however, before he can destroy him. But it does not matter; the monster wishes his own death and promises to place himself on a funeral pile and thus at last secure the spiritual peace for which he has yearned.

I have summarized the book because I suspect that few readers will actually be acquainted with the myth of Frankenstein *as written* by Mary Shelley. For most of us, Frankenstein is Boris Karloff, clumping around stiff, automatic, and threatening: a machine of sorts. We shall have forgotten completely, if ever we knew, that the monster, *cum* machine, is evil, or rather, becomes evil, only because it is spurned by man.

My thesis has been that man is on the threshold of breaking past the discontinuity between himself and machines. In one part, this is because man now can perceive his own evolution as inextricably interwoven with his use and development of tools, of which the modern machine is only the furthest extrapolation. We cannot think any longer of man without a machine. In another part, this is because modern man perceives that the same scientific concepts help explain the workings of himself and of his machines and that the evolution of matter—from the basic building blocks of hydrogen turning into helium in the distant stars, then fusing into carbon nuclei and on up to iron, and then exploding into space, which has resulted in our solar system—continues on earth in terms of the same carbon atoms and their intricate patterns into the structure of organic life.

And now into the architecture of our thinking machines.

It would be absurd, of course, to contend that there are no differences between man and machines. This would be the same *reductio ad absurdum* as involved in claiming that, because he is an animal, there is no difference between man and the other animals. The matter, of course, is one of degree.[17] What is claimed here is that the sharp discontinuity between man and machines is no longer tenable, in spite of the shock to our egos. Scientists, today, know this; the public at large does not, *New Yorker* cartoons to the contrary.[18]

Moreover, this change in our metaphysical awareness, this transcendence of the fourth discontinuity, is essential to our harmonious acceptance of an industrialized world. The alternatives are either a frightened rejection of the "Frankensteins" we have created or a blind belief in their "superhuman virtues" and a touching faith that they can solve all our human problems. Alas, in the perspective I have suggested, machines are "mechanical, all too mechanical," to paraphrase Nietzsche. But, in saying this, I have already also said that they are "all too human" as well. The question, then, is whether we are to repeat the real Frankenstein story and, turning from the "monsters" we have created, turn aside at the same time from our own humanity or, alternatively, whether we are to accept the blow to our egos and enter into a world beyond the fourth discontinuity?

REFERENCES

[1] After finishing the early drafts of this article, I secured unexpected confirmation of my "fantasy" concerning an analyst-machine (which is not, in itself, critical to my thesis). A story in the *New York Times,* March 12, 1965, reports that "a computerized typewriter has been credited with remarkable success at a hospital here in radically improving the condition of several children suffering an extremely severe form of childhood schizophrenia. . . . What has particularly amazed a number of psychiatrists is that the children's improvement occurred without psychotherapy; only the machine was involved. It is almost as much human as it is machine. It talks, it listens, it responds to being touched, it makes pictures or charts, it comments and explains, it gives information and can be set up to do all this in any order. In short, the machine attempts to combine in a sort of science-fiction instrument all the best of two worlds—human and machine. It is called an Edison Responsive Environment Learning System. It is an extremely sophisticated 'talking' type-

writer (a cross between an analogue and digital computer) that can teach children how to read and write. . . . Dr. Campbell Goodwin speculates that the machine was able to bring the autistic children to respond because it eliminated humans as communication factors. Once the children were able to communicate, something seemed to unlock in their minds, apparently enabling them to carry out further normal mental activities that had eluded them earlier."

[2] Ernest Jones, *The Life and Work of Sigmund Freud* (3 vols.; New York, 1953–1957), II, 224–226.

[3] Ernst Cassirer, *The Problem of Knowledge: Philosophy, Science, and History since Hegel,* trans. William H. Woglom and Charles W. Hendel (New Haven, Conn., 1950), p. 160.

[4] Jones, III, 304.

[5] For Bruner's views, see his "Freud and the Image of Man," *Partisan Review,* XXIII, No. 3 (Summer 1956), 340–347. In place of both Bruner's sixth-century Greek physicists and Freud's Copernicus, I would place Galileo as the breaker of the discontinuity that was thought to exist in the material world. It was Galileo, after all, who first demonstrated that the heavenly bodies are of the same substance as the "imperfect" earth and subject to the same mechanical laws. In his *Dialogue on the Two Principal World Systems* (1632), he not only supported the "world system" of Copernicus against Ptolemy but established that our "world," i.e., the earth, is a natural part of the other "world," i.e., the solar system. Hence, the universe at large is one "continuous" system, a view at best only implied in Copernicus. Whatever the correct attribution, Greek physicists, Copernicus, or Galileo, Freud's point is not in principle affected.

[6] "Tools and Human Evolution," *Scientific American,* CCIII, No. 3 (September 1960), 63–75.

[7] E.g., see Charles Darwin, *The Descent of Man* (New York, n.d.), pp. 431–432, 458.

[8] Italics mine; Karl Marx, *Capital,* trans. Eden and Cedar Paul (2 vols.; London, 1951), I, 392–393, n. 2.

[9] René Descartes, *Discourse on Method,* trans. Laurence J. Lafleur (Indianapolis, 1956), p. 24. The rest of the quotations are also from this translation, pp. 29, 35–36, and 36–37.

[10] See G. Rattray Taylor, "The Age of the Androids," *Encounter* (November 1963), p. 43. On p. 40 Taylor gives some of the details of the work of W. K. Taylor and others.

[11] For details, see J. Bronowski and Bruce Mazlish, *The Western Intellectual Tradition: From Leonardo to Hegel* (New York, 1960), pp. 233–241.

[12] See Stephen Toulmin, "The Importance of Norbert Wiener," *New York Review of Books,* September 24, 1964, p. 4, for an indication of La Mettrie's importance in this development. While Toulmin does not put his material in the context of the fourth discontinuity, I find we are in fundamental agreement about what is afoot in this matter.

[13] See Philip and Emily Morrison (eds.), *Charles Babbage and His Calculating Engines* (New York, 1961).

[14] The quotations that follow are from Samuel Butler, *Erewhon* (Baltimore, 1954), pp. 161, 164, 167–168, and 171.

[15] "Science and Technology in Popular Culture," *Daedalus* (Winter 1965), p. 163.

[16] The quotations that follow are from Mary Shelley, *Frankenstein* (New York, 1953), pp. 30–33, 36–37, 85, and 160–161.

[17] In semi-facetious fashion, I have argued with some of my more literal-minded friends that what distinguishes man from existing machines, and probably will always so distinguish him, is an *effective* Oedipus complex: *vive la différence!* For an excellent and informed philosophical treatment of the difference between man and machines, see J. Bronowski, *The Identity of Man* (Garden City, N.Y., 1965).

[18] As in so much else, children "know" what their parents have forgotten. As O. Mannoni tells us, in the course of explaining totemism, "children, instead of treating animals as machines, treat machines as living things, the more highly prized because they are easier to appropriate. Children's appropriation is a virtual identification and they play at being machines (steam-engines, motor cars, aeroplanes) just as 'primitive' people play at being the totem [animal]" (*Prospero and Caliban, The Psychology of Colonization,* trans. Pamela Powesland [New York, 1964], p. 82). In *Huckleberry Finn,* Mark Twain puts this "identification" to work in describing Tom Sawyer's friend, Ben Rogers: "He was eating an apple, and giving a long, melodious whoop, at intervals, followed by a deep-toned ding-dong-dong, ding-dong-dong, for he was personating a steamboat. As he drew near, he slackened speed, took the middle of the street, leaned far over to starboard and rounded to ponderously and with laborious pomp and circumstance—for he was personating the *Big Missouri,* and considered himself to be drawing nine feet of water. He was boat and captain and engine-bells combined, so he had to imagine himself standing on his own hurricane-deck giving the orders and executing them! . . . 'Stop the stabboard! Ting-a-ling-ling! Stop the labboard! Come ahead on the stabboard! Stop her! Let your outside turn over slow! Ting-a-ling! Chow-ow-aw! Get out that head-line! *Lively* now! Come—out with your spring-line—what're you about there! Take a turn round that stump with the bight of it! Stand by that stage, now—let her go! Done with the engines, sir! Ting-a-ling! *sh't! sh't! sh't!*' " See the analysis of this passage in Erik H. Erikson, *Childhood and Society* (2d ed.; New York, 1963), pp. 209 ff.

Transporting Sixty-Ton Statues in Early Assyria and Egypt

C. St. C. Davison

In these days of bulldozers, heavy trucks, and huge cranes which easily transport and lift building materials, we often marvel at how the ancients moved stone blocks weighing over sixty tons for building pyramids and other structures, for they had little else to assist them in this task than muscle and brawn, aided by strong ropes and simple wooden frames. The earliest—and most difficult—method of transporting large blocks of building stone was by dragging them over the ground. Eventually the sledge was utilized for this purpose, followed by the wheeled wagon. However, historians have seemed unwilling to accept this simple evolution in transportation for which there is ample and incontrovertible evidence; they reason that rollers (loose) *must* have preceded the wheel, and they have eagerly looked for proof to bear this out.

At Nineveh between 1845–51, Sir A. H. Layard discovered some important bas-reliefs (Sculptures 124820, 124822, and 124823 in the Nineveh Room of the British Museum) which seemed to provide evidence as to how King Sennacherib (705–681 B.C.) had moved his huge sculptures from the quarries at Mosul, down the river in boats for a stretch of twenty miles, and then on sledges over the Kouyunjik Mounds to the palace and library.[1] Layard said: "The only mechanical powers possessed by the ancients were ropes, rollers and levers." This dictum became firmly established among archaeologists from then on; it is still repeated in many scholarly and widely read historical

Dr. Davison, of the Science Museum (London), is the author of The History of Steam Road Vehicles *and* Historic Books on Machines. *This article was published in* Technology and Culture *(Vol. 2, No. 1): 11–16.*

works,[2] and illustrations showing how the Trojan horse was drawn on a sledge with a series of long, loose, wooden rollers underneath are a commonplace in textbooks. Despite the wide acceptance of this notion, there is absolutely no foundation for Layard's surmise about rollers.

Transporting a Colossus to King Sennacherib's Palace

The bas-reliefs described by Layard show an Assyrian colossus being carried from the river to the palace (seventh century B.C.). Workmen can be seen placing small pieces of wood in front of the sledge, either in order to reduce friction between the sledge and the ground or to raise the bottom above the sharp rocks. The form and function of these pieces of wood as interpreted by archaeologists require close examination. Layard referred to them as "rollers." In support of Sir Henry's theory it is claimed that the sculptors lacked knowledge of perspective; hence they showed the "rollers" under the sledge with their long axes in the direction of motion.[3] Two questions come to mind which cast doubt upon this explanation: since the Assyrian artists had the skill correctly to draw such minute details as the strands of the ropes, the engravings on the hafts of the overseers' swords, and the axles on the carts, can it not justifiably be assumed that the pieces of wood were also correctly drawn? Further, since they could depict an end-view of an axle, can one doubt that they could also draw the end-view of the so-called "rollers"?

In fact, the "rollers" are portrayed as being neither round nor straight; they clearly are trimmed tree-branches, and it is obvious that they are far from being cylindrical rollers. Indeed, should a number of rollers have been placed under the sledge, they would have had to be exactly parallel to each other; otherwise they would have taken their own different directions of motion and collided. When this happens with two rollers, they jam because they rotate in opposite directions where they touch at their peripheries. The ground also would have had to be leveled with an accuracy which it was impossible for the ancients to achieve.

Despite the arguments militating against the employment of rollers by the ancients, archaeologists have searched for sledge-rollers to prove that they were thus employed. They found only a small "roller" at Deir el-Bahri near the pyramids. It is approximately nine inches long and of a diameter varying from about two inches at the ends to about three inches in the middle. This specimen is now on exhibition at the British Museum and is dated about 2100 B.C.[4] The only justification for claiming that it had been utilized for carrying building stones is that it bears marks like those caused by heavy weights. Yet it could equally have served as a fulcrum for an oscillating beam or lever. In spite of the inadequate evidence this "roller" has been accepted as additional "proof" that rollers were employed by stonemasons in transporting colossal weights.[5]

In his book Layard explained how he transported the winged bulls and lions now in the British Museum from the ruins to the banks of the Tigris, stating that this was done on a four-wheeled vehicle with very strong roller-like wheels (about four ft. in diameter by three ft. wide) pulled by 300 men. This, he maintained, was proof enough that rollers were employed under sledges. He obviously failed to see that the action of the rollers on the cart axle was entirely different from that of axleless "rollers" placed haphazardly between rough ground and a sledge. In an embossed bronze band from the palace gate of Shalmaneser III, king of Assyria (859–824 B.C.), one can see the large wheel felloes on the Assyrian chariots of about 850 B.C.; it stands to reason that the wider the felloes of the wheels, the less likelihood there was for them to sink into soft earth or sand. This representation of an early scene of chariots and a bridge of boats also proves that the Assyrians were capable of making an orthographic drawing *par excellence,* even though they were unacquainted with the laws of perspective.

Herodotus, the historian (c. 484–424 B.C.), said that rollers were employed for transporting boats. To this day we have a device for winching up boats from the sea on top of a series of wooden rollers, each one with an axle fitting into a hole in a stationary framework. One may, therefore, accept the existence of Herodotus' rollers. Moreover, Vitruvius, Roman architect, engineer, and superintendent of machines of war, described how large stone columns for a building had each been fitted with an axle and framework; they were then drawn like wagons to the site of a new building (c. 25 B.C.).[6] But the combination of loose rollers and sledges has no counterpart in modern technology—another fact which serves to cast doubt on their employment by the ancients.

Transport of a Colossus in Ancient Egypt

It is clear from Sir A. H. Layard's words that a different method was employed in Egypt for transporting enormous weights: "The absence of levers and rollers is remarkable, as the Egyptians must have been well acquainted with the use of both, and no doubt employed them for heavy weights."[7] Credit must be given to Sir Henry for stating that rollers were not found in Egypt, but the statement indicates that he strongly believed the Assyrian tree branches to be rollers. It also shows that his knowledge of the kinematics of rollers was insufficient.

In referring to a painting in the Egyptian grotto at El Bersheh (XII dynasty ca. 1880 B.C.) where the transport of an enormous statue is portrayed, Layard wrote: "On the statue as in the Assyrian bas-reliefs stands an officer who claps his hands in measure-time to regulate the motion of the men and from the pedestal another pours some liquid, probably grease, on the ground to facilitate the progress of the sledge, which would scarcely be needed were rollers used." Such an unscientific conclusion as this can only be rivaled by one made in 1930 by Clarke and Engelbach: "It is incredible that Assyria should have known the roller and not Egypt. A nation which used blocks and which never deduced the value of a roller for reducing friction from such homely occurrences as slipping on a walking-stick left on the floor would be subhuman in intellect, which the Egyptians were not."[8] Have the authors never slipped on a banana skin? Sliding objects sometimes can have less friction than rolling ones, and there would have been no advantage gained with rollers.

Did the Egyptians Invent Lubricated Sliding Bearings?

Upon examination of the ancient Egyptian painting at El Bersheh referred to by Layard, we see 172 men pulling a statue weighing some sixty tons and a man pouring a liquid on the ground in front of the sledge. If the ground were earthen it would become more slippery, but if it were soft, flat, and sandy, as it is in Egypt, the liquid would filter through the sandy soil with no effect at all. Should, however, large flat pieces of wood, like the one carried by the three men on their shoulders, be placed under the sledge and covered with oil, the work of transporting the statue would become easier. Although the Egyptians are known to have employed 100,000 men each year to transport heavy stone blocks,[9] it was obviously to their advantage to make this operation as efficient as possible, if only with the sole object of having fewer mouths to feed. Although the illustration does not show the flat boards being placed under the sledge, the following calculation leads one to conclude that this is the reason why the board is shown in the picture and why the artist showed likewise three men going before the board and carrying six pots of lubricant. Assuming that:

    (a) The alabaster colossus and sledge weighed[10] about 60 tons (1 ton = 2240 lb.)

    (b) Each board measured about 192 in. × 18 in. (from the drawing)

(If these boards were placed end to end under each side of the sledge there would be a bearing pressure of about 20 lb. per square inch, this being a pressure employed on modern tracked vehicles.)

    (c) The coefficient of sliding friction (μ) was 0.16 for hard wood on hard wood well lubricated with oil at 20 lb. per square inch bearing pressure[11]

    (d) The average pull of a man was 120 lb.

Then, the number of men required to pull the sledge can be arrived at by the following equation:

Frictional force $= \mu R$, Thus we obtain:

$$\text{Number of men} = \frac{0.16 \times 60 \times 2240}{120} = 179.$$

The number of men arrived at, i.e., 179, is very close to the 172 men pulling the sledge as portrayed in the mural painting.

If the boards had not been lubricated it would have taken about three times as many men to pull the sledge. The economic value of such a method of transport becomes obvious.

It is obvious, too, that tree branches placed under a sledge on anything but a very hard ground would sink below the surface and would remain immovable both in translation and rotation. Also, flat pieces of wood, if made of sufficient area, would prevent a sledge from sinking in the sand. Some caterpillar tracks on tanks in World War II, which the author helped to design whilst at the Admiralty, prevented the tanks from sinking into the sand when the ground pressure was about 20 lb. per square inch; this is about the same pressure as would exist on the flat boards under the Egyptian sledge. It thus appears most likely that lubricated boards were employed, and we again find the beginnings of a modern technique, namely, lubricated flat machine surfaces.

REFERENCES

[1] Sir A. H. Layard, *Nineveh and Babylon* (London, 1867).

[2] Charles Singer, E. J. Holmyard, and A. R. Hall (eds.), *A History of Technology,* Vol. I (Oxford, 1954), pp. 452, 470, 710; I. E. S. Edwards, *The Pyramids of Egypt* (Penguin Books, 1955), pp. 216, 229.

[3] S. M. Cole, "Land Transport Without Wheels: Roads and Bridges," in Singer *et al.* (ed.), *A History of Technology,* Vol. I, p. 710.

[4] The British Museum description of this "roller" is as follows: "Exhibit No. 43229. Wooden roller for moving building-stones. The ends of these rollers are generally pointed. From Deir el-Bahri, XI Dynasty, c. 2100 B.C. Presented by the Egyptian Exploration Fund, 1906."

[5] S. Clarke and R. Engelbach, *Ancient Masonry in Egypt* (London, 1930), pp. 89, 90.

[6] Marcus Vitruvius Pollio, *De Architectura,* translated by M. H. Morgan (Cambridge, Mass., 1914), p. 288.

[7] Sir A. H. Layard, *Discoveries in Nineveh and Babylon* (London, 1853), p. 115.

[8] Clarke and Engelbach, *op. cit.,* pp. 89, 90.

[9] Edwards, *op. cit.,* pp. 216, 229.

[10] Although the reader may doubt the existence of such a heavy piece of alabaster, the Keeper of the Egyptology Section of the British Museum informs me that there was a description beside the painting which stated that the material was alabaster. The description is quoted in Edwards, *op. cit.,* p. 216. The painting and description were later destroyed in an earthquake.

[11] E. Oberg and F. D. Jones (eds.), *Machinery Handbook* (New York, 1940), p. 507.

Eilmer of Malmesbury: An 11th-Century Aviator

Lynn White, Jr.

Recent histories of aviation have either entirely overlooked, or else have garbled, the name, the probable date, and the accomplishment of the first Occidental who can be shown to have flown: Eilmer, an Anglo-Saxon Benedictine monk of the Wiltshire Abbey of Malmesbury in the eleventh century.[1] This neglect is doubtless rooted in the fact that the standard, and fairly impressive, *History of Aeronautics in Great Britain,* by John E. Hodgson, dismisses Eilmer's flight as "hardly less legendary" than that of the Anglo-Saxon King Bladud,[2] which is very legendary indeed and which has no source earlier than the notorious Geoffrey of Monmouth.[3]

The evidence about Eilmer is quite different, and comes from his own abbey at no great interval after his death. In his *De gestis regum Anglorum,* William of Malmesbury tells us that shortly before the Norman invasion of England in 1066:

> A comet, a star foretelling, they say, change in kingdoms, appeared trailing its long and fiery tail across

Professor of history and director of the Center for Medieval and Renaissance Studies at the University of California, Los Angeles, Dr. White is the author of Medieval Technology and Social Change, *a landmark both in the history of technology and in the interpretations of medieval history. Dr. White was formerly president of Mills College and was one of the founders of the Society for the History of Technology.*

This article, published in Technology and Culture *(Vol. 2, No. 2): 97–111, was Dr. White's presidential address at the Third Annual Meeting of the Society for the History of Technology, held in New York City in December 1960.*

the sky. Wherefore a certain monk of our monastery, Eilmer by name, bowed down with terror at the sight of the brilliant star, sagely cried, "Thou art come! A cause of grief to many a mother art thou come; I have seen thee before; but now I behold thee much more terrible, threatening to hurl destruction on this land."

He was a man learned for those times, of ripe old age, and in his early youth had hazarded a deed of remarkable boldness. He had by some means, I scarcely know what, fastened wings to his hands and feet so that, mistaking fable for truth, he might fly like Daedalus, and, collecting the breeze on the summit of a tower, he flew for more than the distance of a furlong. But, agitated by the violence of the wind and the swirling of air, as well as by awareness of his rashness, he fell, broke his legs, and was lame ever after. He himself used to say that the cause of his failure was his forgetting to put a tail on the back part.[4]

William of Malmesbury was the best informed and most reliable historian in twelfth-century England.[5] The comet which provided the occasion for him to mention Eilmer's escapade was Halley's, the appearance of which in April, 1066, caused such a stir that it is pictured in the Bayeux Tapestry.[6] Since William was born about 1080 and entered the Abbey of Malmesbury in his boyhood, he was almost certainly acquainted with monks who had known the halt and aged Eilmer in their own youth and who had heard from him and from others the account of his flight and his diagnosis of the cause for his crash. William's slightly condescending reference to Eilmer's erudition may have been based on perusal of astrological works from his pen which seem still to have existed in the sixteenth century[7] and which would account for Eilmer's special concern over the comet. William completed the first three sections of his *De gestis regum* shortly after 1120.[8] We have no reason to doubt the essential accuracy of his account of Eilmer.

What sort of contraption did Eilmer construct? Since it specifically had wings ("pennae"), we may rule out a balloon, even though the Middle Ages from the ninth century at least were familiar with hot-air aerostats used as military signals.[9] The words "collectaque e summo turris aura" evoke the tragic image of the Saracen who in 1162 stood upon a column in the Hippodrome of Constantinople equipped with some sort of sail-like cloak and gathered the air for flight, only to crash to his death.[10] But William says that Eilmer flew "spatio stadii et plus," or more than 600 feet,[11] before falling, and this could not have been done with any loose quasi-parachute apparatus. We must conclude that Eilmer flew with rigid wings of considerable size, since they were attached both to his arms and legs. Probably they were intended to flap like those of a bird, but were hinged in such a way that they would not fold upward but would soar like a glider.

Can Eilmer's flight be dated? In 1066 he claimed to have seen the same comet before, although it was now "much more terrible." Between the appearance of Halley's comet in the spring of 1066 and its prior appearance late in 989, only one comet, that of 1006, was sufficiently conspicuous to be recorded.[12] But the comet of 1006 would not have been remembered in England as the harbinger of new catastrophe: the years immediately following it were indeed filled with misery; they marked, however, no more than a continuation of the terrible second wave of Danish invasions which in fact had broken a long period of relative peace immediately after the comet of the end of 989. The onslaught of 991, the burning of Ipswich, and the Anglo-Saxon disaster at Maldon opened a generation of turmoil which lasted until the victory of Canute in 1016.[13] Nor can the earlier comet seen by Eilmer be identified as that of 1006 in terms of a provincial view of disaster foretold in the heavens: his county of Wiltshire was ravaged in 1003 and 1005, but did not suffer again from major raids until 1011 and 1015.[14] Eilmer was correct in identifying Halley's comet which he had seen before. Assuming that he must have been at least five years old in 989, he was in his early 80's in 1066. William tells us that he accomplished his flight "in early youth, prima juventute," which would probably mean before he was twenty-five years old. The flight may therefore reasonably be placed in the decade 1000 to 1010 A.D.

What can have given Eilmer the idea of making a device for flying? For reasons already noted, he did not get it from the story of King Bladud, which is a bit of romancing which first appears a century after Eilmer. It is improbable that Eilmer's inspiration came from Suetonius,[15] who describes the fatal fall of an actor taking the part of Icarus in a series of mythological masques presented before Nero: the text indicates that this was a failure of theatrical machinery rather than an abortive attempt at actual flight. Varying versions of the legend of the flight of Simon Magus, and of how St. Peter's prayers brought him tumbling, were circulated throughout the Middle Ages.[16] One of them, to be sure, indicates that he had flapping wings which were entangled in Peter's ascending supplications[17]; but otherwise Simon Magus is thought to have been carried aloft by demons rather than by any mechanical device. And, as another English churchman, Bishop Wilkins, was to point out six centuries after Eilmer, although demons or angels may be quite serviceable for flight, this method of aviation must be sharply distinguished from "natural and artificial" means.[18] It should be emphasized that William of Malmesbury's account of Eilmer breathes an atmosphere of approval and admiration:[19] his adventure was purely mechanical, without supernatural involvement.

Were any of Eilmer's contemporaries experimenting with flight? If our dating of his tower-jump between 1000 and 1010 A.D. is correct, there is a curious coincidence. In either 1003 or 1008 A.D. the great Iranian student of Arabic philology, al-Jauharī, met his death attempting flight by some apparatus from the roof of the old mosque of Nishapur in Khorosan.[20] However, by about 985 nomadic incursions had severed the trade routes from the North Sea-Baltic region down the Volga to the Caspian and on to Persia, if we may judge by the Scandinavian numismatic evidence.[21] It is therefore improbable that Eilmer had heard of al-Jauharī, or the latter of Eilmer.

Perhaps some word had reached Malmesbury along the great pilgrimage road from Compostella of a more successful attempt at flight which appears to have been made in Andalusia about 875 A.D. The Moroccan historian al-Maqqarī, who died in 1632 A.D. but who used many early sources no longer extant,[22] tells of a certain Abū'l-Qāsim 'Abbās b. Firnās who lived in Cordoba in the later ninth century. Ibn Firnās was a polymath: a physician, a rather bad poet, the first to make glass from stones (quartz?), a student of music, and inventor of some sort of metronome. In his house, Al-Maqqarī tells us, he constructed a room in which, thanks to mechanisms hidden in the basement, spectators saw stars and clouds, and were astonished by thunder and lightning. "Among other very curious experiments which he made," continues al-Maqqarī, "one is his trying to fly. He covered himself with feathers for the purpose, attached a couple of wings to his body, and, getting on an eminence, flung himself down into the air, when, according to the testimony of several trustworthy writers who witnessed the performance, he flew a considerable distance, as if he had been a bird, but, in alighting again on the place whence he had started, his back was very much hurt, for not knowing that birds when they alight come down upon their tails, he forgot to provide himself with one."[23]

This sounds as specific as William of Malmesbury's account of Eilmer's adventure, and it is amusing that in each case the crackup is ascribed to failure to provide a tail. No modern historian can be satisfied with a source written 750 years after the event, and it is astonishing that, if indeed several eyewitnesses recorded b. Firnās's flight, no mention of it independent of al-Maqqarī has survived. Yet al-Maqqarī cites a contemporary poem by Mu'min b. Saīd, a minor court poet of Cordoba under Muhammad I (d. 886 A.D.), which appears to refer to this flight and which has the greater evidential value because Mu'min did not like b. Firnās: he criticized one of his metaphors and disapproved his artificial thunder. Mu'min wrote of b. Firnās, "He flew faster than the phoenix in his flight when he dressed his body in the feathers of a vulture."[24] Although the evidence is slender, we must conclude that b. Firnās was the first man to fly successfully, and that he has priority over Eilmer for this honor.

But it is not necessary to assume that Eilmer needed foreign stimulus to build his wings. Anglo-Saxon England in his time provided an atmosphere conducive to originality, perhaps particularly in technology. Until the invention of the mechanical clock in the early fourteenth century, the pipe organ was the most complex machine known to Europe. About 950 Bishop Aelfeg (d. 951) of Winchester broke all precedents by installing in his cathedral the first giant organ of the Middle Ages: 400 pipes were supplied with wind from 26 bellows operated by 70 men.[25]

The respect for manual operation guided by high intelligence, of which both the Winchester organ and Eilmer's gliding device are symptomatic, is reflected in a basic Anglo-Saxon mutation of the iconography of the Creator God. In earlier representations God is depicted in passive majesty, creating the cosmos Platonically, by power of thought. But in the Eadiwi Gospel Book, produced at Winchester shortly after 1000 A.D., the text of *Sapientia* XI:21, "Omnia in mensura et numero et pondere disposuisti," is construed literally, and the hand of God—now the master craftsman —is shown holding both scales and a pair of compasses.[26] This concept spread; perhaps under the influence of *Proverbia*, VIII:27, "certa lege et gyro vallabat abyssos," the scales were eliminated and the compasses alone—the supreme medieval and Renaissance symbol of the engineer

—are held in God's hand: a tradition culminating in William Blake's "Ancient of Days."[27]

Meyer Schapiro[28] has pointed to still another Anglo-Saxon iconographic innovation which displays a curious interest in the levitation of the human body. A sermon on Christ's Ascension preached in 971 A.D.[29] shows concern about the mechanics of the event, and, in expounding the words "et nubes suscepit eum ab oculis eorum," explains that it was not the cloud which carried Jesus up, but Jesus the cloud. By 1000 A.D. this sort of speculation had produced a new, realistic, and almost irreverent way of representing the Ascension. The earliest depictions show Christ majestically rising, often carried by angels. Later, Christ is taking some initiative by climbing the clouds as one would climb a mountain, and occasionally the hand of God the Father emerges from the topmost cloud to give him help up. But the Anglo-Saxons of Eilmer's days were beginning to show Christ almost jet-propelled, zooming heavenward so fast that only his feet appear at the top of the picture, while the garments of his astounded disciples flutter in the air currents produced by his rocketing ascent. Schapiro rightly remarks that the daring originality of such a break with iconographic tradition "is a sign of advanced conditions in England . . . and of the progressive character of English culture at the end of the tenth century."[30]

Eilmer's society, then, helps us to understand his feat, but the fact of a favorable context does not diminish his glory: he is the first European (in the cultural sense) about whom we have certain information that he flew successfully. Nor was he forgotten. Quite the contrary, although only the most confused references to him are found in modern histories of aviation, he was remembered throughout the Middle Ages and Renaissance, all later accounts stemming, usually indirectly, from that given by William of Malmesbury. In his *Chronicon* written before 1229, Helinand, a monk of Froidmont near Beauvais,[31] quotes William verbatim on Eilmer's flight, as, before 1241, does Alberic of the Cistercian abbey of Trois-Fontaines near Châlons.[32] Vincent of Beauvais, the chief scholar at the court of St. Louis in the 1250's, recalls him in the famous *Speculum*,[33] the favorite encyclopedia of the next three centuries. Before 1352 Ralph Higden, a Benedictine of Chester, tells about Eilmer in his *Polychronicon*,[34] which was the standard work on English history for the next two hundred years.[35] Unfortunately Higden misread a manuscript, and by fusing the

initial letters *Ei* into *O,* and fissioning the triple strokes of
m into *iv,* he rebaptized Eilmer as "Oliver." Before 1367
Henry Knighton incorporated Higden's account into his
own *Chronicon*[36] and when, in 1385, John of Trevisa, and
a little later still another Englishman, translated Higden
into English, "Oliver's" prowess became known to the less
educated public.[37] Clearly, at least in England, Eilmer was
long a personage alive in the communal recollection. In
1648 John Wilkins cited Eilmer as the prime example of
flight with wings[38]; in 1670 John Milton recorded him in
his *History of Britain Collected out of the Antientest and
Best Authorities.*[39] The first major work on flying to be
published in America, John Wise's *System of Aeronautics,
Comprehending the Earliest Investigations and Modern
Practice* (Philadelphia, 1850), knows Eilmer[40] through
Bishop Wilkins.

Roger Bacon would seem to have been aware of Eilmer,
but his thinking about the mechanical possibilities of flight
had gone much further: by about 1260 he was insisting
that "flying machines can be constructed so that a man sits
in the midst of the machine revolving some engine by which
artificial wings are made to beat the air like a flying bird.
. . . Such devices have long since been made, as well as in
our own day, and it is certain that there is a flying machine.
I have not seen one, nor have I known anyone who has
seen one. But I know a wise man who has designed one."[41]

There is, however, no evidence that memory of Eilmer's
feat helped to stimulate the new burst of speculation and
experiment about aviation which occurred in Italy in the
later fifteenth century. Even before 1449 the engineer Giovanni da Fontana rejects the idea of ascent by hot-air
balloons as too hazardous of fire, but expresses entire confidence that human flight can be achieved with mechanical
wings. Indeed, he has thought of making some himself,
"sed aliis distractus occupationibus non perfeci."[42] Leonardo da Vinci's sketches of parachutes and flying devices
are well known[43] but none appears to have been constructed. The evidence that in the 1490's Giovanni Battista
Danti of Perugia flew in a glider over Lake Trasimeno has
as yet emerged in no document earlier than 1648.[44] But we
must assume that when, in October 1507, an Italian named
Giovanni Damiani, who in 1504 had been appointed Abbot
of Tungland, a Premonstratensian monastery in Galloway,
garbed himself in wings made of feathers, took flight from
the walls of Stirling Castle, plummeted, and broke his leg,[45]

he was inspired by experiments in his native land rather than by Eilmer's example. Damiani sardonically announced that his error had been to include hens' feathers in his wings, since hens have more affinity for scratching in dunghills than for soaring to the heavens.[46]

Yet despite perpetual catastrophe, interest in flight continued endemic: it is symptomatic that the oration of the Tübingen humanist Friedrich Hermann Flayder, *De arte volandi, cujus ope quivis homo sine periculo facilius quam ullum volucre, quocumque lubet, semetipsum promovere potest*, first issued in 1627 at Tübingen, was reprinted not only the following year, but again at Frankfurt in 1737.[47] Even more significant is John Wilkins' detailed and enthusiastic discussion of the possibility of flight in his *Mathematicall Magick* of 1648. Wilkins was Cromwell's brother-in-law and at that time Master of Wadham College. He was already deeply immersed in the scientific activities which were to make him the first secretary of the Royal Society and eventually Bishop of Chester. On the page following his reference to "a certain English monk called *Elmerus . . .* that did by such wings fly from a tower above a furlong," he urges that experiment with flight be started on the gliding principle, that is, that one "use his wings in running on the ground as an Estrich or tame Geese will doe, touching the earth with his toes; and so by degrees learn to rise higher till he shall attain unto skill and confidence. I have heard it from credible testimony," he continues, "that one of our own Nation hath proceeded so far in this experiment, that he was able by the help of wings in such a running pace to step constantly ten yards at a time."[48]

It is doubtful, however, whether any great progress could have been made in human flight with heavier-than-air devices in Europe so long as the kite was unknown: preliminary observation of its reaction to air currents would seem to have been essential to our success in the twentieth century. Yet historians of aviation have largely neglected the kite.

Kites had been used in China since Han times at least.[49] There is no evidence of kites—as distinct from dragon-shaped aerostats—in either classical antiquity or most of the Middle Ages.[50] The meager literature on the subject seems to have overlooked the fact not only that the first word about kites to reach Europe comes where one might reasonably expect it—in Marco Polo's book—but also that Polo describes the East Asian kite in a way almost designed

to stimulate ideas about aviation: he speaks of a man-carrying kite, and lays great stress on how the kite reacts to the
wind. Unfortunately, however, this passage (although it is
almost certainly authentic) is found only in one of the
rarest versions of Polo: a Latin translation which exists in
two manuscripts, the earlier and better being of the fifteenth century. Polo tells us that when a merchant ship is
about to start on a voyage, a kite is sent up for augury:

> The men of the ship will have a wicker framework,
> that is a grate of switches, and to each corner and side
> of that framework will be tied a cord, so that there are
> eight cords and all of these are tied at the other end
> to a long rope. Next they will find some fool or drunk-
> ard and lash him to the frame, since no one in his right
> mind or with his wits about him would expose himself
> to that peril. This is done when the wind is high; then
> they raise the framework into the teeth of the wind and
> the wind lifts up the framework and carries it aloft,
> and the men hold it by the long rope. If, while it is up
> in the air, the frame tips in the direction of the gale,
> they haul in the rope a bit, and then the frame straight-
> ens up and they pay out the rope and the frame goes
> higher. And if again it tips, once more they pull in the
> rope until the frame is upright and climbing, and then
> they yield rope again, so that by this means it might go
> up until it could no longer be seen, if only the rope
> were long enough.[52]

Probably because this particular version of Polo had so
little circulation, no one followed up his description of a
man-carrying kite for nearly three centuries. In 1589 Giovanni Battista della Porta published at Naples the expanded
edition of his *Magia naturalis*,[53] which describes the making
of a kite and its action in relation to the flow of air. Della
Porta adds: "Hence may an ingenious man take occasion
to consider how to make a man fly with huge wings bound
to his elbows and breast."[54] This idea was spread not only
by several Latin editions of the *Magia naturalis* but also by
versions in Dutch (Leiden, 1655) and English (London,
1658). Moreover, the expanded version of Johann Jacob
Wecker, *De secretis libri XVII variis authoribus collecti*
(Basel, 1613), Lib. XIII and XVII, twice quotes verbatim
della Porta's passage on the kite as a possible basis for
human flight.[55] In addition to at least three reprints of the

Latin (Basel, 1642, 1662, 1701), a French translation appeared at Rouen in 1627 and 1633, and an English in London in 1660 and 1661.[56]

Clearly, during the later sixteenth and seventeenth centuries thousands of curious minds pondered the idea that the secret of human flight might be found not in the bird but in the kite. Yet toward the end of the seventeenth century two notable publications diverted the attention of European technologists away from the heavier-than-air tradition of experiment, of which Eilmer of Malmesbury was then recognized as the founder, toward the idea of flight by lighter-than-air apparatus. In 1670 the Jesuit Francesco Lana-Terzi[57] made the suggestion—brilliant in theory, even though unfeasible in practice—of an airship to be raised by means of evacuated metal spheres. An age greatly impressed by experiments with vacua conducted by von Guericke, Papin, and many others, was ready for some new departure. Moreover, ten years later, in 1680, Gian Alfonso Borelli's *De motu animalium*[58] demonstrated that the human musculature is quite insufficient to make any man-powered winged device fly on the analogy of birds. Thereafter, for two centuries, while there remained a residual interest in winged flight,[59] the focus of speculation and experiment shifted to aerostats, a development culminating and terminating in the gigantism of the Zeppelins of the 1930's.

The tradition of Eilmer was revived in the 1880's by two men. The technical priority must be granted to John Joseph Montgomery, a native of Yuba City, California, who became a professor at the University of Santa Clara. By 1884 he appears to have built three gliders and made many flights.[60] The effective father of modern aviation, however, was a Pomeranian, Otto Lilienthal. In 1889 he declared the balloon to be a blind alley—"eine falsche Bahn"—on the journey toward human flight, and lamented that for so long attention had been distracted by aerostats from the real problems of aviation.[61] In 1891 he built his first glider: a wicker framework covered with waxed cotton cloth, and without a tail.[62] Wilbur and Orville Wright's experiments were directly inspired by Lilienthal's glider experiments.[63] They decided, however, that the best way to get experience with air currents was not with a free glider but rather with a glider flown like a kite. It may be noted that their first model had no tail. By 1900 they were able to fly a kite-glider with a pilot aboard, presumably neither an idiot nor a drunkard. At last they felt that the point had been reached where the tug of the kite string should be replaced by motor-driven propellers.[64] Eilmer of Malmesbury's tradition of experiment with heavier-than-air devices was vindicated on December 17, 1903, at Kitty Hawk.

REFERENCES

[1] The undervaluation of Eilmer's achievement is the more curious because it was noted in detail by W. Hunt, "Oliver of Malmesbury," *Dictionary of National Biography,* XLII (London, 1895), p. 140, and by M. Massip, "Un victime d'aviation au onzième siècle," *Mémoires de l'Académie des Sciences, Inscriptions et Belles-lettres de Toulouse,* sér. 10, X (1910), pp. 199–217. Nor is his memory dead locally: a recent window in the Abbey church depicts him holding the model of a batwinged device, the iconography of which stems from Leonardo da Vinci.

[2] (London, 1924), p. 55.

[3] *Historia regum Britanniae* ed. A. Griscom (London, 1929), p. 262; J. S. P. Tatlock, *The Legendary History of Britain* (Berkeley, 1950), p. 47.

[4] Translation modified from that of J. A. Giles, in William of Malmesbury, *Chronicle of the Kings of England,* Book II, Chap. 13 (London, 1847), pp. 251–252. "Is erat litteris, quantum ad id temporis, bene imbutus, aevo maturus, immanem audaciam prima juventute conatus: nam pennas manibus et pedibus haud scio qua innexuerat arte, ut Daedali more volaret, fabulam pro vero amplexus, collectaque e summo turris aura, spatio stadii et plus volavit; sed venti et turbinis violentia, simul et temerarii facti conscientia, tremulus cecidit, perpetua post haec debilis, et crura effractus. Ipse ferebat causam ruinae quod caudam in posteriori parte oblitus fuerit"; *Gesta regum Anglorum,* ed. T. D. Hardy (London, 1840), I, p. 380; ed. W. Stubbs (London, 1887), I, pp. 276–277.

[5] Cf. M. R. James, *Two Ancient English Scholars: St. Aldhelm and William of Malmesbury* (Glasgow, 1931), pp. 8, 15.

[6] *The Bayeux Tapestry,* ed. F. Stenton (New York, 1957), pl. 35; cf. p. 169.

[7] John Bale, *Illustrium maioris Britanniae scriptorum, hoc est Angliae, Cambriae ac Scotiae summarium* (Wesaliae [imprint fictitious], 1548), fol. p. 72v, credits "Oliver" of Malmesbury ("Elmerum externi scriptores vocant") with three works, *Astrologorum dogmata, Eulogium historiarum,* and *De planetarum signis.* In his revised and expanded version, *Scriptorum illustrium maioris Brytannie . . . catalogus* (Basel, 1557), p. 163, Bale substitutes a book *De geomantia* for the *Eulogium historiarum.* His working notes, *Index Britanniae scriptorum,* ed. R. L. Poole (Oxford, 1902), offer no explanation of the change. T. Wright, *Biographia Britannica literaria* (London, 1846), II, p. 18, could find no trace of such works. F. Saxl and H. Meier,

Catalogue of Astrological and Mythological Illuminated Manuscripts of the Latin Middle Ages, III: Manuscripts in English Libraries (London, 1953), I, p. 291, connect only one, dubiously, with Malmesbury.

⁸ Giles, *op cit.*, p. viii, note.

⁹ Cf. R. Hennig, "Beiträge zur Frühgeschichte der Aeronautik," *Beiträge zur Geschichte der Technik und Industrie*, VIII (1918), pp. 105–108, 110–114; F. M. Feldhaus, *Technik der Vorzeit, der geschichtlichen Zeit und der Naturvölker* (Leipzig, 1914), pp. 653–659; J. Duhem, "Les aérostats du moyen-âge d'après les miniatures de cinq manuscrits allemands," *Thalès*, II (1935), pp. 106–114; these overlook a dragon-shaped aerostat of 1327 on the end of a cord held by three soldiers, flying over a besieged city and dropping fire-bombs on it, in Walter de Milimete, *De nobilitatibus . . . regum*, ed. M. R. James (Oxford, 1931), pl. p. 154; and another in D. Schwenter, *Deliciae physicomathematicae* (Nuremberg, 1636), I, p. 472.

¹⁰ Niketas Akominatos, Bk. III, in Migne, *Patrologia graeca*, CXXXIX (Paris, 1894), p. 458, written about 1206. For the dates, cf. F. Dölger, *Regesten der Kaiserurkunden des oströmischen Reiches*, II (Munich, 1925), nr. 1446.

¹¹ C. du Cange, *Glossarium mediae et infimae latinitatis* (Niort, 1886), s.v. "stadium," despairingly defines it as "mensurae species, sed ignota prorsus." In J. H. Baxter and C. Johnson, *Medieval Latin Word-list from British and Irish Sources* (London, 1934), "stadium" does not appear as a measure of length, but in the school Latin of modern Britain, which presumably reflects medieval usage, it means "furlong"; cf. J. MacFarlane, *New Pocket Dictionary of Latin and English Languages* (London, n.d.), p. 635. The Roman *stadium* was 606 feet, 9 inches. The medieval furlong (or normal long furrow of Northern Europe) varied, but was roughly 660 feet.

¹² J. B. Galle, *Verzeichnis der Elemente der bisher berechneten Cometenbahnen* (Leipzig, 1894), p. 153.

¹³ The military history of the second Danish Wars from 991 to 1016 is conveniently summarized by F. Stenton, *Anglo-Saxon England* (Oxford, 1943), pp. 371–387.

¹⁴ R. R. Darlington, "Anglo-Saxon Wiltshire," in *The Victoria History of the Counties of England: Wiltshire*, II (London, 1955), p. 15.

¹⁵ *Nero*, XII, ed. J. C. Rolfe (London, 1930), II, p. 104. The effort of Hennig, *op. cit.*, pp. 103–105 and "Zur Vorgeschichte der Luftfahrt," *ibid.*, XVIII (1928), pp. 91–92, to identify this episode with the legend of the flight of Simon Magus lacks all probability.

¹⁶ L. Thorndike, *History of Magic and Experimental Science*, I (New York, 1929), pp. 400–427.

¹⁷ "Et statim in voce Petri implicatis remigiis alarum quas sumserat, corruit"; *ibid.*, p. 426, n. 1.

[18] J. Wilkins, *Mathematicall Magick, or the Wonders That May Be Performed by Mechanicall Geometry* (London, 1648), p. 201.

[19] The statement of Dom Aelred Watkin, "The Abbey of Malmesbury" in *The Victoria County History: Wiltshire*, III (London, 1956), p. 214, that "The abbot . . . forbade a repetition of the experiment" is embroidery for which the record does not provide thread.

[20] Ahmed Zéki, "L'aviation chez les Arabes," *Bulletin de l'Institut Égyptien, Alexandrie*, 5th ser., IV (1910), pp. 93–95; for the date, C. Brockelmann, *Geschichte der arabische Literatur*, 2nd ed. (London, 1943), I, p. 133.

[21] A. R. Lewis, *The Northern Seas: Shipping and Commerce in Northern Europe, A.D. 300–1100* (Princeton, 1958), pp. 437–439.

[22] B. Carra de Vaux, *Les penseurs de l'Islam* (Paris, 1921), I, p. 163.

[23] Al-Makkari, *History of the Muhammadan Dynasties in Spain*, Bk. 2, ch. 3, tr. Pascual de Gayangas (London, 1840), I, p. 148.

[24] In al-Makkari, *loc. cit.* (and cf. p. 426, n. 34), the poem is mistranslated by Gayangas; cf. Ahmed Zéki, *op. cit.*, p. 100. I owe the present translation to Dr. Wilhelm Hoenerbach of the University of Bonn.

[25] *Frithegodi monachi Breuiloqium vitae beati Wilfredi, et Wulfstani cantoris Narratio metrica de sancto Swithuno*, ed. Alistair Campbell (Zurich, 1950), pp. 69–70, lines 141–170.

[26] E. Panofsky and F. Saxl, *Dürer's "Melencolia I"* (Leipzig, 1923), p. 67, n. 3.

[27] A. Blunt, "Blake's 'Ancient of Days': the symbolism of the compass," *Journal of the Warburg Institute*, II (1938–39), pp. 53–63.

[28] M. Schapiro, "The image of the disappearing Christ: the Ascension in English art around the year 1000," *Gazette des beaux-arts*, XXIII (1943), pp. 135–152.

[29] *The Blicking Homilies*, ed. R. Morris (London, 1880), p. 121.

[30] *Op. cit.*, p. 152.

[31] In Migne, *Patrologia latina*, CCXII (Paris, 1855), p. 953.

[32] *Chronicon*, in *Monumenta Germaniae historica, Scriptores*, XXIII (1874), p. 795.

[33] *Speculum historiale*, Lib. XXVI, cap. 35 (Strassburg, c. 1473), folios unnumbered. I am grateful to Dr. William Bark of Stanford University for confirming this reference in the Stanford copy of the *Speculum*.

[34] *Ranulphi Higden monachi Cestrensis Polychronicon, together with the English Translation of John of Trevisa and of an Unknown Writer of the Fifteenth Century*, ed. Joseph Rowson Lumby (London, 1879), VII, p. 222.

[35] James Westfall Thompson, *History of Historical Writing* (New York, 1942), I, p. 399.

[36] *Chronicon Henrici Knighton, vel Cnitthon,* Lib. I, cap. 15, ed. J. R. Lumby (London, 1889), p. 49.

[37] *Ed. cit.,* VII, p. 223.

[38] *Op. cit.,* p. 204.

[39] Book VI, in *Prose Works,* ed. C. Symmons (London, 1806), IV, p. 252. The story also appears in Sir Roger Twysden, *Historiae Anglicanae scriptores decem* (London, 1652), col. 2338.

[40] P. 25.

[41] *De secretis operibus,* cap. 4, in *Opera quaedam hactenus inedita,* ed. J. S. Brewer (London, 1859), p. 533. For the date, cf. S. C. Easton, *Roger Bacon and His Search for a Universal Science* (New York, 1952), p. 111.

[42] L. Thorndike, *op. cit.,* IV (New York, 1934), pp. 173–174, n. 99; for the date, p. 156.

[43] The essential evidence is assembled by F. M. Feldhaus, *Leonardo der Techniker und Erfinder* (Jena, 1913), pp. 140–150.

[44] Hennig, *op. cit.,* VIII (1918), p. 115; XVIII (1928), pp. 93–94.

[45] The episode is mentioned in two contemporary satirical poems by William Dunbar, *Poems,* ed. J. Small, I (Edinburgh, 1884), pp. 139–143, 150. Damiani was, or claimed to be, a physician, given to alchemy and tinkering at a blacksmith's forge. The available information is gathered by Æ. J. G. Mackay, *ibid.,* III (1889), pp. xlvi, cxviii, ccxiv–v.

[46] Jhone Leslie, *Historie of Scotland,* ed. E. G. Cody and W. Murison, II (Edinburgh, 1895), p. 125; this work was written in 1568–1570.

[47] Cf. G. Bebermeyer, *Tübinger Dichterhumanisten: Bebel, Frischlin, Flayder* (Tübingen, 1927), p. 88; H. F. Flayder, *Ausgewählte Werke,* ed. G. Bebermeyer (Leipzig, 1925), pp. 168–169.

[48] *Op. cit.,* p. 205.

[49] Wang Ch'ung, *Lun-hêng,* tr. A. Forke (Berlin, 1907), I, p. 499; B. Laufer, *The Prehistory of Aviation* (Chicago, 1928), pp. 34–35; J. Duhem, *Histoire des idées aéronautiques avant Montgolfier* (Paris, 1943), p. 194.

[50] Feldhaus, *Technik der Vorzeit,* col. 650, fig. 442, mistakes for a kite a toy consisting of a small fluttering pennant at the end of a string pictured on a Greek vase in Naples; *ibid.,* col. 651, fig. 443, dated about 1450, appears to be a dragon-shaped aerostat such as had been known for centuries; cf. *supra* n. 9.

[51] Cf. Marco Polo, *Description of the World,* ed. A. C. Moule and P. Pelliot (London, 1938), I, pp. 47–52.

[52] Translation modified from those of Moule, *op. cit.,* I, pp. 356–357, and R. Latham, *Travels of Marco Polo* (London, 1958), p. 215. "Homines vero navis habebunt unam cratem, id est unum graditum de viminibus, et in quolibet angulo et latere

ipsius cratis erit ligata una funis ita quod erit octo funes, et omnes ligate erunt ab alio capite cum una sartia longa. Item invenient aliquem stultum vel ebrium et ipsum ligabunt super cratem, quia nullus sapiens nec sincerus ad periculum illud se exponeret. Et hoc fit quando ventus regnat intensus. Ipsi quidem errigunt cratem in opositum venti, et ventus cratem elevat et portat ipsam in altum, et homines per sartiam longam tenent. Et si cratis dum est in aere versus cursum venti declinet, ipsi aliquantulum sartiam ad se trahunt, et tunc cratis erigitur et ipsi sartie cedunt et cratis ascendit. Et si iterum declinet, tanto sartiam trahunt donec cratis erigitur et ascendit, et ipsi sartie cedunt, ita quod per hunc modum tantum ascenderet quod videri non posset, dummodo foret sartia tam longa"; Moule and Pelliot, *ed. cit.,* II, p. lviii. Marco Polo, *Il milione,* ed. L. F. Benedetto (Florence, 1928), pp. 162–163, gives a slightly less satisfactory text from the eighteenth century manuscript.

[53] The first edition, Naples, 1558, contained only 4 books; that of 1589 had 20.

[54] Bk. 20, ch. 10; facsimile of the English tr. (London, 1658), ed. D. J. Price (New York, 1957), p. 409.

[55] In the edition of Basel, 1613, which I have used, the passages appear on pp. 483 and 651.

[56] Cf. J. Ferguson, *Bibliographical Notes on Histories of Inventions and Books of Secrets* (London, 1959) I, ii, p. 62; index, pp. 34–35; II, index, p. 30.

[57] *Prodromo overo saggio di alcune inventioni* (Brescia, 1670); cf. Arturo Uccelli, *Storia della tecnica dal medio evo ai nostri giorni* (Milan, 1945), p. 891; G. Boffito, *Biblioteca aeronautica italiana* (Florence, 1929), pp. 233–237; supplement (1937), p. 309. For anticipations of this idea, cf. *supra,* n. 42 and Feldhaus, *Technik,* p. 647.

[58] (Rome, 1680), Pars I, Prop. 204, pp. 322–326: "Est impossible ut homines propriis viribus artificiose volare possint."

[59] It is curious to note that in 1759 Samuel Johnson incorporates into a conversation in his *History of Rasselas, Prince of Abissinia,* ch. 6, ed. R. W. Chapman (Oxford, 1927), p. 29, a statement by a craftsman in mechanics which formulates quite adequately the principle of heavier-than-air flight: "You will be necessarily upborn by the air, if you can renew any impulse upon it, faster than the air can recede from the pressure."

[60] For a brief biography and bibliography of Montgomery's writings, cf. *National Cyclopaedia of American Biography,* XV (New York, 1916), pp. 338–339. In 1893 he participated in the Aeronautical Congress in Chicago presided over by Octave Chanute, and thereafter was in correspondence with Chanute. Since in the early stages of his experiments Montgomery was secretive, J. Milbank, Jr., *The First Century of Flight in America* (Princeton, 1943), pp. 183–185, treats his claims with some reserve. But there seems to have been no skepticism in California among those most interested in flight, and in a position

to get verbal information not only from Montgomery but also from his colleagues and students at Santa Clara; cf. the lecture by Prof. Joseph Hidalgo of the University of California, Berkeley, *History of Aerial Navigation* (San Francisco, 1910), p. 46, delivered before the Pacific Aero Club on 1 January 1910. Since no adequate study of Montgomery exists, it should be noted that the library of the University of California, Los Angeles, contains a 36-page pamphlet by one of his students, L. A. Redman, *Professor Montgomery's Discoveries in Celestial Mechanics* (San Francisco, 1919) which is not related to his aeronautical work.

[61] O. Lilienthal, *Der Vogelflug als Grundlage der Fliegekunst* (Berlin, 1889), pp. 155–158.

[62] A. E. Berriman, *Aviation* (London, 1913), p. 111.

[63] W. Wright, "Some Aeronautical Experiments," *Journal of the Western Society of Engineers,* VI (1901), p. 494.

[64] Berriman, *op. cit.,* pp. 121–131.

Advances in American Agriculture: The Mechanical Tomato Harvester As a Case Study

Wayne D. Rasmussen

Technological advance in agriculture is not, in the United States today, the result of adopting some one tool or technic. Rather, it is adopting what has been called a "package" of agricultural technology, which usually includes machines, improved (often hybrid) seeds and breeding stock, careful tillage, fertilizer, productive use of water through irrigation or drainage, and the application of chemicals to control weeds, fungi, and insects. The package idea is actually a systems approach to the problems of increasing agricultural productivity. Some of this appears in the histories of such major technological changes as hybrid corn, the cotton picker, and the mechanization of sugar-beet production.[1]

The American farmer today is the most efficient, in terms of production per man-hour, that the world has ever seen. This efficiency is not entirely new, for agriculture over the years has made a massive contribution to American economic growth. However, productivity in the past twenty-five years has advanced at a rate greater than ever before. One hundred years ago, drawing upon a crude but dramatic measure of productivity, one American farm worker was producing the food and fiber used by 5 people. Twenty-

Dr. Rasmussen, of the Economic Research Service of the U.S. Department of Agriculture, has been president of the Agricultural History Society and has written many articles and books on the history of American agriculture. This article, published in Technology and Culture *(Vol. 9, No. 4): 531–43, was derived from a paper presented by Dr. Rasmussen at the Ninth Annual Meeting of the Society for the History of Technology in Washington, D.C., December 1966.*

seven years ago, in 1941, the figure was 12 people; in 1956, 22 people; and in 1966, 37 people.[2]

The rise in productivity has depended upon the desire and ability of American farmers to adopt new technological advances and upon the desire and ability of others to provide technology. Inventors, manufacturers, the state agricultural colleges and experiment stations, and the U.S. Department of Agriculture have provided the new technology. The economic conditions accompanying two great wars—the Civil War and World War II—gave farmers temporary high prices and a seemingly unlimited demand for their products, combined with a shortage of labor, which induced them to adopt the labor-saving and productive technology then available.[3] Farmers have continued since World War II to adopt new technology because price supports have, in general, ensured a floor for farm prices; because nearly all farmers are producing for the commercial market; and because competition among farmers, even with price supports, is forcing the marginal producers out of farming. Producers, except for a hard core of poor, nonproductive subsistence farmers, adopt every bit of technology which promises to reduce costs of production.[4] Thus, incongruous as it may seem, both high prices (combined with a scarcity of agricultural labor) and low prices (combined with rising labor costs) have motivated the adoption of new technology in American agriculture.

The cotton picker is often mentioned as a technological innovation which has brought about major changes in Southern farm life. It might also be cited as an example of the traditional invention. Many attempts had been made by individuals and implement companies to invent a practical cotton picker before 1927. In that year, John D. Rust had the simple idea of moistening the spindles which twisted the cotton fiber from the bolls—the key to a successful, commercial cotton picker. Recently, plant breeders, chemists, and engineers have brought other technological changes to the cotton fields, but they came long after the picker was a practical reality.[5]

The invention of the mechanical tomato harvester contrasted decidedly with the development of the cotton picker. The tomato harvester resulted from the "systems approach." A team made up of an engineering group and a horticultural group, with advice and assistance from agronomists and irrigation specialists, developed suitable plants and an efficient harvester at the same time. The necessary changes in planting, cultivation, and irrigating were developed concurrently. This is one reason, among others, that growers adopted the mechanical tomato harvester within a few years after its development, while twenty years elapsed between the invention of a successful cotton picker and its large-scale manufacture and adoption.

Many developments and inventions—for example, the virtual elimination in the United States of screw worms by sterilizing male screw worm flies with irradiation, the storage of apples in inert gases, the mechanical harvesting of lettuce by a machine which tests each head for firmness— draw upon complicated and sophisticated research. In other instances, costs have been reduced by combining various aspects of what I have called the package of agricultural technology into something new. The mechanical tomato harvester is such a device.

At first glance, the invention of a mechanical tomato harvester might seem simple; at a second glance, impossible. Tomatoes grow on low vines, and the ripe fruit is readily identified by its color. The yield of fruit per acre is large, and the vines can be grown in rows with space for machine operations between them. On the other hand, the fruit is usually thin-skinned and bruises easily. The vines tend to flower and set fruit over a period of weeks, which means that the fruit normally ripens in the same way. The usual method of hand picking is to go over the field three or four times at intervals of one or two weeks, picking the ripe and near-ripe tomatoes.

The first major impetus for the development of machines to pick tomatoes came during World War II—a period of high prices, strong demand, and labor shortages. The county agent at Lancaster, Pennsylvania, whose job as an employee of the State Extension Service was to help farmers apply the latest research results, reported in 1943 that a Mr. Garber had operated a tomato-picking machine in 1941 and 1942. This machine was a large wing extension from the side of a truck. Hammocks were suspended from the wing. The pickers, usually youngsters, were carried down the rows of tomatoes in the hammock, picking as they went. The truck was geared down to move very slowly. The youngsters placed the picked tomatoes on a conveyor belt in front of them, which carried the fruit to the truck.[6]

It was also reported in 1943 that a blacksmith in Holt, California, was building a tomato picker for a canning firm in Stockton. This picker would strip all of the fruit from the vines in a once-over operation. The tomatoes would then be sorted by color—red, pink, green—and be handled accordingly. We have no record of the operation of the machine.[7] However, the basic idea of going "once down the row" has been adopted in all successful pickers.

It is understandable that California led in mechanizing the harvesting of tomatoes for processing since, at about the time the first field experiments with mechanical pickers began, California grew approximately 60 per cent of the nation's processing tomatoes. These nearly 2.5 million tons of fruit were valued at $60 million. California was followed by Indiana, New York, and Ohio.

Another part of the package of technology necessary for a successful picker began to receive attention about the same time. Evidence had accumulated that research to develop a tomato with special new qualities would be necessary for mechanical harvesting of tomatoes. In 1942, A. M. Jongeneel, a California tomato grower, suggested to G. C. Hanna of the Department of Vegetable Crops, University of California, Davis, that the university develop a tomato that could be harvested by machine. Hanna began a search in 1943 in the breeding stocks to find a firm-fruited type that would withstand rough handling. Little progress was made, except to note that pear-shaped fruits would withstand rough treatment much better than any of the round types. During the late 1940's, a pear-shaped tomato particularly adapted to machine harvesting was developed and released to growers.[8]

Pear-shaped tomatoes, for no very good reason, are difficult to market fresh or canned whole. They are used mostly for paste, juice, catsup, and similar products. Usually, tomatoes for processing or canning are processed within twenty-four hours of picking. This means that bruises which would disqualify tomatoes for the fresh market do not eliminate them for processing. It is not practical to incur added expense to avoid bruising in tomatoes which are going to be used for catsup. Consequently, mechanical picking, which has often resulted in more bruising than hand picking, has been restricted to tomatoes for processing, including canning.

Even as the first breeding and mechanical work began in earnest, some experts were doubtful of the prospects of success, while others experimented with measures short of complete mechanization. In 1952, a California farm paper quoted a man from the University of California's College of Engineering "who has made a study of tomato harvesting" as saying "we aren't close at all" to mechanical harvesting.[9]

During the 1951 and 1952 harvesting seasons in California, a number of growers experimented with conveyor belts. Some years later, in 1958, E. S. Holmes and others at the Florida Agricultural Experiment Station developed a conveyor belt which carried hand-harvested vegetables to a machine for packaging. It was reported that tomatoes suffered internal bruising, which might be reduced by better padding and reducing the distance the tomatoes dropped from the belt.[10]

There can be little doubt that the persistence of one man, G. C. Hanna, led to the development of the mechanical tomato picker in California. Hanna's success in developing a pear-shaped tomato, all the fruit of which ripened at about the same time, persuaded Coby Lorenzen of the Department of Agricultural Engineering, University of California, Davis, that a successful mechanical picker could be built. Serious work to this end began in 1949. The undertaking was limited to developing a harvester for the pear-shaped tomato, with the understanding that, after success here, attention would move to the round type.[11]

After exploring several alternatives, it seemed to Lorenzen and Hanna that a successful machine must be based upon a few fundamental ideas. There was no practical alternative to the "once down the row" idea. That is, a machine would go over the field of tomatoes only one time. In that one trip, the vines would probably be destroyed. It also seemed to Lorenzen that a successful harvester might be built by combining several techniques already in use in agricultural machines. This proved to be true, but adjusting the techniques to the peculiar requirements of the tomato took both ingenuity and patience.

Lorenzen and others at the University of California experimented with various devices for doing each essential part of the job. They worked with devices to cut the vines, to lift the plants onto the machine, to separate the tomatoes from the vines, to sort and clean the tomatoes, and to convey the fruit to a suitable container. They then put these together and had a machine in field operation in 1959. The University of California then patented the machine and licensed the Blackwelder Manufacturing Company to undertake its commercial manufacture. The company built fifteen machines in 1960, although all of them were not in use. Apparently single machines of five other types were in operation for at least short periods. About 1,200 tons of pear-shaped tomatoes were harvested by machine—a minute part of the total crop. By 1961 Blackwelder had twenty-five machines in the hands of growers. However, the harvesting operations with the Blackwelder and five other types of machines were still experimental. In 1961 only ½ of 1 per cent of the total California tonnage of tomatoes for processing was harvested mechanically.[12]

Meanwhile, another team, composed of an agricultural engineer, B. A. Stout, and a horticultural specialist, S. K. Ries, of Michigan State University, had built a crude machine in 1958, a more elaborate model in 1959, and still a third in 1960. The 1960 model was tested in both Michigan and Florida. In 1961, approximately four acres of tomatoes were harvested, and the fruit was carried satisfactorily in bulk containers to a commercial processing plant. The Chisholm-Ryder Company of Niagara Falls, New York, according to S. K. Ries, built harvesters based on the Michigan State University design.[13]

Other commercial firms had at least experimental machines in operation by 1961. The H. D. Hume Company of Mendota, Illinois, demonstrated one of its machines in Erie County, New York. Other harvesters included the FMC (Food Machinery Corporation), the Peto Ayala, and the Ziegenmeyer, all in California. The same year, Robert L. Button, Jr., a grower at Winters, California, built a machine. He turned over the manufacturing rights to the Benner Nawman Company. However, in 1965, other arrangements were made for manufacturing the Button harvester, and Benner Nawman marketed its own model. The Massey-Ferguson Company had a model developed by Frank Gill in the fields by 1962.[14]

The years 1960–62 were crucial in the development of the mechanical picker. Several machines, many experimental, were being tested. All of them worked on the same general principles. When most of the fruit was fully ripe, harvesters moved down the row, cutting the plants off and lifting them onto an elevator. The plants were carried to a shaking or other device for separating the tomatoes from the vines. After the tomatoes had been removed, the plants were discarded. The fruit was carried along a sorting area, where green and spoiled fruit and clods of earth were removed by hand. The ripe tomatoes were then moved to a suitable container, often a bulk box holding 500 to 1,000 pounds. The containers were carried by a trailer alongside the harvester until filled and were then hauled to a processing plant.[15]

The different machines, of course, varied in a number of respects. Some, such as the University of California–Blackwelder machine, were self-propelled and sold for about $15,000. The FMC machines offered in 1961 were of the pull type and thus cost less.

The cutting and pickup devices varied, partly because of particular problems involved in harvesting ripe tomatoes. By the time the tomatoes were ripe enough for harvesting, some had fallen from the vines or did so at the first touch of the machine. If the machine cut the vines above ground level, the fallen tomatoes were lost. If cut at ground level, some were recovered. Cutting below ground level and picking up a thin layer of dirt saved nearly all of the tomatoes at the cost of having more dirt to separate from the fruit. The more dirt the machine picked up, the greater the possibility of contamination through a mixture of juice and earth adhering to the fruit.

The Hume machine cut the vines at ground level with a slowly operating sickle. The FMC machine featured self-sharpening counter-rotating disks which cut the vines slightly below ground level and picked up virtually all of the tomatoes. Several machines cut the vines at or just below ground level with a v-shaped blade.[16]

A number of devices were used to separate the tomatoes from the vine, with a shaker being the most usual. The University of California people used rubber-covered walkers on their machine. The mechanism was operated by a crankshaft. The Hume harvester removed the fruit with six belts with reciprocating rubber fingers. The Michigan State University model was equipped with a shaker bed which reciprocated with a four-inch stroke at 175 to 200 cycles a minute. The Ziegenmeyer machine was equipped with a rotary shaker operated on pneumatic tires.[17]

Various belts and blowers were used to remove trash and dirt from the tomatoes after they were separated from the vines. The fruit then moved to sorting belts. The first machines did not have space for enough sorters, who sometimes had to lift the discards rather far, to do a rapid, efficient job. The sorting belts carried the tomatoes to an elevator, which placed them in boxes, bulk bins, or tanks on a trailer pulled parallel to the harvester. During the first years the mechanical pickers were used, much experimenting was done with different containers, but bulk bins are now widely used.

Searching criticisms were made of the early harvesters. All of the early machines had a limited capacity, and tomatoes tended to pile up at one point or another. The major problem, however, was that too much dirt was picked up with the tomatoes. At the same time, some growers were doubtful about the efficiency of the pickers because many loose tomatoes were left on the ground. Many of the tomatoes were cracked and bruised. Too many green or poor-quality tomatoes were getting by the sorters. Some of the machines were not sturdy enough.[18] As late as 1964, there were complaints of crop losses because of machine failures. This problem, however, appeared to have been solved, at least for California, by 1966.[19]

While these criticisms had validity, later developments took care of many of them. Larger machines were built. Various devices were developed to eliminate much of the dirt. Spray washers were used in some cases. Better arrangements of conveyor belts eliminated some of the cracking and bruising. Some of the machines were rebuilt to pick up more of the tomatoes. However, the engineer who invented the University of California picker has said many farmers came to realize that the 15 to 20 per cent loss with the machine was no more than the loss suffered by hand picking the fields three or four times. Space on the harvester for more sorters reduced the number of poor-quality tomatoes getting by, particularly after the sorters had some experience. Some growers preferred women as sorters; others, men. One report suggested that men were more subject to motion sickness. On the other hand, a grower told me that because women were always arguing, he preferred men.[20]

Sorting, still done with hand labor, will doubtless be mechanized. One suggestion is that less shaking would detach the ripe fruit, but would leave most of the green fruit on the vines. An automatic color detection-rejection system, mounted on the harvester, could then discard the comparatively few green tomatoes. An alternative, now in an experimental state, is to drop all sorting on the machine and sort mechanically at a central point. Perhaps the sorting could be done at less expense as a part of the processing operations.[21]

The key to the success of the harvester lay in the breeding of varieties of tomatoes which could be handled by machine. This point was made again and again in early accounts. Later, in 1965, a Virginia economist stated that the tomato harvester was practical to use only under specific conditions. The yield should be thirty or more tons to the acre. The tomatoes should be resistant to bruising and nearly all should reach optimum maturity at the same time. Such conditions were so rare that complete or significant moves to mechanized harvest, other than in California, were several years away. This analysis was supported by an estimate made in 1966 that there would be 774 machines available for use in that year. Of these, only 14 were outside of California and 3 of those were outside the United States, leaving 760 in California. However, an expert in the U.S. Department of Agriculture disagreed with this analysis. He stated in 1967 that mechanization in the East and Midwest depended largely on cultural practices, particularly measures to ensure feasibility of direct seeding and weed control. He predicted that acceptable varieties would be grown on a very large scale in 1968–69.[22]

Research workers agreed on the major characteristics of varieties suitable for mechanical harvesting. The plants should blossom and set fruit at the same time and only once. All the fruit should ripen at about the same time. Plants should not have excessive foliage. Stem or pedical should be jointless to prevent puncture of the fruit in handling. The fruit should be firm and crack-resistant. The tomato should have good vine-holding ability at maturity. Finally, quality standards must be maintained.[23]

Generally, it takes ten generations to get new, fixed characteristics in tomatoes, which is why many breeders grow a second crop each year in Mexico or Puerto Rico. While the present round varieties now used for mechanical harvesting have most of the essential characteristics, further work is needed, particularly if tomatoes are ever to be harvested mechanically for the fresh market. Hanna has said that the greatest problem lies in overcoming the backyard garden philosophy of growing tomatoes. For example, why should consumers insist upon a round tomato? Cylindrical tomatoes would be more efficient in a mechanical age. Perhaps such tomatoes could be sliced and canned, and thus win consumer acceptance.[24]

Growing tomatoes for successful mechanical harvest requires very careful attention to cultural practices. The small size of the plants means that 12,000 to 18,000 can be grown per acre. This in turn means that the plants must be direct seeded rather than transplanted. Planting in straight rows, with the plants evenly spaced, facilitates machine harvesting and encourages even ripening of the fruit.

Fertilizer and irrigation management is another key to the mechanization of tomato picking. The plants need enough fertilizer and water to produce a full crop, but excessive or incorrectly timed applications of either cause irregular blossoming and setting of fruit, which leads to irregular ripening.[25]

It is evident that the successful mechanization of tomato picking depends upon a package of technology, which includes effective machines, specially bred tomatoes, careful irrigation and fertilization, and particular planting techniques. The engineer, the botanist, and the agronomist have to bring their specialties together to ensure successful development. This dependence upon a package of technology is true today for virtually every advance in farming.

It is also true that advances in technology mean little unless they are adopted. The first mechanical picker was in California fields on an experimental basis in 1959. In 1965, 262 mechanical harvesters picked 25 per cent of the state's crop of tomatoes for processing. In 1966, an estimated 460 harvesters picked at least 60 and perhaps 80 per cent of the crop. The mechanical tomato picker has been adopted, at least in California, at a much faster rate than such devices were before World War II.

In 1965, the California Agricultural Experiment Station made an economic study of sixty-three harvesters operated in fifteen California counties.[26] The researchers considered all equipment and all equipment costs involved in the mechanical harvesting system. These included the tomato harvesting machine, trailers to carry the fruit from the harvester to the loading area, tractors to pull the trailers, forklifts to load the fruit onto trucks, and washing equipment.

The study showed that the average harvester crew picked .35 ton of tomatoes per hour per worker. The good hand picker averaged .19 ton per hour. A total of 61 man-hours of labor were required to harvest an acre of tomatoes yielding 21.5 tons using a machine, compared with 113 hours using good hand workers. The machine saved 52 man-hours of labor per average acre. The total average cost of harvesting a ton of tomatoes with a machine was $9.84. The total average costs of hand harvest in the two major tomato-producing counties in California were $17.04 and $17.19, respectively. This reported savings of $7 per ton in harvest costs might cause any efficient tomato grower to consider making the shift from hand to mechanical picking.

Although we do not have definite information on this point, it appears that the producers received several dollars less per ton for their tomatoes in 1966 than in 1965. This would have the effect of forcing growers using high-cost hand harvesters to shift to lower-cost machine harvesters.[27]

The fear of a lack of labor to handle the tomato crop, whether justified or not, and the fear of organized labor from the late 1950's on, were major influences in encouraging the invention of a satisfactory picker, research in essential breeding and crop management, and the purchase of machines by growers. Local newspapers, growers, and scientists are virtually unanimous on this point. Whether or not such fears were justified is beyond the scope of this paper. However, it might be mentioned that during World War II, the federal government began a program of bringing farm workers from Mexico to the United States. A majority of the workers were employed in California. When the wartime program ended in 1947, it was followed by a series of laws which first permitted employers themselves, and later the secretary of labor, to recruit contract laborers under specified conditions. Attacks on this program were intensified in the late 1950's and early 1960's. Finally, in 1964, the last of the laws expired. Mexican laborers were thenceforth to enter under contract only in exceptional cases.[28]

Some local reports in 1966 suggested that there was no shortage of labor for hand harvesting and machine operations. Similar reports indicated that machine workers were making higher wages than the hand pickers. However, the situation is now so fluid that any long-term effects on labor are difficult to determine. Perhaps the most that one might venture is that, in general terms, there will be fewer workers engaged in the tomato harvest, but those remaining in it, because of the higher levels of skills required and greater productivity, will receive better pay than the hand pickers have received over the years.

The changes under way also have had definite implications for the growers. They have led to investments of larger amounts of capital in machines as compared with growing for hand harvest. This, in turn, has led to large-scale plantings in order to utilize the machines to the fullest extent possible. The trend is toward fewer growers with larger acreages.

The mechanical tomato harvester, which depended for

its success upon the combined efforts of the engineers, plant breeders, and agronomists, may be considered rather typical of recent advances within the traditional framework of farming. The effectiveness of one factor or practice in increasing agricultural productivity usually has been enhanced because of its use in conjunction with a complex of other improved practices. At the same time, one improvement often made another possible. Hybrid corn, bred for a stronger root system and stalk, could better utilize fertilizer and was more suited for mechanical harvesting than earlier types. The semidwarf wheat in the Pacific Northwest is spectacularly productive because it was selected to make use of irrigation and fertilization. More abundant and better quality livestock are a result, at least partly, of more abundant feed grains as well as of better feeding and breeding practices and the control of insect pests and diseases.

There is, as of now, virtually no theoretical limit to possible increases in American agricultural productivity over the next several years. Such advances will provide the American people with all the food and fiber that they need and can use in the foreseeable future and will leave some surpluses for export to disaster-stricken or less-developed areas. As a result, we should maintain our position in the United States of enjoying a better diet at lower real cost than ever before in history. Further, we can hope that American agricultural technology, particularly the systems or package approach to problems, will aid developing nations to improve their agriculture.

REFERENCES

[1] Henry A. Wallace and William L. Brown, *Corn and Its Early Fathers* (East Lansing, Mich., 1956); James H. Street, *The New Revolution in the Cotton Economy: Mechanization and Its Consequences* (Chapel Hill, N.C., 1957); and Wayne D. Rasmussen, "Technological Change in Sugar Beet Production," *Agricultural History*, XLI (January 1967), 31–35.

[2] U.S. Department of Agriculture, *Agriculture and Economic Growth* (Agricultural Economic Report 28, March 1963), pp. 10–25; U.S. Department of Agriculture, *Changes in Farm Production and Efficiency* (Statistical Bull. 233, June 1966), p. 34.

[3] Wayne D. Rasmussen, "The Impact of Technological Change on American Agriculture, 1862–1962," *Journal of Economic History*, XXII (December 1962), 578–591.

[4] Ralph A. Loomis and Glen T. Barton, *Productivity of Agriculture, United States, 1870–1958* (U.S. Department of Agriculture, Technical Bull. 1238, April 1961), p. 63; Harold F.

Breimyer, "Sources of Our Increasing Food Supply," *Journal of Farm Economics,* XXXVI (May 1954), 228–242.

[5] James H. Street, "Mechanizing the Cotton Harvest," *Agricultural History,* XXXI (1957), 12–22.

[6] F. S. Buchen, "Letter to the Editor," *New Agriculture* (San Francisco), XXV, No. 8 (May 1943), 4.

[7] Letter from F. Hal Higgins to F. S. Buchen, May 9, 1943, in the F. Hal Higgins Library of Agricultural Technology, University of California, Davis. I wish to acknowledge the assistance of F. Hal Higgins in carrying out the research for this study.

[8] G. C. Hanna, "Development of Tomato Varieties for Mechanical Harvesting" (unpublished manuscript, Agricultural History Branch, U.S. Department of Agriculture, 1966).

[9] "Mechanical Tomato Harvesting," *California Farmer,* February 9, 1952.

[10] "Conveyor Belt Harvesting for Cannery Tomatoes the Answer?" *Western Canner and Packer,* XLV, No. 3 (March 1953), 23, 24, 26, 28, 30; *Florida State Horticultural Society, Proceedings,* LXXII (1959), 153–155.

[11] Interview with Coby Lorenzen, September 29, 1966; Coby Lorenzen, "Tomato Harvester," *American Vegetable Grower,* IV, No. 10 (October 1956), 20.

[12] Robert C. Pearl, "1961 Tomato Mechanical Harvest in California," *Food Technology,* XVI, No. 7 (July 1962), 54–56; John H. Battin, "Tomato Harvesters Show Promise in Imperial Valley Tests," *Western Grower and Shipper,* XXXII, No. 8 (August 1961), 15, 17.

[13] S. K. Ries, "A Summary of 1960 Mechanical Tomato Harvesting Research at Michigan State University," National Canners Association, *Information Letter,* No. 1813 (January 31, 1961), 62; S. K. Ries, "They Said It Couldn't Be Done!" *American Vegetable Grower,* IX, No. 5 (May 1961), 10–12.

[14] F. Leland Elam, "Bob Button's Tomato Picker," *American Vegetable Grower,* X, No. 11 (November 1962), 12, 26; H. B. Peto, "Seedman Endorses Mechanical Harvesting," *California Tomato Grower,* V, No. 10 (November 1962), 9; "New Button Tomato Harvester Will Be Available Next Year," *California Farmer,* December 19, 1964.

[15] Joe Marks, "Tomato Harvesters," *Nation's Agriculture,* XL, No. 5 (May 1965), 12–13.

[16] "Mechanization in Tomato Harvesting," *Farm Technology,* XXII, No. 5 (May 1966), 12–13.

[17] Don M. Taylor, "Mechanized Tomato Harvest Nears," *Western Crops and Farm Management,* IX, No. 10 (October 1960), 16, 18; S. K. Ries, "They Said It Couldn't Be Done!" B. A. Stout and S. K. Ries, "Mechanical Asparagus and Tomato Harvester Research," *Horticultural News* (New Jersey Agricultural Experiment Station), XLIII (March 1962), 145; F. Leland Elam, "Giving 'em the Bumps!" *American Vegetable Grower,* IX, No. 10 (October 1961), 10.

¹⁸ H. B. Peto (see n. 14); Max D. Reeder, "Problems Encountered in the Use of Mechanical Harvesters for Tomatoes," National Canners Association, *Information Letter*, No. 1813 (January 31, 1961), pp. 60–62; S. K. Ries et al., "Mechanical Harvesting and Bulk Handling Tests with Processing Tomatoes," Michigan Agricultural Experiment Station, *Quarterly Bulletin*, XLIV (November 1961), 282–300.

¹⁹ *Bakersfield* (Calif.) *Californian*, August 22, 1964; (Calif.) *Enterprise*, October 7, 1966.

²⁰ "Tomatoes via Machine," *California Farmer*, September 2, 1961; John H. MacGillivary, "Reduction of Sorting Costs on the Tomato Harvester," *California Tomato Grower*, V, No. 8 (September 1962), 4–5; S. K. Ries and B. A. Stout, "Status of Tomato Harvester," *Canner/Packer*, CXXXII, No. 3 (March 1963), 34–35; G. K. York et al., "Sanitation in Mechanical Harvesting and Bulk Handling of Canning Tomatoes," *Food Technology*, XVIII (January 1964), 97–100.

²¹ K. Q. Stephenson, "Selective Fruit Separation for Mechanical Tomato Harvester," *Agricultural Engineering*, XLV (May 1964), 250–253; interview with Coby Lorenzen, September 29, 1966.

²² J. M. Johnson, "Harvest Headaches," *Virginia Farm Economics* (Virginia Cooperative Extension Service), No. 190 (March–April 1965), p. 7; "Here Is the Latest on Tomato Picker Production," *California Farmer*, August 6, 1966; memorandum, Raymon E. Webb, Agricultural Research Service, USDA, to Wayne D. Rasmussen, February 13, 1967.

²³ Max D. Reeder, "Tomato Harvesters—Needed Improvements," *Horticultural News* (New Jersey Agricultural Experiment Station), XLII (May 1961), 52; W. A. Gould et al., "Automated Tomato Harvesting," *Ohio Report on Research and Development* (Ohio Agricultural Experiment Station, Wooster), L, No. 2 (March–April 1965), 27.

²⁴ Interview with G. C. Hanna, September 27, 1966.

²⁵ "What's the Verdict?" *American Vegetable Grower*, X, No. 4 (April 1962), 14–15, 44, 46; H. B. Peto, "Mechanical Harvesting of Tomatoes: A Progress Report," *Seed World*, XCIII, No. 9 (November 8, 1963), 14–15; John C. Lingle, "Once-Over Harvest—What It Means to the Tomato Grower," *American Vegetable Grower*, XIII, No. 1 (January 1965), 16–17, 26–29.

²⁶ Philip S. Parsons, *Costs of Mechanical Tomato Harvesting Compared to Hand Harvesting*, University of California Agricultural Extension Service Publ. AXT-224, May 1966.

²⁷ Interview with grower, September 26, 1966.

²⁸ Wayne D. Rasmussen, *A History of the Emergency Farm Labor Supply Program, 1943–47* (Department of Agriculture, Agriculture Monograph 13, Washington, D.C., 1951), pp. 199–233; Ellis W. Hawley, "The Politics of the Mexican Labor Issue, 1950–1965," *Agricultural History*, XL (July 1966), 157–176.

Part IV
Invention and Innovation

Introduction

The comic strips provide us with a graphic description of how invention takes place: an electric bulb lights up over the head of the inventor, signalizing a flash of inspiration leading to a new discovery. The patent laws for a long time embodied this concept of invention as "flash of insight"— as if this sudden inspiration encompassed the entire innovative process. Yet the history of technology demonstrates quite the contrary: while the flash of inventive insight is an important element of the innovation process, it is but one step in a complex series of developments which leads from the concept of a new or improved process, product, or device to its ultimate application, which makes it an innovation.

Invention is an act of creativity, and the investigation of technical creativity is analogous to similar studies in fields such as literature, art, science, or other aspects of human endeavor. We want to know what psychological traits stimulate creativity, the sociocultural matrix which might encourage creative concepts, the ways in which creativity can be taught or learned—if it can be taught or learned at all. Is creativity solely dependent upon the inventor, or is it a function of the milieu, or is it a product of the interaction of personal and social forces of many different kinds?

Because technology caters to human wants and social needs, there is reason to believe that while the individual inventor might be motivated by various kinds of personal incentives, the successful introduction and application of his invention is controlled more by social, economic, and cultural forces. That is why such great interest is displayed in the steps between invention and innovation, that is, between the first concept or model or a technical device or process and its eventual introduction as part of the technological workings of society.

271

Study of the inventive enterprise is not simply an exercise in intellectual curiosity; it has important practical consequences. Although some nay-sayers decry continual technological advance and advocate a moratorium on technical changes, most people recognize that it would be impossible to dam the wellsprings of the human creative imagination and that society is badly in need of further innovations— if only to mitigate the antisocial effects of previous technological applications! If we are to cope successfully with problems of the environment, poverty, and disease, and to raise the standards of living of people throughout the world, we must know what are the incentives to invention and the conditions which foster innovation. How are technical advances in one field transferred to another, and perhaps more importantly, can study of the innovative process shed light on the transfer of industrial technology from the Western "have" nations to the less-developed "have-not" countries which desperately need industrial growth in order to take care of their fast-rising populations?

Our knowledge of the act of invention is limited, as Lynn White points out: we are ignorant of technological history and we know little about the origins and impact of many inventions that we take for granted. The historical record shows us many cases of blindness to innovation, where the basic inventions already existed but there was a failure to exploit their possibilities. Yet, as White points out, a simple idea transferred out of its original context may have a vast expansion; but we still know little about the ways in which technical devices have been transferred from one field to another.

The complexity of the innovative process is well demonstrated by F. M. Scherer's case study of the development of James Watt's steam engine. Scherer uses the history of the steam engine to test the conceptual formulations of Abbott Payson Usher and Joseph Schumpeter regarding the process of innovation. Even the reader who is not interested in such theoretical models will be impressed by the time taken and the difficulties encountered in translating Watt's original idea and model of the steam engine into a workable and commercially salable innovation. Scherer shows the role of Matthew Boulton, the entrepreneur, in the innovation process, and also how a series of complementary inventions was necessary to take full advantage of the original idea of the separate condenser, as well as the economic variables involved in the steps from the original inventive idea to the practical innovation.

The difficulties in assigning credit for an invention are illustrated in the selection by Robert S. Woodbury. In one of the most famous articles ever published in *Technology and Culture,* Woodbury destroys the legend—imbedded in American folklore and perpetuated in scholarly historical works—that Eli Whitney invented the system of interchangeable parts. This article gains its importance not because of its iconoclastic treatment of Whitney but because it provides us with a truer perspective on one of the most important developments in American economic and social history. Interchangeable parts were among the essential components of the "American system of manufactures," the system of mass production, which is one of the foundations of America's industrial supremacy in the twentieth century.

Norman B. Wilkinson also strikes a blow at another American myth, namely, the notion that American ingenuity and inventiveness made our country technologically independent of the rest of the world from the earliest stages of our history. Looking at the development of American industries in the lower Brandywine Valley during the period 1791–1816, he shows that America was "an apprentice nation—learning, imitating, and sometimes improving upon European technology." In showing how this early American industrial development was indebted to Europe for methods, machinery, and "know-how," Wilkinson also illuminates the problem of the transfer of technology from the more advanced industry of Europe to the then-underdeveloped United States.

Reynold M. Wik tells the amusing story of Henry Ford's attempts to introduce a scientific technology to American agriculture in the 1920's and the 1930's. Ford's naïve belief that scientific and technological advance automatically meant progress is characteristic of much public thought today; and his failure or only partial successes in scientific experiments to produce useful products demonstrate the complexities and nebulosity of the relationships between scientific investigations and technological applications.

The Act of Invention:
Causes, Contexts, Continuities, and Consequences

Lynn White, Jr.

The rapidly growing literature on the nature of technological innovation and its relations to other activities is
largely rubbish because so few of the relevant concrete facts
have thus far been ascertained. It is an inverted pyramid of
generalities, the apex of which is very nearly a void. The
five plump volumes of *A History of Technology*,[1] edited
under the direction of Charles Singer, give the layman a
quite false impression of the state of knowledge. They are
very useful as a starting point, but they are almost as much
a codification of error as of sound information.[2] It is to be
feared that the physical weight of these books will be widely
interpreted as the weight of authority and that philosophers,
sociologists, and others whose personal researches do not
lead them into the details of specific technological items
may continue to be deceived as to what is known.

Since man is a hypothesizing animal, there is no point in
calling for a moratorium on speculation in this area of
thought until more firm facts can be accumulated. Indeed,
such a moratorium—even if it were possible—would slow
down the growth of factual knowledge because hypothesis
normally provokes counter-hypotheses, and then all factions
adduce facts in evidence, often new facts. The best that we
can do at present is to work hard to find the facts and then
to think cautiously about the facts which have been found.

In view of our ignorance, then, it would seem wise to discuss the problems of the nature, the motivations, the con-

These comments by Professor White appeared in Technology and Culture *(Vol. 3, No. 4): 486–500, which was
devoted to the proceedings of the Encyclopaedia Britannica
Conference on the Technological Order (March 1962). For
a note on Dr. White, see p. 239.*

ditioning circumstances, and the effects of the act of invention far less in terms of generality than in terms of specific instances about which something seems to be known.

1. The beginning of wisdom may be to admit that even when we know some facts in the history of technology, these facts are not always fully intelligible, i.e., capable of "explanation," simply because we lack adequate contextual information. The Chumash Indians of the coast of Santa Barbara County built plank boats which were unique in the pre-Columbian New World: their activity was such that the Spanish explorers of California named a Chumash village "La Carpintería."[3] A map will show that this tribe had a particular inducement to venture upon the sea: they were enticed by the largest group of off-shore islands along the Pacific Coast south of Canada. But why did the tribes of South Alaska and British Columbia, of Araucanian Chile, or of the highly accidented Eastern coast of the United States never respond to their geography by building plank boats? Geography would seem to be only one element in explanation.

Can a plank-built East Asian boat have drifted on the great arc of currents in the North Pacific to the Santa Barbara region? It is entirely possible; but such boats would have been held together by pegs, whereas the Chumash boats were lashed, like the dhows of the Arabian Sea or like the early Norse ships. Diffusion seems improbable.

Since a group can conceive of nothing which is not first conceived by a person, we are left with the hypothesis of a genius: a Chumash Indian who at some unknown date achieved a break-away from log dugout and reed balsa to the plank boat. But the idea of "genius" is itself an ideological artifact of the age of the Renaissance when painters, sculptors, and architects were trying to raise their social status above that of craftsmen.[4] Does the notion of genius "explain" Chumash plank boats? On the contrary, it would seem to be no more than a traditionally acceptable way of labeling the great Chumash innovation as unintelligible. All we can do is to observe the fact of it and hope that eventually we may grasp the meaning of it.

2. A symbol of the rudimentary nature of our thinking about technology, its development, and its human implications, is the fact that while the *Encyclopaedia Britannica* has an elaborate article on "Alphabet," it contains no discussion of its own organizational presupposition, alphabetization. Alphabetization is the basic invention for the classification and recovery of information: it is fully comparable

in significance to the Dewey decimal system and to the new electronic devices for these purposes. Modern big business, big government, big scholarship are inconceivable without alphabetization. One hears that the chief reason why the Chinese Communist regime has decided to Romanize Chinese writing is the inefficiency of trying to classify everything from telephone books to tax registers in terms of 214 radicals of ideographs. Yet we are so blind to the nature of our technical equipment that the world of Western scholars, which uses alphabetization constantly, has produced not even the beginning of a history of it.

Fortunately, Dr. Sterling Dow of Harvard University is now engaged in the task. He tells me that the earliest evidence of alphabetization is found in Greek materials of the third century B.C. In other words, there was a thousand-year gap between the invention of the alphabet as a set of phonetic symbols and the realization that these symbols, and their sequence in individual written words, could be divorced from their phonetic function and used for an entirely different purpose: an arbitrary but very useful convention for storage and retrieval of verbal materials. That we have neglected thus completely the effort to understand so fundamental an invention should give us humility whenever we try to think about the larger aspects of technology.

3. Coinage was one of the most significant and rapidly diffused innovations of late antiquity. The dating of it has recently become more conservative than formerly: the earliest extant coins were sealed into the foundation of the temple of Artemis at Ephesus c. 600 B.C., and the invention of coins, i.e., lumps of metal the value of which is officially certified, was presumably made in Lydia not more than a decade earlier.[5]

Here we seem to know something, at least until the next archaeological spades turn up new testimony. But what do we know with any certainty about the impact of coinage? We are compelled to tread the slippery path of *post hoc ergo propter hoc*. There was a great acceleration of commerce in the Aegean, and it is hard to escape the conviction that this movement, which is the economic presupposition of the Periclean Age, was lubricated by the invention of coinage.

If we dare to go this far, we may venture further. Why did the atomic theory of the nature of matter appear so suddenly among the philosophers of the Ionian cities? Their notion that all things are composed of different arrangements of identical atoms of some "element," whether water, fire, ether, or something else, was an intellectual novelty of the first order, yet its sources have not been obvious. The psychological roots of atomism would seem to be found in the saying of Heraclitus of Ephesus that "all things may be reduced to fire, and fire to all things, just as all goods may be turned into gold and gold into all goods."[6] He thought that he was just using a metaphor, but the metaphor had been possible for only a century before he used it.

Here we are faced with a problem of critical method. Apples had been dropping from trees for a considerable period before Newton discovered gravity:[7] we must distinguish cause from occasion. But the appearance of coinage is a phenomenon of a different order from the fall of an apple. The unprecedented element in the general life of sixth-century Ionia, the chief stimulus to the prosperity which provided leisure for the atomistic philosophers, was the invention of coinage: the age of barter was ended. Probably no Ionian was conscious of any connection between this unique new technical instrument and the brainstorms of the local intellectuals. But that a causal relationship did exist can scarcely be doubted, even though it cannot be "proved" but only perceived.

4. Fortunately, however, there are instances of technological devices of which the origins, development, and effects outside the area of technology are quite clear. A case in point is the pennon.[8]

The stirrup is first found in India in the second century B.C. as the big-toe stirrup. For climatic reasons its diffusion to the north was blocked, but it spread wherever India had contact with barefoot aristocracies, from the Philippines and Timor on the east to Ethiopia on the west. The nuclear idea of the stirrup was carried to China on the great Indic culture wave which also spread Buddhism to East Asia, and by the fifth century the shod Chinese were using a foot stirrup.

The stirrup made possible, although it did not require, a new method of fighting with the lance. The unstirrupped rider delivered the blow with the strength of his arm. But stirrups, combined with a saddle equipped with pommel and cantle, welded rider to horse. Now the warrior could lay his lance at rest between his upper arm and body: the blow was delivered not by the arm but by the force of a charging stallion. The stirrup thus substituted horsepower for manpower in battle.

The increase in violence was tremendous. So long as the blow was given by the arm, it was almost impossible to impale one's foe. But in the new style of mounted shock combat, a good hit might put the lance entirely through his body and thus disarm the attacker. This would be dangerous if the victim had friends about. Clearly, a baffle must be provided behind the blade to prevent penetration by the shaft of the lance and thus permit retraction.

Some of the Central Asian peoples attached horse tails behind the blades of lances—this was probably being done by the Bulgars before they invaded Europe. Others nailed a piece of cloth, or pennon, to the shaft behind the blade. When the stirrup reached Western Europe c. 730 A.D., an effort was made to meet the problem by adapting to military purposes the old Roman boar spear which had a metal crosspiece behind the blade precisely because boars, bears, and leopards had been found to be so ferocious that they would charge up a spear not so equipped.

This was not, however, a satisfactory solution. The new violence of warfare demanded heavier armor. The metal crosspiece of the lance would sometimes get caught in the victim's armor and prevent recovery of the lance. By the early tenth century Europe was using the Central Asian cloth pennon, since even if it got entangled in armor it would rip and enable the victor to retract his weapon.

Until our dismal age of camouflage, fighting men have always decorated their equipment. The pennons on lances quickly took on color and design. A lance was too long to be taken into a tent conveniently, so a knight usually set it upright outside his tent, and if one were looking for him, one looked first for the flutter of his familiar pennon. Knights riding held their lances erect, and since their increasingly massive armor made recognition difficult, each came to be identified by his pennon. It would seem that it was from the pennon that distinctive "connoissances" were transferred to shield and surcoat. And with the crystallization of the feudal structure, these heraldic devices became hereditary, the symbols of status in European society.

In battle, vassals rallied to the pennon of their liege lord. Since the king was, in theory if not always in practice, the culmination of the feudal hierarchy, his pennon took on a particular aura of emotion: it was the focus of secular loyalty. Gradually a distinction was made between the king's two bodies,[9] his person and his "body politic," the state. But a colored cloth on the shaft of a spear remained the primary symbol of allegiance to either body, and so remains even in polities which have abandoned monarchy.

The grimly functional rags first nailed to lance shafts by Asian nomads have had a great destiny. But it is no more remarkable than that of the cross, a hideous implement in the Greco-Roman technology of torture, which was to become the chief symbol of the world's most widespread religion.

In tracing the history of the pennon, and of many other technological items, there is a temptation to convey a sense of inevitability. However, a novel technique merely offers opportunity; it does not command. As has been mentioned, the big-toe stirrup reached Ethiopia. It was still in common use there in the nineteenth century, but at the present time Muslim and European influences have replaced it with the foot stirrup. However, travelers tell me that the Ethiopian gentleman, whose horse is equipped with foot stirrups, rides with only his big toes resting in the stirrups.

5. Indeed, in contemplating the history of technology, and its implications for our understanding of ourselves, one is as frequently astonished by blindness to innovation as by the insights of invention. The Hellenistic discovery of the helix was one of the greatest of technological inspirations. Very quickly it was applied not only to gearing but also to the pumping of water by the so-called Archimedes screw.[10] Somewhat later the holding screw appears in both Roman and Germanic metal work.[11] The helix was taken for granted thenceforth in Western technology. Yet Joseph Needham of Cambridge University assures me that, despite the great sophistication of the Chinese in most technical matters, no form of helix was known in East Asia before modern times: it reached India but did not pass the Himalayas. Indeed, I have not been able to locate any such device in the Far East before the early seventeenth century when Archimedes screws, presumably introduced by the Portuguese, were used in Japanese mines.[12]

6. Next to the wheel, the crank is probably the most important single element in machine design, yet until the fifteenth century the history of the crank is a dismal record of inadequate vision of its potentialities.[13] It first appears in China under the Han dynasty, applied to rotary fans for winnowing hulled rice, but its later applications in the Far East were not conspicuous. In the West the crank seems to have developed independently and to have emerged from the hand quern. The earliest querns were fairly heavy, with a handle, or handles, inserted laterally in the upper stone, and the motion was reciprocating. Gradually the stones

grew lighter and thinner, so that it was harder to insert the peg-handle horizontally: its angle creeps upward until eventually it stands vertically on top. All the querns found at the Saalburg had horizontal handles, and it is increasingly clear that the vertical peg is post-Roman.

Seated before a quern with a single vertical handle, a person of the twentieth century would give it a continuous rotary motion. It is far from clear that one of the very early Middle Ages would have done so. Crank motion was a kinetic invention more difficult than we can easily conceive. Yet at some point before the time of Louis the Pious the sense of the appropriate motion changed; for out of the rotary quern came a new machine, the rotary grindstone, which (as the Latin term for it, *mola fabri,* shows) is the upper stone of a quern turned on edge and adapted to sharpening. Thus, in Europe at least, crank motion was invented before the crank, and the crank does not appear before the early ninth century. As for the Near East, I find not even the simplest application of the crank until al-Jazari's book on automata of 1206 A.D.

Once the simple crank was available, its development into the compound crank and connecting rod might have been expected quite quickly. Yet there is no sign of a compound crank until 1335, when the Italian physician of the Queen of France, Guido da Vigevano, in a set of astonishing technological sketches, which Rupert Hall has promised to edit,[14] illustrates three of them.[15] By the fourteenth century Europe was using crankshafts with two simple cranks, one at each end; indeed, this device was known in Cambodia in the thirteenth century. Guido was interested in the problem of self-moving vehicles: paddlewheel boats and fighting towers propelled by windmills or from the inside. For such constricted situations as the inside of a boat or a tower it apparently occurred to him to consolidate the two cranks at the ends of the crankshaft into a compound crank in its middle. It was an inspiration of the first order, yet nothing came of it. Evidently the Queen's physician, despite his technological interests, was socially too far removed from workmen to influence the actual technology of his time. The compound crank's effective appearance was delayed for another three generations. In the 1420's some Flemish carpenter or shipwright invented the bit-and-brace with its compound crank. By c. 1430 a German engineer was applying double compound cranks and connecting rods to machine design: a technological event as significant as the Hellenistic invention of gearing. The idea spread like wildfire, and European applied mechanics was revolutionized.

How can we understand the lateness of the discovery, whether in China or Europe, of even the simple crank, and then the long delay in its wide application and elaboration? Continuous rotary motion is typical of inorganic matter, whereas reciprocating motion is the sole movement found in living things. The crank connects these two kinds of motion; therefore we who are organic find that crank motion does not come easily to us. The great physicist and philosopher Ernst Mach noticed that infants find crank motion hard to learn.[16] Despite the rotary grindstone, even today razors are whetted rather than ground: we find rotary motion a bar to the greatest sensitivity. Perhaps as early as the tenth century the hurdy-gurdy was played with a cranked resined wheel vibrating the strings. But by the thirteenth century the hurdy-gurdy was ceasing to be an instrument for serious music. It yielded to the reciprocating fiddle bow, an introduction of the tenth century which became the foundation of modern European musical development. To use a crank, our tendons and muscles must relate themselves to the motion of galaxies and electrons. From this inhuman adventure our race long recoiled.

7. A sequence originally connected with the crank may serve to illustrate another type of problem in the act of technological innovation: the fact that a simple idea transferred out of its first context may have a vast expansion. The earliest appearance of the crank, as has been mentioned, is found on a Han-dynasty rotary fan to winnow husked rice.[17] The identical apparatus appears in the eighteenth century in the Palatinate,[18] in upper Austria and the Siebenbürgen,[19] and in Sweden.[20] I have not seen the exact channel of this diffusion traced, but it is clearly part of the general Jesuit-inspired *Chinoiserie* of Europe in that age. Similarly, I strongly suspect, but cannot demonstrate, that all subsequent rotary blowers, whether in furnaces, dehydrators, wind tunnels, air conditioning systems, or the simple electric fan, are descended from this Han machine which seems, in China itself, to have produced no progeny.

8. Doubtless when scholarship in the history of technology becomes firmer, another curious device will illustrate the same point. To judge by its wide distribution,[21] the fire piston is an old invention in Malaya. Dr. Thomas Kuhn, now of Princeton University, who has made careful studies of the history of our knowledge of adiabatic

heat, assures me that when the fire piston appeared in late eighteenth-century Europe not only for laboratory demonstrations but as a commercial product to light fires, there is no hint in the purely scientific publications that its inspiration was Malayan. But the scientists, curiously, also make no mention of the commercial fire pistons then available. So many Europeans, especially Portuguese and Netherlanders, had been trading, fighting, ruling, and evangelizing in the East Indies for so long a time before the fire piston is found in Europe, that it is hard to believe that the Malayan fire piston was not observed and reported. The realization of its potential in Europe was considerable, culminating in the diesel engine.

9. Why are such nuclear ideas sometimes not exploited in new and wider applications? What sorts of barriers prevent their diffusion? Why, at times, does what appeared to be a successful technological item fall into disuse? The history of the faggoted forging method of producing sword blades[22] may assist our thinking about such questions.

In late Roman times, north of the Alps, Celtic, Slavic, and Germanic metallurgists began to produce swords with laminations produced by welding together bundles of rods of different qualities of iron and steel, hammering the resulting strip thin, folding it over, welding it all together again, and so on. In this way a fairly long blade was produced which had the cutting qualities of steel but the toughness of iron. Although such swords were used at times by barbarian auxiliaries in the Roman army, the Roman legions never adopted them. Yet as soon as the Western Empire crumbled, the short Roman stabbing sword vanished and the laminated slashing blade alone held the field of battle. Can this conservatism in military equipment have been one reason for the failure of the Empire to stop the Germanic invasions? The Germans had adopted the new type of blade with enthusiasm, and by Carolingian times were manufacturing it in quantities in the Rhineland for export to Scandinavia and to Islam where it was much prized. Yet, although such blades were produced marginally as late as the twelfth century, for practical purposes they ceased to be used in Europe in the tenth century. Does the disappearance of such sophisticated swords indicate a decline in medieval metallurgical methods?

We should be cautious in crediting the failure of the Romans to adopt the laminated blade to pure stupidity. The legions seem normally to have fought in very close formation, shield to shield. In such a situation, only a stabbing sword could be effective. The Germans at times used a "shield wall" formation, but it was probably a bit more open than the Roman and permitted use of a slashing sword. If the Romans had accepted the new weapon, their entire drill and discipline would have been subject to revision. Unfortunately, we lack studies of the development of Byzantine weapons sufficiently detailed to let us judge whether, or to what extent, the vigorously surviving Eastern Roman Empire adapted itself to the new military technology.

The famous named swords of Germanic myth, early medieval epic, and Wagnerian opera were laminated blades. They were produced by the vast patience and skill of smiths who themselves became legendary. Why did they cease to be made in any number after the tenth century? The answer is found in the rapid increase in the weight of European armor as a result of the consistent Frankish elaboration of the type of mounted shock combat made possible by the stirrup. After the turn of the millennium a sword in Europe had to be very nearly a club with sharp edges; the best of the earlier blades was ineffective against such defenses. The faggoted method of forging blades survived and reached its technical culmination in Japan[23] where, thanks possibly to the fact that archery remained socially appropriate to an aristocrat, mounted shock combat was less emphasized than in Europe and armor remained lighter.

10. Let us now turn to a different problem connected with the act of invention. How do methods develop by the transfer of ideas from one device to another? The origins of the cannon ball and the cannon may prove instructive.[24]

Hellenistic and Roman artillery was activated by the torsion of cords. This was reasonably satisfactory for summer campaigns in the Mediterranean basin, but north of the Alps and in other damper climates the cords tended to lose their resilience. In 1004 A.D. a radically different type of artillery appeared in China with the name *huo p'ao*. It consisted of a large sling beam pivoted on a frame and actuated by men pulling in unison on ropes attached to the short end of the beam away from the sling. It first appears outside China in a Spanish Christian illumination of the early twelfth century, and from this one might assume diffusion through Islam. But its second appearance is in the northern Crusader army attacking Lisbon in 1147 where a battery of them were operated by shifts of one hundred men for each. It would seem that the Muslim defenders were quite unfamiliar with the new engine of destruction and soon capitulated. This invention, therefore, appears to have reached the West from China not through Islam but directly across Central Asia. Such a path of diffusion is the more credible because by the end of the same century the magnetic needle likewise arrived in the West by the northern route, not as an instrument of navigation but as a means of ascertaining the meridian, and Western Islam got the compass from Italy.[25] When the new artillery arrived in the West it had lost its name. Because of structural analogy, it took on a new name borrowed from a medieval instrument of torture, the ducking stool or *trebuchetum*.

Whatever its merits, the disadvantages of the *huo p'ao* were the amount of manpower required to operate it and the fact that since the gang pulling the ropes would never pull with exactly the same speed and force, missiles could not be aimed with great accuracy. The problem was solved by substituting a huge counterweight at the short end of the sling beam for the ropes pulled by men. With this device a change in the weight of the caisson of stones or earth, or else a shift of the weight's position in relation to the pivot, would modify the range of the projectile and then keep it uniform, permitting concentration of fire on one spot in the fortifications to be breeched. Between 1187 and 1192 an Arabic treatise written in Syria for Saladin mentions not only Arab, Turkish, and Frankish forms of the primitive trebuchet, but also credits to Iran the invention of the trebuchet with swinging caisson. This ascription, however, must be in error; for from c. 1220 onward oriental sources frequently call this engine *magribī,* i.e., "Western." Moreover, while the counterweight artillery has not yet been documented for Europe before 1199, it quickly displaced the older forms of artillery in the West, whereas this new and more effective type of siege machinery became dominant in the Mameluke army only in the second half of the thirteenth century. Thus the trebuchet with counterweights would appear to be a European improvement on the *huo p'ao*. Europe's debt to China was repaid in 1272 when, if we may believe Marco Polo, he and a German technician, helped by a Nestorian Christian, delighted the Great Khan by building trebuchets which speedily reduced a besieged city.

But the very fact that the power of a trebuchet could be so nicely regulated impelled Western military engineers to seek even greater exactitude in artillery attack. They quickly saw that until the weight of projectiles and their friction with the air could be kept uniform, artillery aim would still be variable. As a result, as early as 1244 stones for trebuchets were being cut in the royal arsenals of England calibrated to exact specifications established by an engineer: in other words, the cannon ball before the cannon.

The germinal idea of the cannon is found in the metal tubes from which, at least by the late ninth century, the Byzantines had been shooting Greek fire. It may be that even that early they were also shooting rockets of Greek fire, propelled by the expansion of gases, from bazooka-like metal tubes. When, shortly before 673, the Greek-speaking Syrian refugee engineer Callinicus invented Greek fire, he started the technicians not only of Byzantium but also of Islam, China, and eventually the West in search of ever more combustible mixtures. As chemical methods improved, the saltpeter often used in these compounds became purer, and combustion tended toward explosion. In the thirteenth century one finds, from the Yellow Sea to the Atlantic, incendiary bombs, rockets, firecrackers, and fireballs shot from tubes like Roman candles. The flame and roar of all this has made it marvelously difficult to ascertain just when gunpowder artillery, shooting hard missiles from metal tubes, appeared. The first secure evidence is a famous English illumination of 1327 showing a vase-shaped cannon discharging a giant arrow. Moreover, our next certain reference to a gun, a "pot de fer à traire garros de feu" at Rouen in 1338, shows how long it took for technicians to realize that the metal tube, gunpowder, and the calibrated trebuchet missile could be combined. However, iron shot appear at Lucca in 1341; in 1346 in England there were two calibers of lead shot; and balls appear at Toulouse in 1347.

The earliest evidence of cannon in China is extant examples of 1356, 1357, and 1377. It is not necessary to assume the miracle of an almost simultaneous independent Chinese invention of the cannon: enough Europeans were wandering the Yuan realm to have carried it eastward. And it is very strange that the Chinese did not develop the cannon further, or develop hand guns on its analogy. Neither India nor Japan knew cannon until the sixteenth century when they arrived from Europe. As for Islam, despite several claims to the contrary, the first certain use of gunpowder artillery by Muslims comes from Cairo in 1366 and Alexandria in 1376; by 1389 it was common in both Egypt and Syria. Thus there was roughly a forty-year lag in Islam's adoption of the European cannon.

Gunpowder artillery, then, was a complex invention

which synthesized and elaborated elements drawn from diverse and sometimes distant sources. Its impact upon Europe was equally complex. Its influences upon other areas of technology such as fortification, metallurgy, and the chemical industries are axiomatic, although they demand much more exact analysis than they have received. The increased expense of war affected tax structures and governmental methods; the new mode of fighting helped to modify social and political relationships. All this has been self-evident for so long a time that perhaps we should begin to ask ourselves whether the obvious is also the true.

For example, it has often been maintained that a large part of the new physics of the seventeenth century sprang from concern with military ballistics. Yet there was continuity between the thought of Galileo or Newton and the fundamental challenge to the Aristotelian theory of impetus which appeared in Franciscus de Marchia's lectures at the University of Paris in the winter of 1319–20,[26] seven years before our first evidence of gunpowder artillery. Moreover, the physicists both of the fourteenth and of the seventeenth centuries were to some extent building upon the criticisms of Aristotle's theory of motion propounded by Philoponus of Alexandria in the age of Justinian, a time when I can detect no new technological stimulus to physical speculation. While most scientists have been aware of current technological problems, and have often talked in terms of them, both science and technology seem to have enjoyed a certain autonomy in their development.

It may well be that continued examination will show that many of the political, economic, and social as well as intellectual developments in Europe which have traditionally been credited to gunpowder artillery were in fact taking place for quite different reasons. But we know of one instance in which the introduction of firearms revolutionized an entire society: Japan.[27]

Metallurgical skills were remarkably high in Japan when, in 1543, the Portuguese brought both small arms and cannon to Kyushu. Japanese craftsmen quickly learned from the gunsmiths of European ships how to produce such weapons, and within two or three years were turning them out in great quantity. Military tactics and castle construction were rapidly revised. Nobunaga and his successor, Hideyoshi, seized the new technology of warfare and utilized it to unify all Japan under the shogunate. In Japan, in contrast to Europe, there is no ambiguity about the consequences of the arrival of firearms. But from this fact we must be careful not to argue that the European situation is equally clear if only we would see it so.

11. In examining the origins of gunpowder artillery, we have seen that its roots are multiple, but that all of them (save the European name *trebuchet*) lie in the soil of military technology. It would appear that each area of technology has a certain self-contained quality: borrowings across craft lines are not as frequent as might be expected. Yet they do occur, if exceptionally. A case in point is the fusee.

In the early fifteenth century clock makers tried to develop a portable mechanical timepiece by substituting a spring drive for the weight which powered stationary clocks. But this involved entirely new problems of power control. The weight on a clock exerted equal force at all times, whereas a spring exerts less force in proportion as it uncoils. A new escapement was therefore needed which would exactly compensate for this gradual diminution of power in the drive.

Two solutions were found, the stackfreed and the fusee, the latter being the more satisfactory. Indeed, a leading historian of horology has said of the fusee: "Perhaps no problem in mechanics has ever been solved so simply and so perfectly."[28] The date of its first appearance is much in debate, but we have a diagram of it from 1477.[29] The fusee equalizes the changing force of the mainspring by means of a brake of gut or fine chain which is gradually wound spirally around a conical axle, the force of the brake being dependent upon the leverage of the radius of the cone at any given point and moment. It is a device of great mechanical elegance. Yet the idea did not originate with the clock makers: they borrowed it from the military engineers. In Konrad Keyser's monumental, recently republished treatise on the technology of warfare, *Bellifortis,* completed c. 1405, we find such a conical axle in an apparatus for spanning a heavy crossbow.[30] With very medieval humor, this machine was called "the virgin," presumably because it offered least resistance when the bow was slack and most when it was taut.

In terms of eleven specific technological acts, or sequences of acts, we have been pondering an abstraction, the act of technological innovation. It is quite possible that there is no such thing to ponder. The analysis of the nature of creativity is one of the chief intellectual commitments of our age. Just as the old unitary concept of "intelligence" is giving way to the notion that the individual's mental capacity consists of a large cluster of various and varying factors mutually affecting each other, so "creativity" may well be a lot of things and not one thing.

Thirteenth-century Europe invented the sonnet as a poetic form and the functional button[31] as a means of making civilized life more nearly possible in boreal climes. Since most of us are educated in terms of traditional humanistic presuppositions, we value the sonnet but think that a button is just a button. It is doubtful whether the chilly northerner who invented the button could have invented the sonnet then being produced by his contemporaries in Sicily. It is equally doubtful whether the type of talent required to invent the rhythmic and phonic relationships of the sonnet pattern is the type of talent needed to perceive the spatial relationships of button and buttonhole. For the button is not obvious until one has seen it, and perhaps not even then. The Chinese never adopted it: they got no further than to adapt the tie cords of their costumes into elaborate loops to fit over cord-twisted knobs. When the Portuguese brought the button to Japan, the Japanese were delighted with it and took over not only the object itself but also its Portuguese name. Humanistic values, which have been cultivated historically by very specialized groups in quite exceptional circumstances, do not encompass sufficiently the observable human values. The billion or more mothers who, since the thirteenth century, have buttoned their children snugly against winter weather might perceive as much of spirituality in the button as in the sonnet and feel more personal gratitude to the inventor of the former than of the latter. And the historian, concerned not only with art forms but with population, public health, and what S. C. Gilfillan long ago identified as "the coldward course" of culture,[32] must not slight either of these very different manifestations of what would seem to be very different types of creativity.

There is, indeed, no reason to believe that technological creativity is unitary. The unknown Syrian who, in the first century B.C., first blew glass was doing something vastly different from his contemporary who was building the first water-powered mill. For all we now know, the kinds of ability required for these two great innovations are as different as those of Picasso and Einstein would seem to be.

The new school of physical anthropologists who maintain that *Homo* is *sapiens* because he is *faber,* that his biological differentiation from the other primates is best understood in relation to tool making, are doubtless exaggerating a provocative thesis. *Homo* is also *ludens, orans,* and much else. But if technology is defined as the systematic modification of the physical environment for human ends, it follows that a more exact understanding of technological innovation is essential to our self-knowledge.

REFERENCES

[1] (Oxford, 1954–58).

[2] Cf. the symposium in *Technology and Culture,* I (1960): 299–414.

[3] E. G. Gudde, *California Place Names,* 2nd ed. (Berkeley and Los Angeles, 1960), p. 52; A. L. Kroeber, "Elements of Culture in Native California," in *The California Indians,* ed. R. F. Heizer and M. A. Whipple (Berkeley and Los Angeles, 1951), p. 12–13.

[4] E. Zilsel, *Die Entstehung des Geniebegriffes* (Tübingen, 1926).

[5] E. S. G. Robinson, "The Date of the Earliest Coins," *Numismatic Chronicle,* 6th ser., XVI (1956): 4, 8, arbitrarily dates the first coinage c. 640–630 B.C. allowing "the Herodotean interval of a generation" for its diffusion from Lydia to the Ionian cities. But, considering the speed with which coinage appears even in India and China, such an interval is improbable.

D. Kagan, "Pheidon's Aeginetan Coinage," *Transactions and Proceedings of the American Philological Association,* XCI (1960): 121–136, tries to date the first coinage at Aegina before c. 625 B.C. when, he believes, Pheidon died; but the argument is tenuous. The tradition that Pheidon issued a coinage is late, and may well be no more than another example of the Greek tendency to invent culture-heroes. The date of Pheidon's death is uncertain: the belief that he died c. 625 rests solely on the fact that he is not mentioned by Strabo in connection with the war of c. 625–600 B.C.; but if Pheidon, then a very old man, was killed in a revolt of 620 (cf. Kagan's note 21) his participation in this long war would have been so brief and ineffective that Strabo's silence is intelligible.

[6] H. Diels, *Fragmente der Vorsokratiker* (6th ed. Berlin, 1951), 171 (B. 90).

[7] The story of the apple is authentic: Newton himself told William Stukeley that when "the notion of gravitation came into his mind [it] was occasion'd by the fall of an apple, as he sat in a contemplative mood"; cf. I. B. Cohen, "Newton in the Light of Recent Scholarship," *Isis,* LI (1960): 490.

[8] The materials on pennons, and other baffles behind the blade of a lance, are found in L. White, Jr., *Medieval Technology and Social Change* (Oxford, 1962), pp. 8, 33, 147, 157.

[9] See the classic work of Ernst Kantorowicz, *The King's Two Bodies* (Princeton, 1957).

[10] W. Treue, *Kulturgeschichte der Schraube* (Munich, 1955), pp. 39–43, 57, 109.

[11] F. M. Feldhaus, *Die Technik der Vorzeit, der Geschichtlichen Zeit und der Naturvölker* (Leipzig, 1914), pp. 984–987.

[12] E. Treptow, "Der älteste Bergbau und seiner Hilfsmittel," *Beiträge zur Geschichte der Technik und Industrie*, VIII (1918): 181, fig. 48; C. N. Bromehead, "Ancient Mining Processes as Illustrated by a Japanese Scroll," *Antiquity*, XVI (1942): 194, 196, 207.

[13] For a detailed history of the crank, cf. White, *op. cit.*, pp. 103–115.

[14] A. R. Hall, "The Military Inventions of Guido da Vigevano," *Actes du VIIIe Congrès International d'Histoire des Sciences* (Florence, 1958), pp. 966–969.

[15] Bibliothèque Nationale, MS latin 11015, fols. 49ʳ, 51ᵛ, 52ᵛ. Singer, *op. cit.*, II, figs. 594 and 659, illustrates the first and third of these, but with wrong indications of folio numbers.

[16] H. T. Horwitz, "Uber die Entwicklung der Fahigkeit zum Antreib des Kurbelmechanismus," *Geschichtsblätter fur Technik und Industrie*, XI (1927): 30–31.

[17] White, *op. cit.*, p. 104 and fig. 4. For what may be a slightly earlier specimen, now in the Seattle Art Museum, see the catalogue of the exhibition *Arts of the Han Dynasty* (New York, 1961), No. 11, of the Chinese Art Society of America.

[18] I am so informed by Dr. Paul Leser of the Hartford Theological Foundation.

[19] L. Makkai, in *Agrártörténeti Szemle*, I (1957): 42.

[20] P. Leser, "Plow Complex; Culture Change and Cultural Stability," in *Man and Cultures: Selected Papers of the Fifth International Congress of Anthropological and Ethnological Sciences*, ed. A. F. C. Wallace (Philadelphia, 1960), p. 295.

[21] H. Balfour, "The Fire Piston," in *Anthropological Essays Presented to E. B. Tylor*, (Oxford, 1907), pp. 17–49.

[22] É. Salin, *La Civilisation Mérovingienne*, III (Paris, 1957): 6, 55–115.

[23] C. S. Smith, "A Metallographic Examination of Some Japanese Sword Blades," *Quaderno II del Centro per la Storia della Metallurgia* (1957): 42–68.

[24] White, *op. cit.*, pp. 96–103, 165.

[25] *Ibid.*, p. 132.

[26] A. Maier, *Zwei Grundprobleme der scholastischen Naturphilosophie*, 2nd ed. (Rome, 1951), p. 165, n. 11.

[27] D. M. Brown, "The Impact of Firearms on Japanese Warfare, 1543–98," *Far Eastern Quarterly*, VII (1948): 236–253.

[28] G. Baillie, *Watches* (London, 1929), p. 85.

[29] Singer, *op. cit.*, III, fig. 392.

[30] Göttingen University Library, Cod. phil. 63, fol. 76ᵛ; cf. F. M. Feldhaus. "Uber den Ursprung vom Federzug und Schnecke," *Deutsche Uhrmacher Zeitung*, LIV (1930): 720–723. See also Götz Quarg, ed., Konrad Keyser aus Eichstätt, *Bellifortis*, 2 vols. (Dusseldörf, 1967).

[31] Some buttons were used in antiquity for ornament, but apparently not for warmth. The first functional buttons are found c. 1235 on the "Adamspforte" of Bamberg Cathedral, and in 1239 on a closely related relief at Bassenheim; cf. E. Panofsky, *Deutsche Plastik des 11. bis 13. Jahrhundert* (Munich, 1924), pl. 74; H. Schnitsler, "Ein unbekanntes Reiterrelief aus dem Kreise des Naumburger Meisters," *Zeitschrift des Deutschen Vereins für Kunstwissenschaft,* 1 (1935): 413, fig. 13.

[32] In *The Political Science Quarterly* XXXV (1920): 393–410.

Invention and Innovation in the Watt-Boulton Steam-Engine Venture

F. M. Scherer

It is well known how the Watt-Boulton steam engine freed England and then all nations from the geographical and climatic vagaries of water power and how it permitted man for the first time to concentrate great quantities of efficient motive power in one location. But how did this important transformation come about? What lessons about the economic characteristics of technological invention and innovation can we learn from the steam engine's history?

The terms "invention" and "innovation" suggest the conceptual formulations of Abbott Payson Usher and Joseph A. Schumpeter. Crucial to Usher's conception of invention is an "act of insight" going beyond the exercise of normal technical skill, even though additional activities (perception of a problem, setting the stage, and critical revision) are also recognized.[1] Schumpeter, on the other hand, defined innovation as "the carrying out of new combinations."[2] For the case of new technology, this can be identified with reducing an invention to practice and exploiting it commercially. Schumpeter emphatically distinguished his concept of innovation from that of invention.

> Economic leadership in particular must be distinguished from "invention." As long as they are not carried into practice, inventions are economically irrelevant. And to carry any improvement into effect is a

---

Dr. Scherer, professor of economics at Michigan State University, is the author of books and articles on technological change in civilian and military fields, including The Weapons Acquisition Process: Economic Incentives. *This article was published in* Technology and Culture *(Vol. 6, No. 2): 165–87.*

task entirely different from the inventing of it, and a task, moreover, requiring entirely different kinds of aptitudes. Although entrepreneurs of course *may* be inventors just as they may be capitalists, they are inventors not by nature of their function but by coincidence and vice versa. . . . It is, therefore, not advisable, and it may be downright misleading, to stress the element of invention as much as many writers do.[3]

Between the Usherian and Schumpeterian conceptions of technological change there are significant differences in emphasis and, to some extent, substance. Unfortunately, a dialogue to resolve or focus these differences did not take place before Schumpeter's death in 1950. Through an analysis of the Watt-Boulton steam engine's history, this paper assesses the ability of the two partly conflicting theories to integrate key behavioral variables in the process of technological change and identifies a sector of the problem underemphasized by both authors. Particular attention will be devoted to the roles played by insight, special aptitudes, risk-bearing, and motivation. The impact of the patent system on invention and innovation in the steam-engine venture will also be explored. An attempt is made to reiterate familiar historical details only to the extent that they are directly germane to the Usher and Schumpeter schemata.

The Separate Condenser:
A Classic Example of Invention

Let us begin with Watt's first and most fundamental invention in the field of steam-engine technology: the separate condenser. We shall find it to be an unusually clear example of Usherian invention.

In 1712 Thomas Newcomen erected a steam engine that combined for the first time a piston-in-cylinder arrangement and a basic motive principle involving the formation of a vacuum within the cylinder through the induced condensation of steam. This "atmospheric" engine saw extensive use, especially for pumping water from the English coal mines. Even though many minor improvements were made in its design during the following fifty years, substantially increasing the engine's operating efficiency, no one successfully challenged Newcomen's general approach.

In the winter of 1763–64 John Anderson, professor of natural philosophy at Glasgow University, brought a small model of the Newcomen engine for repair to James Watt, twenty-eight-year-old mathematical instrument maker to the University. Watt was perplexed by several aspects of the model's operation, especially by the unexpectedly large quantity of steam consumed. At first he attributed this to conduction of heat through the cylinder walls, but tests on a larger model with a wooden cylinder, which conducted less heat than the University model's brass cylinder, showed that to be only a partial explanation. Further experiments occupied most of the time Watt could spare from his regular duties for more than a year and led him to recognize a paradoxical deficiency in the Newcomen engine's operating concept: to utilize steam efficiently, the cylinder had to be kept at 100° C. so there would be no condensation during the piston's unpowered stroke; but to form an effective vacuum for the power stroke, considerable cooling water had to be injected into the cylinder, and this tended to cool the cylinder below 100°.[4] This was, in Usher's terminology, the perception of an unsatisfactory pattern. Then, presumably while Watt was strolling on the Green of Glasgow "on a fine Sabbath afternoon" early in 1765, came his famous insight: he would condense the engine's steam not in the operating cylinder, as Newcomen and all his followers had done, but rather in a separate condensing vessel to which it would be drawn by a pump or by other means.

The following day Watt began work on a small (1¾-inch cylinder diameter) model to test his separate-condenser idea, and in his own words, "in three days, I had a model at work nearly as perfect . . . as any which have been made since that time."[5] In another narrative Watt said that with the testing of this model: "excepting the non-application of the steam-case and external covering, the invention was complete, in so far as regarded the savings of steam and fuel. A large model, with an outer cylinder and wooden case, was immediately constructed, and the experiments made with it served to verify the expectations I had formed, and to place the advantage of the invention beyond the reach of doubt."[6] There is some reason to question whether Watt actually accomplished decisive results in only three days.[7] We know from a letter written by Watt to Dr. James Lind on April 29, 1765, that he was working on a new engine and soon hoped "to have the decisive trial." The first surviving contemporary reference to successful results is a letter from Watt to Dr. John Roebuck dated August 23, 1765. Nevertheless, there is no doubt that invention of the separate condenser in the Usherian sense had occurred by the summer of 1765, somewhat more than a year after Watt first received the Newcomen engine model for repair. Usher sets the date for Watt's strategic invention more precisely. "There is an unusually clear basis for recognizing that the solution of the problem was achieved during this Sunday afternoon walk. If the concept then achieved had been less adequate, we might properly set a later date for the solution of the problem, but the actual task proved to be the realization in actual mechanism of the apparatus conceived at that time. The concept itself did not require revision."[8]

The Support of Watt's Work by Roebuck and Boulton: Investment or Schumpeterian Innovation?

When the principle of Watt's separate condenser was successfully demonstrated with a test model, the new steam engine was far from ready for commercial exploitation. There remained much technical effort to scale it up to full size and to make the full-scale version work efficiently and reliably. This effort is logically subsumed under the fourth step—critical revision—of Usher's schema. However, Schumpeter's concept of innovation may also be applied to describe these post-initial-demonstration activities, and here we shall view the steam engine's development initially from the Schumpeterian viewpoint. Let us for the moment ignore the specific technical activities pursued and consider only how these activities were supported.

This support was financial in nature. Watt's funds were quite limited, but he was able to finance several models and experiments during 1765 and 1766 by borrowing from his friend Professor Joseph Black. From the summer of 1766 to January 1768, Watt found it necessary to engage full time in more remunerative activities (mainly surveying), and thus made little progress on his steam-engine experiments. Neverthleless, by the spring of 1768 Watt's debt to Black and a former partner had accumulated to more than £1,000. At that time Watt entered into a partnership with Dr. John Roebuck, who paid off Watt's debts and the cost of a patent, receiving in return a two-thirds share in the invention. There followed a second period of active experimentation, during which an 18-inch cylinder diameter model was erected and, with somewhat disappointing results, tested at Roebuck's Kinneil residence. However, Roebuck, who was experiencing growing financial embarrassment due to other ventures, was unable to pay all of these costs. Instead Watt defrayed them, partly by borrowing from Black, and from early in 1770 to May 1774 he again found it necessary to take up surveying as a nearly full-time occupation. This decision was explained in 1772 by Watt as follows: "I pursued my experiments till I found that the expense and loss of time lying wholly upon me, through the distress of Dr. Roebuck's situation, turned out to be a burthen greater than I could support; and not having conquered all the difficulties that lay in the way of execution, I was obliged for a time to abandon the project."[9] Thus Watt spent not more than three years' time actively pursuing the development of his steam engine in the nine-year period from 1765, when he first conceived the separate-condenser principle, to 1774. During the remaining time, lack of financial support forced him to accept jobs offering more immediate compensation.

In 1773 Roebuck became bankrupt, and in 1774 Matthew Boulton, a Birmingham manufacturer and creditor of Roebuck who for some time had been interested in Watt's engine, acquired Roebuck's two-thirds share in the invention. The firm of Boulton & Watt was established; facilities connected with Boulton's Soho manufactory were occupied; the 18-inch Kinneil engine was rebuilt there; and in May 1774 Watt turned his full-time attention to the engine business. Under their partnership agreement Boulton financed the firm's developmental expenses and paid Watt an annual salary of £330 plus expenses. From the outset Boulton recognized that the venture would be costly and risky. He wrote to Watt in 1773: "The thing is now a shadow, it is merely ideal, and it will cost money to realize it."[10] Testifying before the House of Commons in 1775, Dr. Roebuck stated that he and Boulton had spent more than £3,000 on early experiments, and that total expenses in view would add at least another £10,000.[11]

This was no small sum, either generally or to Boulton in particular. During the early years of steam-engine work at Soho, skilled craftsmen were paid roughly £50 per year; thus the costs already sunk by 1775 were the equivalent of 60 man-years of skilled labor, and 200 more man-year equivalents were contemplated. The manufactory of Boulton & Fothergill, erected at Soho in 1764 and known throughout England as a leading example of the factory system, initially cost only £9,000. Although he had acquired a landed estate of about £28,000 through two marriages, Boulton's financial position in 1774 was by no means strong. The Boulton & Fothergill firm was apparently on the brink of bankruptcy, and only by selling his wives' properties was he able to keep it solvent.[12] The risks of failure were substantial: the tests at Kinneil had not been notably successful, and of the several steam-engine projects undertaken during the preceding century, only Newcomen's and some improvements of the Newcomen engine by John Smeaton achieved any lasting economic success.[13] Indeed, Boulton's partner Fothergill would have nothing to do with the new steam-engine venture. And with the exception of Boulton, none of Roebuck's creditors placed any value on the engine patent.[14]

What motives led Roebuck and then Boulton to invest in Watt's steam-engine project? For Roebuck, the prospect of a new and more efficient steam engine fitted well with his other business interests. He was principal partner in an iron works at Carron, Stirlingshire, that produced major components for Newcomen atmospheric engines. He had also become committed to a coal-mining venture in Borrowstounness (Bo'ness), but the pits had flooded and the atmospheric engines available were not powerful enough to remove the water. If Watt's steam engine were successful, it might, among other things, save Roebuck's otherwise lost investment at Bo'ness.

Boulton's motives centered not so much on saving an enterprise as on creating one. To be sure, Boulton anticipated that Watt's engine could remedy a problem at his Soho manufactory, for summer water shortages made his water wheels inoperative and necessitated the expense of horses to propel the machinery. But Boulton was not interested in just one or a few engines; in the expansive attitude which characterized his ventures, he wanted to produce for the whole world. As he wrote to Watt on February 7, 1769, explaining his decision not to accept a partnership with Roebuck permitting Boulton to manufacture engines only for the midland counties of Warwick, Stafford, and Derby:

> I was excited by two motives to offer you my assistance—which were, love of you, and love of a money-getting, ingenious project. . . . To . . . produce the most profit, my idea was to settle a manufactory near to my own, by the side of our canal, where I would erect all the conveniences necessary for the completion of engines, and from which manufactory we would

serve all the world with engines of all sizes. . . . It would not be worth my while to make for three counties only; but I find it very well worth my while to make for all the world.[15]

In any event, the availability of financial support was an important determinant of the rate at which Watt perfected his engine. Progress was impeded for long intervals when no investor appeared with the necessary combination of confidence in the engine's principles, available funds, and willingness to bear the risks of development.

The financial support of Watt's work cannot, however, be called an innovative role in the strict Schumpeterian sense. Schumpeter carefully distinguished between the role of the capitalist, or investor of risk capital, and the entrepreneur or innovator. Were Roebuck and (especially) Boulton something more than capitalists in their support of the Watt steam-engine project? Schumpeter provided a number of distinguishing characteristics to help identify his entrepreneur-innovator.

1. New combinations are as a rule embodied in new firms which do not arise out of the old ones but start producing beside them.[16]

2. The entrepreneur-innovator is characterized by "initiative," "authority," and "foresight"; he is the "captain of industry" type.[17]

3. The only man the entrepreneur-innovator has to convince is the banker who is to finance him.[18]

4. The entrepreneur-innovator retires from the arena only when his strength is spent.[19]

5. The entrepreneur-innovator's motivation includes such aspects as the dream to found a private kingdom, the will to conquer and to succeed for the sake of success itself, and the joy of creating and getting things done.[20]

Examination of the evidence against these characteristics suggests that Boulton and, to a lesser extent, Roebuck rather than Watt occupied the innovator's position during most of the steam-engine development. Boulton clearly qualifies under the first (perhaps least important) condition: he set up a wholly new business for the steam-engine venture, since Fothergill would have nothing to do with the scheme.

Classifying the principals under Schumpeter's second criterion is not so simple. Watt's *technical* initiative and foresight were outstanding, particularly in his younger years. Both Roebuck and Boulton also displayed exceptional initiative in establishing several industrial ventures. The quality of their foresight is more debatable, as evidenced by Roebuck's bankruptcy and Boulton's unprofitable experiences in hardware-manufacturing, silver-mining, canal-building, and spelter-manufacturing.[21] J. E. Cule has attributed the unprofitability of Boulton's hardware enterprise in part to mismanagement.[22] But skill in performing the day-to-day management function is not a requisite for Schumpeter's entrepreneur-innovator.[23] And the losses of Boulton & Fothergill probably resulted more directly from overexpansion,[24] due in turn to the very megalomania that made Boulton a "captain of industry" type, than from mismanagement.

In contrast to Roebuck and Boulton, Watt demonstrated little zest for business affairs. It was necessary for Roebuck continually to spur him on when they were partners. During the Watt-Boulton partnership, Watt typically worked at his home while Boulton managed operations at Soho. Watt was perpetually overloaded with work because he tried to do everything himself and would not delegate to others. His letters repeatedly reveal his fear of managing and dealing with other people. For example, listing reasons why Boulton should be taken in as an active partner, Watt wrote to Roebuck on September 24, 1769: "Consider my uncertain health, my irresolute and inactive disposition, my inability to bargain and struggle for my own with mankind; all which disqualify me for any great undertaking." In similar vein was his letter to Dr. William Small on November 24, 1772: "I am extremely indolent, cannot force workmen to do their duty, have been cheated by undertakers and clerks, and am unlucky enough to know it. . . . I would rather face a loaded cannon than settle an account or make a bargain. In short I find myself out of my sphere when I have anything to do with mankind."

Watt also had no aptitude for dealing with bankers. The firm's borrowing was left to Boulton, but their obligations were still a nightmare to him.[25] For instance, even after the engine had proved successful, he wrote to Boulton on June 21, 1781, regarding his part in refinancing their debt, "When I executed the mortgage, my sensations were such as were not to be envied by any man who goes to death in a just cause, nor has time lessened the acuteness of my feelings." And, as his abandonment of the steam-engine project when his friends could provide no more money indicates, Watt was not one to perform the entrepreneurial function of persuading bankers to advance funds.

Boulton's enthusiasm for new ventures never flagged. When one project no longer absorbed his attention and capital, he began searching for another. Watt, on the other hand, was anxious to retire from the arena as soon as his engine's success was assured. The situation in 1785 is typical. With his debts paid off for the first time in a decade, Boulton plunged into a copper-market-stabilization scheme.[26] In contrast, Watt wrote to Boulton on November 5 of that year, "On the whole, I find it now full time to cease attempting to invent new things, or to attempt anything which is attended with any risk of not succeeding, or of creating trouble in the execution. Let us go on executing the things we understand, and leave the rest to younger men, who have neither money nor character to lose." Similarly, when their partnership agreement expired in 1800, Watt retired, at the age of 64, from active business life, while Boulton, at 70, became engrossed in a coin-minting venture.

Finally, we see distinct differences between Boulton and Watt in terms of Schumpeter's set of motivational characteristics. As just one example, Boulton wanted to manufacture engines for all the world; Watt asserted that he would rather have a modest income with fewer risks and troubles.[27]

To sum up, Boulton and Roebuck appear to have met rather well the specifications for Schumpeter's entrepreneur-innovator, as well as providing needed capital. In contrast, it is unlikely that Watt would have persisted in the development of his engine without the business support and drive of Roebuck and Boulton. Watt himself admitted that "without [Boulton] the invention could never have been carried by me to the length it has been."[28]

This does not mean that the roles of Boulton and Roebuck should be deemed more necessary or important than that of Watt. Too many histories of the Watt-Boulton steam-engine venture have shown a distinct "one hero" bias, emphasizing the contribution of one principal while pointedly deprecating the other's. But Watt's inventive genius, the capital of Roebuck and Boulton, and the entrepreneurial contribution of Boulton were all necessary, and none was sufficient in itself for full and timely exploitation of Watt's basic idea. A theory of technological change must recognize that contributions are essentially complementary, just as the theory of conventional production recognizes that land, labor, and capital are complementary.[29]

Technical Development of the Engine:
Invention or Innovation?

Thus far we have viewed the post-1765 effort largely from an economic perspective. But what were the actual technical activities that those expenditures supported? In Usher's theory they are included under the critical-revision step, but do they deserve a position subordinate to the act of insight, as at least some of his writings imply? And where, if at all, do these activities fit into Schumpeter's concept of innovation? To answer these questions, we must examine the technical work Watt and his associates carried out between 1765 and 1780.

In very broad perspective, the effort took the form of building, testing, modifying, and retesting models of increasing scale and sophistication. Although there are ambiguities in the evidence, it would appear that during the 1765–66 period Watt built three models: the original test model with a 1¾-inch brass cylinder but no steam jacket; a 1⅖-inch sheet-iron cylinder model with a steam jacket; and a model with a 5- or 6-inch copper cylinder and a wooden steam jacket.[30] Several more models with 7–8-inch cylinder diameters were built in various configurations during the next two years and tested intermittently. Then, in April 1769, a more or less full-scale engine with an 18-inch cylinder of tin was fabricated. This step up to practical operating scale led to many difficulties. As Watt wrote of Roebuck's plan to perfect the engine without Watt's help in 1772 by experimenting with a smaller and less expensive model, "he had turned his thoughts towards the engine not recollecting that I had been sufficiently successful with an engine of that size formerly, and that it was only in the 18-inch engine that the difficulties appeared. I have, however, dissuaded him from it, as, without flattering myself, I cannot imagine that he can find out in a few days all the difficulties, and the means of avoiding them, which have cost me so much labor."[31] This 18-inch engine was removed from Kinneil to Soho in 1773, where it was redesigned and rebuilt in 1774 and 1775 to serve as a successful prototype for the first commercially erected Watt-Boulton engines: a 50-inch water-pumping model for the Bloomfield Colliery and a 38-inch model for blowing John Wilkinson's blast furnaces, both set to work in March 1776.

The problems encountered with the various early test models were numerous, and Watt proposed and tried out an even greater number of solutions. Watt's comment in

1769 is typical. "I have been trying experiments on the reciprocating engine, and have made some alterations for the better and some for the worse, which latter must return to their former form."[32] Dickinson and Jenkins have observed, "When we reflect on the exhaustive character of these experiments, we can hardly wonder that in after days Watt spoke as if he had explored every possible combination of which the steam engine was capable."[33] Indeed, the development effort may have been complicated by Watt's desire to perfect every facet of the engine's operation—a quest opposed first by Roebuck and then by Boulton, who were anxious to press forward to commercial results with all possible dispatch. Watt himself admitted later in a third-person account of his work: "Had W. been content with the mechanism of steam-engines as they then stood, his machine might soon have been brought before the public; but his mind ran upon making engines *cheap* as well as *good,* and he had a great hankering after inverted cylinders and other modifications of his invention, which his want of experience in the practice of mechanics in great, flattered him would prove more commodious than his matured experience has shown them to be. He tried, therefore, too many fruitless experiments on such variations."[34] Dickinson suggests that Watt's tremendous ability to conceive many alternative solutions to a problem may actually have delayed the engine development effort. "This very fertility of mind . . . may almost be said to have delayed his progress. Of a number of alternatives he does not seem to have had the flair of knowing which was the most practicable, hence he expended his energies on many avenues that led to dead ends. In truth this is the attitude of mind of the scientist rather than that of the craftsman. Still, unless he had explored these avenues he could not be certain that they led nowhere."[35]

One of the most troublesome aspects of Watt's engine was the piston-cylinder arrangement. For the engine to work efficiently, the piston had to fit tightly within the cylinder throughout its stroke so the vacuum would not be dissipated, and yet excessive friction also had to be avoided. Techniques of metal-working at the time were not adequate to secure a tight metal-to-metal fit, and so other means had to be devised. Newcomen had solved the problem for his atmospheric engine by affixing a flexible leather disk atop the piston and by keeping a quantity of water in the cylinder above the piston and disk as a seal. But this method was less suitable for Watt's engine, since Watt wanted to keep the cylinder as hot as possible, and he knew that water used as a seal would invade the cylinder and absorb heat from the working steam.[36] He experimented with one approach after another. He tried tin, copper, wood, and cast iron for his cylinders and pistons to see which materials could be worked to the closest tolerances and still retain sufficient structural strength and durability. (The tin cylinder employed in the 18-inch Kinneil engine was rejected, for example, because it would not retain its shape.) He considered square pistons, round pistons, and flexible pistons (which could be inflated with fluid to make a tight fit). He experimented with piston disks and rings of leather, pasteboard, cloth, cork, oakum, hemp, asbestos, a lead-tin alloy, and a copper-lead alloy. To help seal the piston he tried mercury, oil, graphite, tallow, horse dung, vegetable oil, and a variety of other materials.

The problem was greatly alleviated by an achievement essentially unrelated to the work of Watt. In 1774 John Wilkinson (principal in one of the first two firms to employ the Watt-Boulton engine) patented a new type of boring mill which significantly increased the accuracy to which cylinders could be fabricated. The first bored cast-iron cylinder made by Wilkinson with his new technique arrived at Soho in April 1775, and it was used to resurrect the previously deficient 18-inch engine brought from Kinneil. From then until 1795, Wilkinson was the regularly recommended supplier for Watt-Boulton cylinders and pistons. Only three or four engines were built to Watt's design without Wilkinson cylinders during that twenty-year period.[37]

It is frequently stated that Wilkinson's invention was vital to the success of the Watt-Boulton steam engine.[38] This view must be qualified. Surely the engine could have been operated without Wilkinson's cylinders, as the erection of even the few non-Wilkinson engines implies. Furthermore, there is some evidence that the earlier cylinder problems were not insuperable. Watt wrote to his father on December 11, 1774, several months before the first Wilkinson cylinder arrived at Soho, that the 18-inch engine was "now going, and answers much better than any other that has yet been made, and I expect that the invention will be very beneficial to me." Still earlier, after discussing some of the piston-packing adversities experienced in experiments with the 18-inch engine at Kinneil, Watt wrote that even if no solutions better than those already tried could be found, the engine would probably consume only half as much steam as the atmospheric engines then in common use.[39] It is also likely that the Newcomen solution could have been adopted, if Watt had been willing to settle for the additional efficiency provided by his separate condenser and not worry about a loss of heat through the evaporation of sealing water.[40] Thus, it seems more reasonable to conclude that the Wilkinson invention was essential for the level of technical and economic success actually attained by the Watt-Boulton engine, but that at least moderate success could have been achieved without it.

Perhaps as problematic as the cylinder-piston question was the matter of the design of the condenser. In fact, during the trials of the 18-inch engine in 1769–70 at Kinneil, the condenser proved to be the main source of difficulty. In the course of ten years Watt tried a great number of approaches and designs. Up to 1775 he emphasized the surface-condenser approach, with cooling water circulated outside the chambers within which the steam condensed. But then he changed to Newcomen's water-injection approach to simplify the mechanism, to achieve greater condensing capacity within a given volume, and to prevent gradual deterioration of the condenser's efficiency by accumulation of mineral deposits on the vessel walls. During the period when Watt favored the surface-condensation approach, he considered plate designs, tube designs, worm designs, flooded designs, and unflooded designs and a variety of materials and fabrication methods. Many experiments were required to determine the proper capacity of a condenser for an engine of any given size. After the change to injection condensing in 1775, several different condenser and pump-system designs were tested.

This same pattern was repeated to a greater or lesser extent for each of the other components of the Watt-Boulton steam engine. Only a few examples can be added here. The valves and their working gear were redesigned completely several times, first along the lines of the Newcomen engine and later following different approaches. Iron and copper boiler bottoms were tried, and experiments were conducted to determine the necessary steam outputs for engines of various sizes. The early small models constructed by Watt had no beams and were inverted relative to the Newcomen design (that is, the vacuum was formed above rather than below the piston), but for the 18-inch Kinneil engine Watt shifted to a conventional beam design. He then adhered in general practice to the Newcomen approach, although in 1776 an inverted engine was designed for John Wilkinson.

To determine which designs and materials were best, beams in many different configurations were fabricated and subjected to destructive testing.

In general, extensive testing of every conceivable practical solution seems to have been a major characteristic of the technical activities pursued from 1765 to 1780, and the degree of insight associated with most of the individual solutions tended to be low. New designs and materials were proposed not only to solve strictly mechanical and technical problems, but also to simplify the fabrication of parts, to facilitate the erection and maintenance of engines, to prolong the life of engine components, to avoid expensive materials, to increase the engine's operating efficiency, and even to avoid having to buy components from vendors whose prices were too high.[41] This technical work involved not just the inventor Watt but also Boulton and a number of skilled craftsmen whose assignments included fabrication, erection, testing, and even suggesting new designs. In many ways the Boulton & Watt shops at Soho during the late eighteenth century clearly anticipate our industrial research and development laboratories of the twentieth century.[42]

Once the basic engine design had been more or less proved and emphasis shifted to the design and erection of engines for industrial use, technical activities of a noninventive nature continued. The basic design had to be modified for various specific applications—there were engines in many different sizes and configurations for use in pumping, blowing, and hammering, each requiring special attention to unique problems. For many years Watt did most of this work himself, at such an expenditure of time that he was unable to devote attention to more creative activities. A great deal of the time of Watt, Boulton, and their firm's most skilled workers was also spent away from Soho superintending the erection of new engines and trouble-shooting temperamental engines already erected. Finally, many experiments were conducted to collect engineering data on potentialities of engines and strengths of materials, as well as to determine the savings of coal afforded by the Watt-Boulton engine over atmospheric engines for the purpose of setting royalties.

In sum, it appears that the great bulk of the technical activity applied to the steam engine after Watt's basic work in 1765 does not qualify as invention in the strictest Usherian sense. The problems tended to define themselves through experimental failures instead of requiring significant perception of an unsatisfactory pattern; the level of insight associated with proposed new solutions was typically low; and optimal solutions were selected from a wide variety of alternative approaches through empirical analysis. Such activities may of course be classified as critical revision under a four-step conceptualization of the inventive process. Still, accounting as they did for most of the substantial time and effort expended before Watt's fundamental ideas were translated into an economically advantageous machine, they occupied a position that can scarcely be subordinated to the inventive acts of insight.[43] Technical activities of this nature also hold no special position in the Schumpeterian concept of innovation, and yet their successful prosecution must be one of the entrepreneur's prime concerns. Perhaps the best solution is to apply to such activities the term "development" (from the broader modern term "research and development") and to recognize that development, like investment, is a necessary complement to invention and innovation in the process of technological change.[44]

Improvements

From 1775 through 1780 most of James Watt's energies were absorbed in designing steam engines for specific applications according to the general technical concepts tested successfully at Soho prior to that time, supervising the erection of those engines on the spot, and trouble-shooting. By 1781, however, he found it possible and desirable to concentrate again on more original work. This effort yielded a number of significant improvements, including devices to convert the engine's reciprocating motion into rotary motion, the double-acting engine, expansive working, an automatic centrifugal governor, and a compound (two-cylinder) engine.

No attempt can be made here to describe these improvements. Many were logical extensions of Watt's original idea, requiring relatively little new insight but considerable trial and error. In other instances (especially in the invention of the centrifugal governor and of a parallel-motion mechanism for use with the double-acting engine) a high level of creative insight was revealed. The end result of the invention of the parallel-motion mechanism, however, could have been obtained with less ingenious alternative solutions. The centrifugal governor, an important stimulant to later technological progress in other fields, is even more difficult to assess, since the basic idea was not original with Watt. Although exceptions must be recognized, for the most part the activities during this improvement period appear more like those of the 1765–75 design-development period than like the 1764–65 period of analytical and creative work.

The Role of Patents in the Steam Engine's History

The available material provides some indication of how one economic policy—the granting of patent monopolies—affected the Watt-Boulton invention and development effort. At least since James I accepted the Statute of Monopolies in 1624, the underlying principle of patent systems has been to encourage technological progress by granting to inventors temporary monopolies of their inventions. The objective is often defined in popular discussion as encouraging invention, although, as the foregoing analysis has illustrated, invention strictly defined is a necessary but not a sufficient condition for technological progress.

The existence of a patent system seems to have had little or no influence on Watt's invention of the separate-condenser principle. His introduction to the problem came when he was hired to repair a Newcomen engine model. The experiments which led to Watt's historic insight—costing little more than his spare time—originated in that task and were sustained as a result of scientific curiosity.[45]

It is less clear whether, after the separate-condenser idea was conceived, Watt would have invested his time and his friend's money in reducing the invention to practice. He apparently gave no immediate consideration to obtaining a patent, although the possibility of eventual patent protection could have been in the back of his mind. Not until three years after his strategic insight did Watt, with the encouragement of Roebuck, apply for a patent. By that time Watt had (through borrowing) committed some £1,000 to the development of his invention, relying mainly upon secrecy to protect his investment.

There is reason to believe that application for a patent was a condition of Roebuck's entry into partnership with Watt. Whether or not Roebuck's interest would have been sustained in the absence of patent protection is impossible to determine. We know, however, that due to financial difficulties Roebuck contributed little more to the development than the sum agreed upon in 1768.

Thus we must ask whether Boulton would have participated in the project without patent protection. The status of Watt's patent was clearly of concern to Boulton. When Watt arrived at Soho in May 1774 to begin work on the engine under Boulton's patronage, less than ten years remained of the patent's term. Boulton recognized that the engine venture would be costly; a 1775 estimate of £13,000 has been mentioned earlier. There was clearly a risk that the investment would not be recovered before the patent expired.[46] In any event, a decision to seek extension of the patent must have been made by December 1774, for on January 13, 1775, Watt communicated to Boulton a lawyer's advice "that we might surrender up the present patent, and that he did not doubt a new one would be granted." A petition to Parliament was submitted in February 1775, and, after some heated debate, the petition was enacted into law and received final approval in May.

These actions on the part of Boulton and Watt do not prove that without the extension, Boulton's support would have been withdrawn. Boulton's investment before extension became certain was considerable. By May 1775, the 18-inch Kinneil engine had been completely rebuilt, the first Wilkinson cylinder was on hand, and cumulative development expenditures of £3,000 (including Roebuck's share of more than £1,200) had been recorded. Nevertheless (although the integrity of self-serving declarations must be held suspect) Watt and Boulton (who was a gifted lobbyist, among other things) must have argued convincingly to members of Parliament that their work would be impeded unless the extension were secured, for the enabling bill reads in part as follows:

> AND WHEREAS, in order to manufacture these engines with the necessary accuracy, and so that they may be sold at moderate prices, a considerable sum of money must be previously expended in erecting mills, and other apparatus; and as several years, and repeated proofs, will be required before any considerable part of the publick can be fully convinced of the utility of the invention, and of their interest to adopt the same, the whole term granted by the said Letters Patent may probably elapse before the said JAMES WATT can receive an adequate advantage to his labor and invention:
>
> AND WHEREAS, by furnishing mechanical powers at much less expense, and in more convenient forms, than has hitherto been done, his engines may be of great utility in facilitating the operations in many great works

and manufactures of this kingdom; yet it will not be in the power of the said JAMES WATT to carry his invention into that complete execution which he wishes, and so as to render the same of the highest utility to the publick of which it is capable, unless the term granted by the said Letters Patent be prolonged.[47]

Of the situation in 1775, Watt wrote much later, "Prior to the Act of Parliament we had sold no engines; though we had made some, they were merely to satisfy ourselves,— and, had the Act not passed, the invention had fallen to the ground;—so much did we foresee the moral difficulties before us, and the great necessary expenditure."[48] Again, this appraisal may have been tempered by hindsight. Still the patent extension's importance to Boulton is also indicated by the fact that his partnership with Watt did not begin formally until after the extension had been enacted, and its termination date was set as the extended patent's expiration date. Thus there are reasonable grounds for inferring that Boulton's decision to invest in the project was influenced by the certainty of at least some patent protection, if not by the possibility of extended protection. In this case the grant of a patent monopoly was a probable incentive for investment in technological innovation, although not an incentive for invention.

Had there been no patent protection at all, and had Boulton nonetheless invested in the steam-engine venture, Boulton and Watt certainly would have been forced to follow a business policy quite different from that which they actually followed. Most of the firm's profits were derived from royalties on the use of engines rather than from the sale of manufactured engine components, and without patent protection the firm plainly could not have collected royalties. The alternative would have been to emphasize manufacturing and service activities as the principal source of profits, which in fact was the policy adopted when the expiration date of the patent for the separate condenser drew near in the late 1790's. But establishment of the requisite manufacturing facilities probably would have called for more capital than Boulton was in a position to supply at the time. And the manufacturing which *was* done by Boulton & Watt during the firm's early years was generally unprofitable—due in part to an emphasis on high quality motivated in turn by the desire for maximum engine performance and hence (since royalties were based upon operating cost savings) high royalties.[49] The necessity of charging high prices to cover costs might also have inhibited any substantial expansion of the firm's sales of manufactured goods.[50] Whether in the absence of patent protection a different attitude toward quality would have been adopted and whether emphasis on manufacturing would have led the original partners to implement significant efficiency-increasing production measures, as their sons did during the late 1790's,[51] can only be left to speculation.

It is possible to conclude more definitely that the patent litigation activities of Boulton & Watt during the 1790's did not directly incite further technological progress. The 1769 patent was successfully defended against charges of invalidity, persons exploiting inventions that infringed the separate-condenser principle were prosecuted and enjoined, and users of the Boulton & Watt engine who were in arrears on their royalties were compelled to pay up. The effect of these actions for Boulton & Watt was mainly one of income redistribution: back royalties collected in Cornwall alone after the patent's validity was affirmed amounted to about £ 30,000.[52] No prospective new inventions or developments were at stake in these patent actions. Still two indirect effects on the rate of technological progress may be identified. First, Boulton & Watt's refusal to issue licenses allowing other engine-makers to employ the separate-condenser principle clearly retarded the development and introduction of improvements.[53] On the other hand, it is possible that the handsome profits realized by Boulton and Watt[54] provided a striking success story which encouraged others to accept the risks and burdens of invention, development, and innovation.[55] Naturally, there is no way of demonstrating the existence or strength of such an indirect effect.

Conclusion

It would be presumptuous to suggest modifications in the theories of Usher and Schumpeter on the basis of a single case study. The dangers in generalizing from one example are well known, and the models of both authors are susceptible to a diversity of interpretations. Still, the evidence examined here at least suggests that Usherian invention and Schumpeterian innovation are logically distinguishable and that both invention and innovation, along with investment, are necessary and complementary functions in the advance of technology.[56]

The position of the function "development," as defined in this paper, is more ambiguous. The successful accomplishment of development tasks is a concern of the Schumpeterian innovator, and the activities themselves are an important claimant upon the investor's capital. Development can also be subsumed under Usher's "critical revision." Yet the amount of time and effort devoted to development as opposed to more insightful activities by Watt implies the need for a less subordinate treatment. Nor was the Watt-Boulton case unique in this respect. Ten years of experimental work occurred before Thomas Newcomen had a satisfactory version of the engine that dominated the first half of the eighteenth century.[57] Six years of development preceded successful operation of Jonathan Hornblower's compound expansive engine, a competitor to the Boulton & Watt engine later blocked by Watt's patent.[58] And in extending Watt's basic patent, Parliament in 1775 called attention to "the many difficulties which *always* arise in the execution of such large and complex machines, . . . and the long time requisite to make the necessary trials."[59] In view of this experience, one might wish to assign development a functional status equal to that of invention, innovation, and investment.

Although each of the functions appears to be a necessary condition for technological advance, it is possible that innovation, investment, and development are more sensitive to economic variables than invention is. Inventive acts of insight may follow, as in the steam-engine experience, from scientific curiosity and a fortuitous combination of chance factors without any direct stimulus from profit expectations or such public policies as the granting of patent monopolies. In the case at hand, execution of the innovative, investment, and developmental functions depended much more directly than Usherian invention upon these economic factors. Such a relationship, if valid more generally, may have significant implications for public policy. As Judge Frank observed in 1942, "The controversy between the defenders and assailants of our patent system may be about a false issue—the stimulus to invention. The real issue may be the stimulus to investment. On that assumption, a statutory revision of our patent system should not be too drastic. We should not throw out the baby with the bathwater."[60] Certainly the question of what patent monopolies in fact stimulate deserves further attention by students of technological history.

REFERENCES

[1] A. P. Usher, *A History of Mechanical Inventions* (rev. ed.; Cambridge, Mass., 1954), pp. 60–65.

[2] *The Theory of Economic Development,* trans. Redvers Opie (Cambridge, Mass., 1934), pp. 74–94. See also his *Business Cycles* (New York, 1939), especially pp. 84–98.

[3] *Theory of Economic Development,* pp. 88–89.

[4] In connection with these experiments, it would appear that Watt independently rediscovered the theory of latent heat originally enunciated in 1761 by his friend Joseph Black, although some controversy exists on this point. Cf. Milton Kerker, "Science and the Steam Engine," *Technology and Culture,* II (Fall 1961), 381–390.

[5] Letter, Watt to the Chief Justice of Common Pleas, 1795, cited in T. S. Ashton, *Iron and Steel in the Industrial Revolution* (2d ed.; Manchester, 1951), p. 61. There is no surviving contemporary record by Watt of his earliest experiments.

[6] From an account apparently written by Watt in 1814, cited in J. P. Muirhead, *The Origin and Progress of the Mechanical Inventions of James Watt* (London, 1854), I, lxxiii–lxxiv. Unless otherwise indicated, all of the Watt correspondence cited here is reproduced chronologically in Muirhead, Vols. I and II.

[7] Analysis of the available evidence suggests that there were three early models, not two—the model with a 1¾-inch brass cylinder but no steam jacket described in the text; one with a 1⅖-inch iron cylinder and a steam jacket; and the large model (probably with a five- or six-inch copper cylinder) referred to here by Watt. In *James Watt and the Steam Engine* (Oxford, 1927), pp. 92–96, H. W. Dickinson and Rhys Jenkins propose that the 1¾-inch model and the 1⅖-inch model were one and the same. However, there is good reason to doubt such an inference. What is purported to be the 1⅖-inch model is preserved in the Science Museum at South Kensington, London. Accounts by both Watt and Dr. John Robison cited in Muirhead (*op. cit.*) state quite explicitly that the cylinder of the 1¾-inch model was hastily improvised from a physician's syringe and had no steam jacket. It seems plausible that Watt performed a quick, crude test of his theory with the 1¾-inch syringe model and then built the more elaborate 1⅖-inch model to determine what effect the combination of a separate condenser and steam-jacketing would have on steam consumption.

[8] *History of Mechanical Inventions,* pp. 71–72.

[9] Watt to Dr. William Small, August 30, 1772.

[10] Cited in Paul Mantoux, *The Industrial Revolution in the Eighteenth Century,* trans. Marjorie Vernon (rev. ed.; New York, 1933), p. 332.

[11] Mantoux, *op. cit.,* p. 335, cited from the *Journal of the House of Commons,* XXV, 142. Mantoux reports an estimate by S. Timmins that £47,000 were actually expended for construction, equipment, and other costs, although it is not known

how the estimate was derived from incomplete company records. In an article much more skeptical of Boulton's contribution, J. E. Cule asserts that "the total cost of launching the enterprise was £3,370" ("Finance and Industry in the Eighteenth Century: The Firm of Boulton and Watt," *Economic History, Supplement to Economic Journal,* IV [February 1940]: 320). He does not state what this estimate includes, but it apparently excludes Roebuck's contribution (which was taken over by Boulton in settlement of debts) and the losses borne by the firm in trouble-shooting initial installations and in producing engine components.

The existence of such widely varying estimates of the firm's actual original investment suggests the need for further research in the primary sources. More accurate information would provide a much clearer picture of the degree of risk actually assumed by Boulton.

[12] H. W. Dickinson, *Matthew Boulton* (Cambridge, 1936), pp. 36, 109–110. From 1762 to 1780, Boulton & Fothergill incurred net losses amounting to £11,000 on a total capital of £20,000.

[13] There is also evidence that Newcomen, as the victim of an unfortunate patent-rights situation, did not profit significantly from his invention.

[14] Watt to Dr. Small, July 25, 1773.

[15] As events turned out, Boulton & Watt did relatively little manufacturing of engines up to roughly 1795. Apparently because of the scarcity of capital and skilled labor and because the energies of Watt and Boulton were fully absorbed in design work, experiments, supervision of the erection of engines, and promotion of their venture, manufacturing activities were confined largely to key components such as nozzles and valves. Nearly all of the firm's early profits were derived from royalties paid by engine-users.

[16] *Theory of Economic Development,* p. 66.

[17] *Ibid.,* pp. 75, 78.

[18] *Ibid.,* p. 89.

[19] *Ibid.,* pp. 78, 92.

[20] *Ibid.,* p. 93.

[21] Cule, *loc. cit.,* pp. 320–324.

[22] *Ibid.*

[23] Cf. Schumpeter, *Theory of Economic Development,* especially p. 78.

[24] Cf. Erich Roll, *An Early Experiment in Industrial Organization* (London, 1930), especially pp. 10–11.

[25] Dickinson and Jenkins, *op. cit.,* p. 52. It should be noted that Cule, *loc. cit.,* raises certain points that might appear to deny both the fact and the general interpretation of the firm's borrowing. He argues first that no money was borrowed to finance the Boulton and Watt venture; that in 1772 Boulton sold his Parkington estate to provide all the funds necessary for

the steam-engine enterprise. This analysis of Boulton's intent is dubious, since in 1772 Boulton's negotiations to acquire a part of Roebuck's steam-engine interest were stalemated indefinitely. A more likely reason for the sale was Boulton & Fothergill's urgent need for cash. The partnership, in fact, immediately absorbed £6,000 of the proceeds.

It also seems most difficult to isolate, as Cule has, Boulton's financial transactions as a partner in Boulton & Watt from those as partner in Boulton & Fothergill. He required a certain amount of cash to meet the minimum demands of the two firms together, and this joint demand necessitated both the sale of his personal assets and borrowing, including later refinancing secured additionally by Watt's interest in certain engine royalties. Had Boulton not borrowed as partner in the hardware firm, or had the firm collapsed, the steam-engine firm would have been deprived of capital essential to its initial or continued existence.

Cule also implies that Boulton was "unbusiness-like" for borrowing £7,000 at a 10 per cent interest rate, while Watt was a shrewder business man for urging reduction of the loan through cash payments on the principal. The opposite conclusion might as easily be drawn from the evidence. Perhaps the 10 per cent rate was exorbitant, but it was the best Boulton could do in tight circumstances. The conventional economic theory of profit maximization indicates that funds should be borrowed in an imperfect capital market as long as the additional interest cost is less than the return from investments which would be foregone unless the funds are borrowed. Up to about 1785, the operations of Boulton & Watt were severely constrained by capital scarcity. This condition, plus the high *average* returns to the firm's investment, leads me to suspect that unless there were substantial discontinuities in investment opportunities, the return on marginal investment would have exceeded the 10 per cent cost of capital. By similar reasoning, one might conclude that short-term borrowing through discounting bills of exchange (successfully opposed by Watt, to Cule's evident satisfaction) would have been profitable.

[26] This was not completely unrelated to the steam-engine business, since Boulton recognized that the level of royalties received from engines used in Cornish copper mines depended upon the copper price level.

[27] See, e.g., his letters to Boulton of October 31, 1780, and to Small in August 1772.

[28] Quoted in H. W. Dickinson, *James Watt: Craftsman and Engineer* (Cambridge, 1936), p. 200. Dickinson adds his own opinion that without Boulton, Watt "would never have brought his engine into general use, nor derived any reward for his invention, nor followed it up by those equally brilliant inventions connected with the rotative engine."

[29] One might still wish to stress Watt's contribution because

of its uniqueness, whereas generalized capital and entrepreneurship were present in modest abundance in eighteenth-century England. Economic theory holds that complementary factors are valued on the basis of supply (degree of scarcity) and demand (contribution to society's express wants). In the steam-engine case, while Watt's inventive genius was clearly scarce, capital and entrepreneurship for the specific venture were perhaps equally scarce due to market imperfections.

[30] Cf. n. 7.

[31] Watt to Dr. Small, November 7, 1772.

[32] Watt to Dr. Small, January 28, 1769.

[33] Dickinson and Jenkins, *op. cit.*, p. 102.

[34] Watt, "A Simple Story" (1796), cited in Muirhead, *op. cit.*, p. lxxxix.

[35] *James Watt*, p. 40. For an analysis of analogous problems in twentieth-century technical developments, see M. J. Peck and F. M. Scherer, *The Weapons Acquisition Process: An Economic Analysis* (Boston, 1962), pp. 263–264, 488–492, and 495–501. See also my comment in National Bureau of Economic Research, *The Rate and Direction of Inventive Activity* (Princeton, N.J., 1962), pp. 501–502.

[36] To keep the cylinder hot, Watt also decided to break with the Newcomen approach of using the atmosphere to push the piston into the vacuum formed by steam condensation. He used steam instead of air for this purpose by enclosing the cylinder on both sides of the piston and employing a more complex valving system. This approach later provided the basis for a double-acting engine with steam being condensed on both sides of the piston, forming alternating vacuums to power the piston both upward and downward.

[37] See Roll, *op. cit.*, pp. 25–35, 55–57, and 157–160, and Ashton, *op. cit.*, pp. 62 ff., for accounts of Wilkinson's business relationships with Boulton & Watt, which deteriorated during the 1790's to the point where Boulton & Watt set up their own iron-working shop using Wilkinson's boring-mill technique.

[38] Cf. Dickinson and Jenkins, *op. cit.*, p. 43; Roll, *op. cit.*, p. 25; and Ashton, *op. cit.*, p. 63.

[39] Watt to Dr. Small, September 20, 1769.

[40] This view is held by Dickinson and Jenkins, *op. cit.*, p. 108.

[41] To illustrate the last objective, a major change from using an integral outer cylinder (steam jacket) was decided upon to permit engine-users to buy the outer cylinder from manufacturers other than Wilkinson, about whose prices they had complained (Dickinson and Jenkins, *op. cit.*, p. 193; see also Roll, *op. cit.*, pp. 34–35).

[42] Dickinson and Jenkins (*op. cit.*, p. 1) remark about Watt: "Indeed the business of mechanical engineering may be said to have been set on foot by him."

[43] It would be incorrect to state unequivocally that Usher subordinates critical revision to acts of insight. In his basic work

the tendency to do so is pronounced, but in a later elaboration he distinguishes inventive acts of insight from the broader "process of invention," and he warns against overemphasizing high-level acts of insight while underemphasizing elements of novelty arising from acts of skill (A. P. Usher, "Technical Change and Capital Formation," in the National Bureau of Economic Research conference report, *Capital Formation and Economic Growth* [Princeton, N.J., 1955], pp. 523–550 and esp. pp. 523 and 527). His description of the critical revision phase— "a very intimate interweaving of minor acts of insight and acts of skill"—fits the steam-engine case rather well (p. 528). My only objection to this revised schema is the somewhat confusing double meaning attached to "invention," one definition focusing on acts of insight and the other (process) definition embracing most or all aspects of pioneering technological advance.

⁴⁴ In fact, in the article cited above Usher defines the fourth step of his inventive process as "critical revision and development" rather than critical revision alone, as in his book.

⁴⁵ For his work on the model Watt was paid £5 11s. in June 1766 (Dickinson and Jenkins, *op. cit.,* p. 22). Interestingly, if Watt had been hired by a government agency such as the U.S. National Aeronautics and Space Administration or the Atomic Energy Commission in 1964, he probably would have been unable to secure a patent on his invention, since any patent obtained in connection with contracts from those agencies must normally be assigned to the government.

In assessing the importance of the patent system to Watt's invention, one must not be misled by the vigorous statements Watt made on behalf of stronger patent laws during the 1780's (cf. Roll, *op. cit.,* pp. 145–147). His concept of "invention" appears to have included developmental activities, and in advocating patent protection for the "first introducers" in England of inventions made abroad, he seems to have had in mind the necessity of encouraging the Schumpeterian innovative role. It is hardly necessary to note also that Watt was not without the kind of vested interest that has made all debates on the patent system so acrimonious.

⁴⁶ Most authors indicate that the investment would not have been recouped if the patent expired on schedule in 1783 and if no further revenues were obtained from then on. Cule's paper, however, can be interpreted as saying that cumulative revenues exceeded cumulative costs at an earlier date.

⁴⁷ 15 George III, c. 61 (1775).

⁴⁸ Watt to Mr. T. Parker, June 2, 1797.

⁴⁹ Cf. Roll, *op. cit.,* pp. 121–122.

⁵⁰ *Ibid.,* pp. 56–57.

⁵¹ *Ibid.,* pp. 156–157, 275.

⁵² Dickinson and Jenkins, *op. cit.,* p. 336. This is what was actually obtained through threats and negotiation. In 1799, Boulton & Watt claimed that the back royalties owed them in

Cornwall amounted to more than £162,000 (Dickinson, *Matthew Boulton*, p. 176).

[53] See Dickinson and Jenkins, *op. cit.*, pp. 42 and 69–70; Roll, *op. cit.*, p. 143.

[54] Watt, who started in business by borrowing, left an estate of £60,000 at his death in 1819 (Dickinson and Jenkins, *op. cit.*, p. 79). Boulton's estate in 1809 was £150,000 (Dickinson, *Matthew Boulton*, p. 201).

[55] Cf. Schumpeter, *Capitalism, Socialism, and Democracy* (3d ed.; New York, 1950), pp. 73–74.

[56] For a contrary conclusion see Vernon W. Ruttan, "Usher and Schumpeter on Invention, Innovation, and Technological Change," *Quarterly Journal of Economics*, LXXIII (Nov., 1959), 596–606. Usher also in a very few sentences rejects the Schumpeterian distinction between invention and innovation by what seems to me to be an extension of his own schema (*Capital Formation and Economic Growth*, pp. 527, 534).

[57] H. W. Dickinson, "The Steam Engine to 1830," in *A History of Technology*, ed. Charles Singer *et al.* (London, 1958), IV, 175.

[58] Dickinson and Jenkins, *op. cit.*, pp. 57, 304.

[59] 15 George III, c. 61 (1775) (italics added).

[60] *Picard* v. *United Aircraft Corporation*, 128 F. 2d 632, 643 (1942). See also F. M. Scherer, S. E. Herzstein, *et al.*, *Patents and the Corporation* (2d ed.; Boston, 1959), esp. Chaps. v and xiv.

The Legend of Eli Whitney
and Interchangeable Parts

Robert S. Woodbury

In some legends the story is such that from its very nature we can never establish its truth or falsity; in others patient historical work—usually external to the legend—can ascertain whether the events actually happened or not. The legend of Eli Whitney's part in interchangeable manufacture is, however, unique in that the clues and even much of the evidence for its refutation are part of the legend as customarily recited. It is also unique in that the legend is not merely a popular one nor even a story given "authority" by inclusion in conventional textbooks. This legend has been retold at least twice with all the paraphernalia of historical scholarship—footnotes, elaborate bibliography, discussion of the sources, and even use of archival material.[1] But in both cases we find the same failure to evaluate the evidence critically, to follow leads to other sources, and to question

Professor Woodbury, of the department of humanities of Massachusetts Institute of Technology, is the author of a series of monographs on the history of machine tools, one of which, History of the Lathe to 1850, *was the first volume in the Monograph Series in the History of Technology published jointly by the Society for the History of Technology and the MIT Press. A grant from the Wilkie Foundation made possible the research for this paper, which was presented at a joint session of the Society for the History of Technology and Section L (History and Philosophy of Science) of the American Association for the Advancement of Science in December 1959 in Chicago. This article, published in* Technology and Culture *(Vol. 1, No. 3): 235–54, was awarded the first Abbott Payson Usher Prize by the Society for the History of Technology.*

basic presuppositions. These same faults extend back to the origins of the legend.

Poking back into the beginnings of this legend, one finds evidence to show that it was at least partially created consciously by its hero and uncritically accepted by most of his contemporaries.[2] The *editio princeps* of the legend is equally uncritical; in fact it is frankly an *apologia pro vita sua*. In his *Memoir of Eli Whitney*, Denison Olmstead[3] gives us most of the elements of the legend and claims to have based his account upon conversations with those who knew Whitney, as well as upon examination of his correspondence and Miller's. Yet Blake writing in 1887 said: ". . . there have not been wanting persons who have endeavored to take from Mr. Whitney the credit of originating the uniformity system and making it a great practical success at the beginning of this century, thus leading in the van of progress of the mechanical arts, and laying the foundations for the enormous industry development of the nineteenth century."[4] Evidently some of his contemporaries were not taken in by Whitney's claims, but the scholars have not asked either who these other inventors were or what their contributions may have been. Let us examine the principal parts of this legend in some detail.

I

The Contract

Whitney's contract of June 14, 1798 to manufacture arms for the federal government is the focus of a number of elements of our legend. His motives in this undertaking have been interpreted as those of a prudent businessman doing his patriotic duty and as those of a genius anxious to put into execution a new scheme of manufacture for the good of his country in a time of crisis. His actual motives were quite different.

In 1798 Miller and Whitney had lost all their suits to obtain their cotton-gin patent rights in the courts of the South. What little legal merits these decisions had stemmed from a defect in the Patent Law of 1793; clearly nothing further could be done until Congress corrected this defect. The efforts of Whitney and others did not finally result in a new patent law until 1800. The intervening years could be seen as a lull in the affairs of Miller and Whitney. But Whitney could hardly look forward to any relaxation, for their financial affairs were in desperate straits. Every source of credit had been exhausted by both partners.[5] Certainly Whitney himself was on the verge of a nervous breakdown. Although some have tried to find in this situation a frustrated love for Catherine Greene, a more careful reading of his letter of October 7, 1797 to Miller indicates rather that Whitney's high hopes of financial security, respected position, and prestige have not only come crashing to the ground, but the disgrace of bankruptcy is staring him in the face. All that winter of 1797–98 Whitney brooded alone, half-heartedly carrying on the affairs of Miller and Whitney. He shut himself off from all his old friends and even unjustly accused his partner and friend Miller.

Whitney needed a new opportunity—any opportunity. But, more important, he needed credit—credit to save Miller and Whitney from bankruptcy, credit to enable him to fight for his rightful profit and for his good name lost in the cotton-gin suits. When he heard that the Congress was "about making some appropriations for procuring Arms etc. for the U.S.," here was a heaven-sent opportunity.[6] This would at least keep his manufactory going until he could get his cotton-gin rights. The opportunity was so great and Whitney's situation so desperate that he was willing to promise "ten or Fifteen Thousand stand of Arms," a fantastic proposal! Whitney even promised to begin delivering "in a short time" and he "will come forward to Philadelphia immediately. . . ." New hope for a desperate man!

Why was such a rash proposal not rejected at once by such prudent men as President Adams and Timothy Pickering, the secretary of war? The failure of the Pinckney mission had caused public concern, and French privateers were rumored to be off the coast. Even Washington was called out of retirement to head the armed forces. On the 4th of May 1798 Congress voted $800,000 for the purchase of cannon and small arms. When on the 24th of May Whitney arrived at the seat of government the plum was not only ripe and juicy but begging to be picked. Public sentiment was aroused, and the highest officials must do something— and that right promptly. Both sides could not close the contract quickly enough. Only the purveyor of public supplies, Tench Coxe, seems to have kept his head—"I have my doubts about this matter and suspect that Mr. Whitney cannot perform as to time."[7]

It is not necessary to see "influence" at work here, though it is true that Whitney did have a number of Yale graduates who could help him. Much less is there any evidence that the generous terms of Whitney's contract grew out of the feeling that he had been shabbily used in his cotton-gin suits. But he clearly had a personal friendship with Jefferson arising out of the patent for the cotton gin.

Certain features of the contract deserve closer examination. The legend makes much of the fact that the actual document was wholly handwritten. It says that all the other contractors of this time received printed contracts, and that there was therefore something special about Whitney's contract. Unfortunately an examination of the actual contracts, including Whitney's, in the National Archives shows that this was by no means the only handwritten contract— there were others, such as that of Owen Evans of Providence, Pennsylvania. The fact is that several of the early contracts were handwritten; the later contracts, mostly signed in September, were printed forms. These other contracts, printed or handwritten, were all identical in wording and provisions with Whitney's, except in the terms of the last paragraph. There *was* something special about Whitney's contract—it contained a paragraph six not included in any of the others. It was this paragraph that was crucial for Whitney. Having quickly sized up the situation in which the high officials found themselves, the shrewd Whitney saw his chance, consulted Baldwin as to the form the contract should take,[8] and at one stroke solved all his immediate problems. This paragraph reads:

> 6th. Five thousand dollars shall be advanced to the party of the second part on closing this contract, and on producing satisfactory evidence to the party of the first, that the said advance has been expended in making preparatory arrangements for the manufacture of arms, Five Thousand dollars more shall be advanced. No further advances shall be demanded until One thousand stands of Arms are ready for delivery, at which time the further sum of Five thousand dollars, shall be advanced. After the delivery of One thousand stands of arms, and the payment of the third advance as aforesaid, further advances shall be made at the discretion of the Secretary of the Treasury in proportion to the progress made in executing this contract. It is however understood and agreed by and between the parties to this instrument, that from time to time, whenever the party of the second part shall have the second thousand ready for delivery he shall be intitled to full

payment for the same, so with respect to each and every Thousand until he shall have delivered the said Ten thousand stands.

Here was credit at last! Here was financial standing which assured further credit! Five thousand dollars at once, and five thousand more on terms which could be easily fulfilled by using his cotton-gin laborers and machines. And assured payment for each thousand stands of arms upon delivery— to a total of $134,000. Little wonder that Whitney wrote to his friend Stebbins: "Bankruptcy and ruin were constantly staring me in the face and disappointment trip'd me up every step I attempted to take, I was miserable . . . Loaded with a Debt of 3 or 4000 Dollars, without resources, and without any business that would ever furnish me a support, I knew not which way to turn . . . By this contract I obtained some thousands of Dollars in advance which saved me from ruin."[9]

No wonder that in his eagerness to read paragraph six of the contract Whitney evidently skimmed rapidly over the incredible terms of paragraph one.[10] Whitney had contracted to deliver 4,000 stands of arms by September 30, 1799, and 6,000 more by September 30, 1800. Four thousand stands of arms in fifteen months, from a factory yet to be built, and made by laborers as yet untrained and by methods as yet unknown! And 6,000 more in the following year! In his desperation Whitney had thrown all caution to the winds. He was no experienced manufacturer, as his deliveries of the relatively simple cotton gin indicate. He was aware that he knew nothing of arms making. And a prudent man would have expected at least some of the setbacks with which he fills his later letters to Wolcott, together with requests for further credit, contrary to the provisions of the contract. In short, despite his vague claims of new methods and what could be done by "Machinery moved by water," Whitney had only the vaguest idea of how he would actually fulfill the contract. He was not able to deliver even the first 500 muskets until September 26, 1801, and the contract was not actually completed until January 23, 1809. Further, the records of the Springfield Armory, now in the National Archives, show that even during the period 1815 to 1825, when his plant was fully established, Whitney never delivered muskets at the rate promised in his contract of 1798.[11]

Not only have these facts been forgotten in estimating Whitney's motives in the contract, but also in attempting a proper evaluation of his troubles with Samuel Dexter, who had replaced Wolcott as secretary of the treasury. We are asked to see Dexter as a villain abusing our hero with "malice" by demanding that he perform in accordance with his contract. The other contractors of 1798 had in many cases failed to fulfill their contracts, and some of them had even gone into bankruptcy as a result of their efforts to manufacture arms for the federal government. But Springfield Armory records show that some of them had performed as contracted, a few on time and even more eventually. Yet these men had all ventured into arms making by financing themselves privately. There are no records of their writing the long apologetic letters full of troubles, promises, and requests for further advances, which characterize Whitney's correspondence from 1798 on. Nor had they been given the numerous informal extensions of time with which Wolcott, strongly under the influence of Hamilton's theories of the importance of manufactures, had favored Whitney and which culminated in a formal modification of the contract just before Wolcott left office. Whitney had been given more consideration than any other contractor.

But one might give at least a moment to the position of Dexter. He was a government official sworn to carry out the law and to protect the interests of the government. Whitney had been given every chance and had not performed. Some of the other contractors had. Even had Dexter seen the ultimate interests of the government in this matter in the broad terms that Wolcott and Jefferson did, he had no authority to make the extremely loose interpretation of Whitney's contract that Wolcott had. Actually Wolcott had left office partially as a result of other similar easy exercise of the discretions permitted his high office. Dexter did not deserve such blame.

And had Whitney, for his part, acted in good faith since 1801? We can leave out of our discussion the troubles Whitney so fully related in his numerous lengthy letters to Wolcott. They were real enough, even if recounted in rather unmanly fashion, but they were all of the sort, magnitude, and frequency which a prudent man would expect in an undertaking of this sort. And one could argue that if Whitney had been carelessly optimistic in what he had promised in the contract of 1798, so had the responsible government officials, who had also been warned by the purveyor of public supplies to expect delay in delivery. However, despite Whitney's claims of the exhausting efforts and attention he had devoted to his arms manufacturing, the facts prove otherwise.

It is true that the lull in the affairs of Miller and Whitney from 1798 to 1801, plus the credit advanced him by the federal government, did enable Whitney to devote most of his time in these years to make a beginning on fulfilling his arms contract, and by September of 1801 he did deliver the first 500 muskets. But from this initial delivery until 1807 there is no twelve-month period during which he delivered over 1,000 muskets. During this same period he had been given advances from the Treasury such that he was constantly in debt to the United States. In fact, when Whitney finally completed delivery in January 1809 he received a payment of only $2,450 as final settlement of the total contract of $134,000. Only on this date was his account up to date. The Whitney account in the Springfield Armory records[12] also shows that in 1806 Whitney delivered 1,500 muskets, in 1807 he delivered 2,000, and in 1808 and the first few days of 1809 he delivered 1,500. What is the explanation of these facts?

I do not wish to imply that Whitney was misrepresenting his troubles in his letters to Wolcott, Dexter, and Dearborn; but he most certainly was not telling the whole story, as his other correspondence clearly shows. In April 1800 the Congress revised the patent law which had been the legal means of defeating Whitney's claims to his cotton-gin rights. Under the new law Miller at once started suit against the principal offenders. The "lull" was over. But it became increasingly evident that justice would not be done Miller and Whitney in Georgia under any law. On September 4, 1801, Miller wrote to Whitney of a new possibility—their patent rights were to be purchased by the state legislatures. Here was a greater reward than Whitney could have dreamed of! Miller needs Whitney's help and his "contacts." Whitney can not wait and by November 22, 1801 is dating a letter to Stebbins: "Virginia Nineteen Miles North of the Northern line of North Carolina." He sees a chance that Miller and Whitney may get $100,000 from South Carolina alone for his rights—here was freedom from debt, assured financial security, and a credit reputation of the best. Better still, he will have the fame and prestige of a name officially cleared and full credit for his invention. Is it any wonder that a man of Whitney's ambitions and self-interest rushed off to Columbia and left the troubles and problems of arms manufacturing behind? Whitney had slaved and scrimped to get through Yale that he might become respected and financially secure. Now fortune beckoned, and the arms contract could wait.

From the fall of 1801 until Judge Johnson's decision of December 19, 1806, the ups and downs of cotton-gin affairs were certainly far more important in Whitney's mind than the manufacture of arms.[13] This can be definitely established simply by noting the places where Whitney's and others' letters show him to have been in this five-year period. The contents of his correspondence clearly establish a similar conclusion, as does the mere volume of the lawsuits in which Miller and Whitney were engaged. In the final settlement of the partnership in 1818 Whitney was allowed $11,000 for the expenses of six journeys south on these lawsuits. Certainly he was seldom attending full time to the arms manufactory at New Haven.

In short, from 1801 to 1806 Whitney not only failed to fulfill the contract, he regularly substituted long letters of excuse for honest effort to carry out his obligation, while he chased the richer prize of the rewards he expected from the cotton gin. In the light of these facts Dexter can hardly be blamed for his actions, and Jefferson's intervention seems hardly to have been in the interests of the government, whatever effect it may have had upon the future of American industry.

II

Manufacture by the Uniformity Principle

The shortage of skilled artisans in the formative years of the American republic has been so often repeated as the source of Yankee mechanical ingenuity that it is now taken as axiomatic, without careful examination of the actual numbers as adequate for the needs of the day. This same axiom has served to "explain" Whitney's use of manufacture by interchangeable parts. In fact, Whitney so explains it himself. But let us look at the facts. The Springfield Armory was opened in 1794, and its payroll records from the beginning are to be found in the National Archives. By 1802 the Armory had 76 skilled armorers employed, and by 1814 it had 225. Although the figures for Harper's Ferry have not been preserved and we know it to have been substantially smaller than Springfield, we would be safe in assuming that Harper's Ferry had at least half this number of armorers. This total is impressive and seems hardly to indicate a scarcity of skilled armorers. In addition, we have the records of deliveries by other private contractors of arms to the Springfield Armory. During the whole period which concerns us, either the Springfield Armory, or Asa Waters of Sutton, Mass., or Lemuel Pomeroy of Pittsfield, Mass., delivered at least as many arms in each year as did Whitney. In fact Springfield manufactured 16,120 in the six years from 1795–1801, a much more impressive record than Whitney's 10,000 in ten and a half years. Both started from nothing. Leaving out of account the deliveries of the smaller manufacturers, Springfield, Waters, and Pomeroy certainly had an ample supply of armorers—or are we to believe that they, too, had the principle of interchangeable parts which Whitney claimed was unique in his establishment at New Haven?

But where did Whitney get his ideas for manufacture of arms on this new principle? He always claimed that it was his and his alone, and so the legend says, despite strong evidence to the contrary. There can be no doubt that prior to Whitney other men had actually used the principle of manufacture by interchangeable parts. In the 1720's Christopher Polhem, in Sweden, was manufacturing gears for clocks by using machinery and precision measurement to ensure interchangeability.[14] But there is no evidence that Whitney or anyone in the United States knew of Polhem's work, though it could have influenced Blanc in Europe.

The work of Blanc [sic] was clearly known to Thomas Jefferson; in fact our legend always includes a recital of his letter to John Jay in 1785 describing Blanc's work, and Jefferson's letter to Monroe of November 14, 1801, in which he points out that by 1801 Whitney had not developed the method as far as Blanc had in 1788. But the most amazing thing about the Whitney legend is the failure of scholars to follow up this clear lead to answer two questions of first importance: (1) Who was Blanc and what did he do? (2) Did Blanc's work have any influence upon Whitney?

The sources on Blanc are not only easily available, but are very detailed on his methods and results, for much of his work was done in French government arsenals and created controversies which were the subject of several official investigations and reports.[15] Even a cursory examination of these sources would indicate that Whitney was far from being the first to introduce the principle of interchangeable parts in the manufacture of small arms. It is also quite clear that Blanc had carried the technique much further than we have any evidence for Whitney's doing. Furthermore, Blanc's *Mémoire* of 1790 shows a profound understanding of the nature and probable effects of interchangeable manufacture, of which Whitney had only the barest inkling. Whitney's goal was only a system to use unskilled labor to increase output and reduce cost; whatever interchangeability he achieved was only a by-product of his method.

Blanc had problems to meet that Whitney never did. An entrenched officialdom and a threatened craft labor in long established government arsenals, together with the eclipse of the nascent industrial revolution in France under the revolution and Napoleon, prevented a final fruition and spread of Blanc's ideas and methods in France and on the Continent.

But did the spark fly from Blanc to Whitney? Recent research showed that in the Baldwin Collection of the Sterling Library, Yale University, is a letter from Elizer Goodrich to Simeon Baldwin describing a recent meeting at which Whitney was present with Goodrich and Thomas Jefferson, who led a detailed discussion of Blanc's methods and results. The letter is dated January 8, 1801—over eight months before Eli Whitney's delivery of his first 500 muskets to the Springfield Armory! We must also ask whether Whitney's contemporaries in America may have influenced him, in particular the work being done at the Springfield Armory. It is most significant that after signing the contract in June 1798 Whitney had gone to Springfield to see their methods and to talk with the superintendent.[16] And we have Whitney's letter in the summer of 1799 in which he had originally written, "I might bribe workmen from Springfield to come to make me such tools as they have there." It is clear that Whitney was prepared to copy at least some of the methods already in use at Springfield. What were these machines? Unfortunately a fire in 1801 destroyed many of the records of the Springfield Armory, and the question cannot be answered fully. But we do have one official report[17] that gives clear indications that special machinery was in use by, at the latest, 1799: ". . . the artificers were employed for some time on the buildings, instead of on the manufactory, and in making the necessary pieces of machinery and tools. . . ." [If we take into account the difficulties of opening a new establishment, such as] "unsuccessful attempts in the proper construction of the machinery," [we should be satisfied with the present cost of muskets]. The report also uses such expressions as "The

works now being complete, and labor-saving machines operating to great advantage . . . ," and ". . . improvements in
the machinery and system for carrying on the manufactory."

That these improvements in the machines were effective
is shown by the fact that in the month of September 1798
the armory produced 80 muskets, but the following September 1799 it produced 442 muskets. This was accomplished with the same number of workers on the payroll.
The report goes on to state that it had previously required
21 man-days to produce a musket; with the improved
machinery only 9 man-days were needed. This at a time
when Whitney had not yet delivered a single musket!

The later correspondence between Whitney and Roswell
Lee, then superintendent at Springfield, although lacking
technical details, strongly suggests that, contrary to Whitney's claims, at least a simultaneous development was going
on. And there are patents, contracts, and accounts of
Simeon North that strongly suggest that he, too, was using
interchangeability in making his pistols as early as 1807.[18]

John Hall began work on his rifle designed to be made
by interchangeable parts and on machinery to manufacture
it prior to his patent of 1811 and was installing his methods
and machines in the Harper's Ferry armory by 1817. In
1827 Hall petitioned the government to give him adequate
recompense for his contributions. This resulted in a series
of commissions and investigations to establish the facts, by
which he was finally compensated in 1840. The reports of
these boards are matters of public record. The most significant for our purposes is one of 1827—two years after
Whitney's death.

> In making this examination our attention was
> directed, in the first place, for several days, to viewing
> the operations of the numerous machines which were
> exhibited to us by the inventor, John H. Hall. Captain
> Hall has formed and adopted a system of manufacture
> of small arms, *entirely novel,* and which, no doubt, may
> be attended with the most beneficial results to the coun-
> try, especially if carried into effect on a large scale.
>
> His machines for this purpose . . . are used for cut-
> ting iron and steel, and for executing woodwork . . .
> and *differ materially from any other machines we have
> ever seen in any other establishment . . . By no other
> process known to us (and we have seen most, if not all,*

*that are in use in the United States) could arms be made
so exactly alike as to interchange . . ."*[19]) [Italics mine.]

This report was signed by James Carrington and Luther
Sage, who had been government arms inspectors for years
and were thoroughly familiar with the methods in use at
Springfield and in the manufactories of the private contractors, including Whitney's. That Whitney himself thought
Hall's work new is shown by the fact that he made the long
trip to Harper's Ferry to see the "new system being adapted
there."

A later report indicates that the machinery was especially
desirable for it could manufacture "all other species of arms
identically." This later report also shows that the machinery
had been in use since at least as early as 1819 at Harper's
Ferry: "At Harper's Ferry, and at Springfield, this machinery is believed to be *exclusively* used; and the money
expended upon it, and upon the tools at the former armory
from 1819 to 1834, both inclusive, was within a fraction of
$150,000." The commission stated that since Hall was
employed as an armorer at Harper's Ferry after 1819 he
deserved no compensation in addition to his regular pay
for improvements made after that date. But it recommended
that he be compensated for the work he did from 1811 to
1819.[20]

All this can hardly be said to justify our legend's cate-
gorical statement, "In every way Hall profited by Whitney's work."

III

Manufacture by Machinery

We have thus far taken the term "manufacture by interchangeable parts" to have a clear meaning, based upon
Blanc's, Whitney's, and Hall's dramatic demonstrations in
assembling arms out of parts taken at random. This is a
concept, based upon characteristics of the product. It of
course raises the question of how closely the parts must fit
to be interchangeable. The usual answer is that the tolerances allowed must be sufficiently small for the product to
work as designed and no more, since closer tolerances will
merely increase cost. But this is rather vague. A more significant concept of interchangeable parts results from an
examination of the actual methods by which such parts are
produced. In this sense modern interchangeable parts

require these elements: (1) precision machine tools, (2) precision gauges or other instruments of measurement, (3) uniformly accepted measurement standards, and (4) certain technics of mechanical drawing. We do not, of course, expect Whitney to have all these elements, but we can estimate the contribution he may have made by comparing his work to them.[21]

In what sense were the Whitney firearms interchangeable? A test of a number of known Whitney arms in at least one collection proved that they were *not* interchangeable in all their parts! In fact, in some respects they are not even approximately interchangeable! [22] The answer to this paradox is to be found partly in the actual means of establishment of standards for their manufacture. Each of the contractors of 1798 (and the later contractors as well) was given two or three samples of the Charleville model of 1763, and his contract specified that these were to be followed exactly. This method meant that at best the output of one plant would be interchangeable, but the muskets of a given contractor would not necessarily be interchangeable with those of the other contractors. In short, our third and fourth elements of interchangeable parts—uniform standards of measurement, and working from adequately dimensioned drawings—were absent. In fact, they were not to appear for two more generations.

However, the first steps in this direction were to be taken by John Hall. Writing to Congress February 21, 1840, he says:

> And so in manufacturing a limb of a gun so as to conform to a model, by shifting the points, as convenience requires, from which the work is *gauged* and executed; the slight variations are added to each other in the progress of the work, so as to prevent uniformity. The course which I adopted to avoid this difficulty was, *to perform and gauge every operation on a limb from one point,* called a bearing, so that the variation in any operation could only be the single one from that point[23] [Italics mine.]

What about our second element—use of gauges? Polhem had used these, and so had Blanc. There is clear evidence that gauges were being used at the Springfield Armory by 1801. Hall certainly had used them extensively before he went to Harper's Ferry in 1817, but there is not the slightest evidence that Whitney ever did.

A number of visitors went through the Whitneyville plant in Whitney's lifetime. All were properly amazed, but none wrote an account which tells us what Whitney's actual methods were, except that there were "moulds" and "machines." By putting bits of information together, the "moulds" can be interpreted as what would today be called die forging; Blanc had clearly used this method. But "moulds" may also refer to filing jigs. The legend makes much of: (1) the numerous references by Whitney and others to his "machines," (2) the machine tools listed in the inventory made by Baldwin of Whitney's estate, and (3) Whitney's supposed invention of the milling machine.

Let us examine each of these in detail. First, we may ask what did the term "machine" mean in Whitney's day? It most certainly did not mean what it does today. It included a trip hammer and a water wheel, but it also meant almost any kind of device. What machines did Whitney actually employ? In this connection we have the letter of ten-year-old Philos Blake, Whitney's nephew, written after his visit in September 1801: "Thare is a drilling machine and a boureing machine to bour berels and a screw machine and too great large buildings, one nother shop and a stocking shop to stocking guns in, a blacksmith shop and a trip hammer shop and five hundred guns done." This is the only first-hand evidence we have of Whitney's machines at this time. Yet an official inspection of the Springfield Armory in January 1801 says the following: ". . . the number of Files required at the Factory being so great, some Water Machinery is now preparing which will diminish the demand of this expensive article."[24] Even more advanced machinery was used in the national armories by 1817.[25]

Timothy Dwight, one of Whitney's visitors prior to 1823, says: "Machinery moved by water . . . is employed for hammering, cutting, turning, perforating, grinding, polishing, etc."[26] But by this time we have clear evidence that such machinery was in use at both Springfield and Harper's Ferry.

The list of machine tools in the inventory of Whitney's estate is detailed and tells us much about the tools he had in use at the time of his death in 1825, but lists nothing not already in use at Springfield and Harper's Ferry. In fact, the large number of files listed as on hand would suggest that for much of his work Whitney used only a filing jig or fixture to guide a hand-operated file as his principal means of producing uniform parts for the locks of his muskets. But Polhem had done this two generations earlier.

The Whitney papers at Yale also include a number of drawings, none of them dated, signed, or even identifiable as definitely made by Whitney; these drawings are quite possibly those of Whitney's nephews, for Benjamin Silliman, writing in 1832, says: "The manufactory has advanced, in these respects [machinery], since it has been superintended by Mr. Whitney's nephews, the Messrs. Blake, and to them it is indebted for some valuable improvements."[27] They had been in charge for about five years before Whitney's death.

The legend includes one specific machine—the milling machine discovered in 1912 by Professor Joseph W. Roe of Yale, and now in the collection of the New Haven Colony Historical Society. It was identified by Eli Whitney's grandson of the same name as having been made by his grandfather and as the first one ever made. His authority for this identification was that he remembered having seen it as a boy and having been told this story by workmen in the old Whitneyville plant. Roe dated this machine as of 1818 merely because of a statement in the *Encyclopaedia Britannica* that "the first milling machine was made in a gun manufactory in 1818." All this hardly seems adequate evidence. The first reference we have to the use of milling by Whitney is in his letter to Calhoun of March 20, 1823. But by 1818 we have clear evidence that milling was in common use in both national armories, and by at least Robert Johnson and Lemuel Pomeroy of the private contractors.[28]

In short, we really know practically nothing of what Whitney actually had in his manufactory at Mill Rock; what little we do know of was clearly not an innovation; and we have good evidence to show that all that Whitney claimed as his own contribution was at least independently innovated by others, particularly in the national armories. Whitney's claims of originality seem to have been the exact opposite of the truth. Certainly no one is justified in stating that Whitneyville was the site of "the birth of the machine tool industry," much less the birthplace, even in America, of manufacture by means of interchangeable parts.

IV

"The Birth of American Technology"

There can be no doubt that what became by the 1850's widely known abroad as the "American system of manufacturing" had its origin in this first quarter of the nineteenth century and that its principal features were developed in the northeastern section of the United States. The American system included mass manufacture, by power-driven machinery, by machinery especially designed to serve its particular purpose, and by the use of the principle of interchangeable parts.

The legend says that all this stems from Eli Whitney. We have seen enough to indicate that we actually know very little of what he really did; hence there is no clear beginning from which we can tell what later developed from Whitney's work. It is also clear that other men were working along these very lines in the manufacture of arms at the same time.

The legend also claims that from Whitney stemmed the application of the American system of manufacture to many light metal-working industries—Colt and his revolver, Jerome's clocks, Waltham watches, Yale's locks, Singer's sewing machines, and so on. Even if we knew exactly what Whitney did, there is little evidence to support this application of the-great-man-in-history hypothesis. About all that can be said is that further applications of interchangeable parts would *logically* seem to follow from Whitney's broad *claims*. But this is not the same as proof that Whitney actually had methods similar to those of later innovators, much less that they really did derive their ideas and methods from his. Certainly many other men contributed as much or more than Whitney, and evidence for their work can be found, far more convincing than Whitney's boasting claims. The legend says, for example, that the influence of Whitney was the basis of the Colt Armory methods of manufacture. In fact, it was E. K. Root who was the technical genius behind the manufacture of the Colt revolver, and his work stems directly from that of John Hall at Harper's Ferry. Whitney's influence on the manufacture of clocks, watches, and sewing machines is equally open to question.

We know so little of Whitney's actual methods of manufacture that his contribution to interchangeable parts is difficult to assess. What little we do know indicates, if anything, that Whitney was on the wrong track anyhow. John Hall's methods can be fairly clearly established, at least sufficiently for us to be sure that modern interchangeable manufacture derives far more from his inventive genius at Harper's Ferry than from Eli Whitney's manufactory at Mill Rock.[29] Actually one is led to find the origins of the "American system of manufacturing" in the culmination of a number of economic, social, and technical forces[30] brought to bear on manufacture by several men of genius, of whom Whitney can only be said to have been *perhaps* one.

V

Conclusion

This analysis of the Legend of Eli Whitney and Interchangeable Parts raises more questions than it answers. We have by no means arrived at the truth about the legend, much less about the advent of manufacture by interchangeable parts. However, I hope it is clear that the whole question needs re-examination—a more critical analysis of presuppositions and of the evidence which is known, and a more careful search for other sources.

But why not let this nice convenient legend go on? Were it Whitney alone that concerns us, that might be well enough. But the issue is larger than that. The history of our industrial growth is of first importance to the understanding of our American heritage. That industrial development cannot be properly understood without careful consideration of its technological basis. Therefore the true story of the "Birth of American Technology" is of prime concern to us. We should make certain that the baby is perfect and legitimate.

REFERENCES

[1] J. Mirsky and A. Nevins, *The World of Eli Whitney* (N. Y., 1952); C. M. Green, *Eli Whitney and the Birth of American Technology* (Boston, 1956).

[2] B. Silliman in *American Journal of Science & Arts,* XXI (Jan. 1832), p. 256.

"The private establishment of Mr. Whitney has proved a model for the more extensive manufactories which are the property of the nation. Into them, as the writer of the foregoing article has stated, and as I have been informed by Mr. Whitney, his principal improvements have been transplanted, chiefly by the aid of his workmen, and have now become common property." See also Whitney's numerous letters to government officials, e.g., EW to Wolcott, July 30, 1799; EW to Stebbins, April 26, 1800; Wadsworth to Wolcott, Dec. 24, 1800; Goodrich to Baldwin, Jan. 8, 1801; EW to Stebbins, Oct. 15, 1803; EW memo to War Dept., June 29, 1812; EW to Irvine, Nov. 25, 1813; EW to Calhoun, July 9, 1821; EW to Lee, Aug. 2, 1824.

[3] D. Olmstead, *Memoir of Eli Whitney, Esq.* (New Haven, 1846). First published in *Am. J. of Sci. & Arts*, XXI (Jan. 1832).

[4] New Haven Colony Historical Society Papers, Vol. V, p. 123.

[5] Miller to Whitney, Dec. 2, 1796 and May 11, 1797.

[6] Whitney to Wolcott, May 1, 1798.

[7] *The Report and Estimate of Tench Coxe, Purveyor of Public Supplies,* June 7, 1798.

[8] Whitney to Baldwin, May 27, 1798.

[9] Whitney to Stebbins, Nov. 27, 1798.

[10] Whitney admitted as much in his letter to Wolcott of Oct. 17, 1798, and again June 29, 1799. He had already admitted in a letter to Stebbins: "I have now taken a serious task upon myself and I fear a greater one than is in the power of any man to perform in the given time—but it is too late to go back."

[11] Note that when Whitney had his plant fully established and his contract of 1798 was about to run out, he wrote to Major Rogers (Oct. 28, 1809) looking for a new contract: "I would contract to deliver 2000 or 2500 pr year for seven or ten years." Experience had sobered his promises.

[12] Springfield Armory Records, National Archives.

[13] EW to Stebbins, Oct. 15, 1803.

[14] C. Polhem, *Patriotiska testamente* (1761).

C. H. König, *Inledning til mechaniken och bygningskonsten jämte en beskrifning öfwer atskillige af framledne . . . Hr Polhem opfundne machiner* (1752).

Christopher Polhem Mimesskrift Utgifren af Svenska Teknologföreningen (Stockholm, 1911), pp. 121–169. See also the machines, models, and drawings of Polhem in the collections of the Tekniska Museet, Stockholm.

(The author is preparing a summary of the principal features of Polhem's work in this and other important advances in mass production.)

[15] H. Blanc, *Mémoire Important sur la fabrication des armes de guerre, à l'Assembleé Nationale* (Paris, 1790).

Le Roy, *et al.,* *Rapport fait à l'Académie Royale des Sciences, d'un Mémoire Important de M. Blanc* (Paris, 1791).

Comité Central de l'Artillerie, Archives, 6-c-5: 6th Division, 1st Chap. File 6202. (These records are located at the Laboratoires Central et Ecoles de l'Armament, Fort de Montrouge, Arcueil [Seine] and are quite extensive and complete.)

For a convenient summary see W. J. Durfee, "The First Systematic Attempt at Interchangeability in Firearms," *Cassier's Magazine,* V (Nov. 1893–April 1894): 469–77.

For the later history of Blanc's methods see:

H. Cotty, *Mémoire sur la Fabrication des Armes Portatives de Guerre* (Paris, 1806) (Copies are at the U.S. Military Academy, West Point, N. Y., and the Bibliothèque Nationale.)

and M. Bottet, *La Manufacture d'Armes de Versailles, Boutet Directeur-Artiste* (Paris, 1903).

(Working independently Prof. John E. Sawyer of Yale University, now president of Williams College, had by 1959 already completed a study of Blanc. His project is now being broadened and is expected to result in publication and translation of the significant sources and a detailed analysis of the Blanc record.)

[16] Whitney to Wolcott, June 29, 1799.

[17] *Executive Documents, 6th Congress, 1st Session,* January 7, 1800. "Springfield Armory," passim.

[18] See J. E. Hicks, *United States Ordnance* (Mount Vernon, N. Y., 1940), Vol. II, Chap. VII, where much of the material is reprinted.

[19] *Reports of Committees, No. 375, 24th Congress, 1st Session* (1836), John Hall Petition, Document Number 4, "Extracts from Report of Board of Commissioners to the Colonel of Ordnance in 1827 to examine the machinery," pp. 8-9.

[20] *House Reports No. 453, 26th Congress, 1st Session* (1840), John H. Hall Petition, U.S. Document 671.

[21] For a more complete analysis of the concept of interchangeable parts and for a detailed history of its development see the author's forthcoming monograph *History of Shop Precision of Measurement and Interchangeable Parts.*

[22] C. E. Fuller, *The Whitney Firearms* (1946), p. 156.

[23] *House Reports, No. 453, 26th Congress, 1st Session* (1840), John H. Hall Petition, U.S. Document 671, p. 6.

[24] Lt. Col. John Whiting, *Report on Inspection of Springfield Armories,* to Hon. William Eustis, Esq., Secretary of War, dated Boston, January 13, 1801. Report Number Two.

[25] *American State Papers, Military Affairs, 17th Congress,* No. 246, "Armory at Springfield, 1795-1817."

[26] T. Dwight, *Travels in New England and New York* (Edinburgh, 1823).

[27] Silliman, *loc. cit.*

[28] For further details see R. S. Woodbury, *History of the Milling Machine* (Cambridge, Mass., 1960).

[29] For this story see the author's forthcoming monograph *History of Shop Precision of Measurement and Interchangeable Parts.*

[30] See J. E. Sawyer, "The Social Basis of the American System of Manufacturing," *Journal of Economic History* (1954), especially pp. 369–379.

Brandywine Borrowings
From European Technology

Norman B. Wilkinson

One day in 1797 the French traveler Rochefoucault-Liancourt was shown around a large, modern merchant flour
mill on Brandywine Creek in Wilmington, Delaware, by
Thomas Lea, one of the owners. The visitor's attention was
called to a number of recently installed automatic devices
that moved the grain and processed the flour through successive operations with scarcely a touch of human hand.
Lea commented with satisfaction on these improvements,
the work of the Delawarean Oliver Evans, an "ingenious
mechanician," and he apparently indulged in some quiet
Quaker boasting about the advances being made in American milling technology.

When Liancourt put down his impressions of this visit he
wrote:

> Like a true American patriot, he [Thomas Lea] per-
> suades himself, that nowhere is any undertaking exe-
> cuted so well, or with so much ingenuity, as in Amer-
> ica; that the spirit, invention, and genius of Europe, are
> in a state of decrepitude . . . whilst the genius of
> America, full of vigour, is arriving at perfection.

Such patriotic enthusiasm, characteristic of many Americans he had met, continued Liancourt, did not prevent

*Dr. Wilkinson, of the Hagley Museum of the Eleutherian
Mills-Hagley Foundation in Wilmington, Delaware, pre-
sented this paper at the Fourth Annual Meeting of the
Society for the History of Technology in Washington, D.C.,
December 1961. Published in the Winter 1963 issue of*
Technology and Culture *(Vol. 4, No. 1): 1–13, this article
was awarded the Abbott Payson Usher Prize of the Society
for the History of Technology.*

them, however, from adopting all the good inventions developed by Europeans by which they could improve their mills.[1]

It is our purpose here to illustrate how certain industries of the lower Brandywine Valley, in the quarter century between Hamilton's *Report on Manufactures* of 1791 and the passage of the first protective tariff in 1816, were under obligation to Europe for the methods, the machinery, and the "know-how" in the operations of their mills. This can only be a sampling, for by Liancourt's own estimate, there were some sixty to eighty mills on the stream in the 1790's, "almost all of different descriptions, such as paper, powder, tobacco, sawing, fulling, and flour mills. . . ."[2]

We are inclined to regard this figure of sixty to eighty mills as an exaggeration, but that the four or five miles from Wilmington upstream was one of the nation's most concentrated industrial areas is borne out by the remarks of many other visitors and by the attention given it in such promotional publications as Morse's *Geography* and *Gazeteers* and in Niles' *Register*. In his *American Gazeteer* of 1804 and 1810 Morse noted that "Wilmington and its neighborhood are probably already the greatest seat of manufactures in the United States."[3]

1. Textiles and Textile Machinery

Although we cannot identify him, it seems likely that Samuel Slater had a counterpart on the Brandywine: another English textile workman with a keen memory and mechanical talent, employed by Jacob Broom, a signer of the Constitution from Delaware. Broom was a wealthy entrepreneur, a banker and land owner, with a hand in many commercial ventures; in 1795 he established a carding and spinning mill about four miles north of Wilmington. His factory was described as "very Compleat, all the carding, woofing and spinning is done by water and machines which are excellent in their performance . . . very few excellent Cotton Stuffs are imported as those made here. . . ."[4] All the machines and implements were well made and constructed on Arkwright's plan. Broom employed fifteen workmen, all Englishmen, and fifty more were expected.[5]

This effort to duplicate what Samuel Slater and Moses Brown were achieving at Pawtucket, begun only a few years earlier, was short-lived. Broom's mill was totally destroyed by fire early in February 1797, with a loss of

$10,000. The Delaware legislature granted him the right to conduct a lottery to raise money for rebuilding, but after some unsuccessful efforts Broom abandoned the venture. In 1802 his mill property was sold to E. I. du Pont, and on the site du Pont erected his powder factory between 1802 and 1804. Not long after his own mills began producing powder, du Pont, writing to his father, Pierre Samuel du Pont de Nemours, in France in October 1804, explained that he was compelled to write in code because

> the greatest danger to my business is that of attracting the attention of the English. . . . They employ all possible means to prevent the establishment of manufactures here. They burned my predecessor's cotton mill, and might easily try to do the same to my mills.[6]

If du Pont's charge was true—and we have come across no corroborating evidence in other sources—the English were adding arson and peacetime sabotage to their restrictions on the migration of mechanics and the export of machines and machine drawings in their efforts to retain their industrial leadership and monopoly. But du Pont was an Anglophobe with good cause to dislike the English.

The ease with which English artificers evaded the ban on migration is illustrated in the careers of George and Isaac Hodgson, mechanics from Manchester, who established a cotton spinning machine works a short distance below the du Pont mills in 1811. Declaring themselves to be farmers, an occupation not on the banned list, they wrapped their tools, marked them as fruit trees, and then took passage to Ireland, presumably their new home. From Belfast they sailed to Philadelphia, sending their "fruit trees" by another vessel, however, to lessen the danger of being apprehended by port officers as illegally migrating mechanics. The Hodgsons made it safely, claimed their tools in Philadelphia, and were soon operating the Brandywine Foundry making spinning machinery.[7]

The year 1811 was an opportune time to get into this business, and the Hodgsons quickly earned a reputation as manufacturers of textile machines. John Vaughan of Philadelphia and E. I. du Pont recommended them to General Thomas Pinckney when he was considering entering the textile business in South Carolina in 1812. And Charles Willson Peale, artist, museologist, and manufacturer, at du Pont's suggestion, apprenticed two of his sons, Titian and Franklin, to the Hodgsons to learn the machine trade in preparation for careers as cotton manufacturers in the Peale factory at Belfield, near Philadelphia.[8]

Nearby, a little distance upstream, and using some of the Hodgson machines, was the spinning mill of Duplanty and McCall. This was one of several Brandywine textile mills established at the instigation of E. I. du Pont during the War of 1812. It was erected just outside his lower powder mill yard and was managed by a Frenchman, Raphael Duplanty, a former du Pont accountant, and Robert McCall of the Philadelphia merchant and banking family of that name.

For several years prior to the founding of this business du Pont had contemplated both a cotton mill and a woolen mill as parts of a diversified industrial community he hoped to create in the vicinity of his powder mills. In 1809 he had written to a relative in France, one Pierre Abraham Pouchet-Belmare, owner of a textile mill in Rouen, seeking information about some new French spinning and weaving machines which were reputedly very efficient and economical to operate. He asked Belmare to send him drawings or models of these machines and to give his opinion of them. Du Pont could not afford to experiment; he wanted machines that had already proved satisfactory in the French mills. He assured his cousin that the prospects for going into the textile business here in America were tremendous, particularly for one so well versed in the business as was Belmare. "What would you think of using your own knowledge of this industry for starting a factory here for yourself or one of your children? I am sure that such an establishment would be most successful."[9]

To his father, his confidante and adviser, du Pont remarked that most of the machines used for carding and spinning wool in America were English types, but he wanted to learn more about the newly invented French machines, which might prove preferable. Their adoption in America would not harm the French textile industry, for French textiles did not sell in this country; the preference was for English merchandise, from habit, and because English prices were lower. He believed,

> It is only the English commerce that American manufactures can hurt. This country is now ready for manufacturing; with the industry and activity natural to Americans they will soon make for themselves whatever they most need; if we do not begin others will; in any case the manufactures will be established.[10]

When his cousin replied, Pouchet-Belmare graciously declined the invitation to come to the Brandywine to set up a textile mill on French principles. His father had built up the family business in Rouen over twenty years; Belmare was now sole owner, and "When a man is forty-five years old he should not leave his own country or start new enterprises." But he did give a careful opinion of some of the machines about which du Pont had inquired, negative and discouraging. He recommended other types, widely used in both France and England, and offered to send several to du Pont. Also, he would try to find a good experienced textile man willing to go to America as the foreman in du Pont's contemplated mill. "And," he offered, "if you . . . want to send any of your children to France to learn our methods of manufacturing and you will trust them to me, be quite sure that I will do all in my power to help them. . . ."[11]

One outcome of this correspondence, delayed some years, was the founding of the Duplanty and McCall cotton factory. A number of French textile workers were employed at spinning fine yarn, from number 80 thread upwards to 110, and a French couple was placed under contract to construct, operate, and teach others to operate French stocking frame machines. The firm planned to do its own bleaching and for this purpose tried to engage persons in France to come to America "to take charge and establish a Berthaleinne laundry," which we believe meant the installation of Berthollet's bleaching process, first announced in 1785.[12] Postwar conditions and the competition from imported textiles forced Duplanty and McCall into bankruptcy in 1819, owing over $82,000.[13]

2. The Tanning Industry

A long-established Delaware industry that E. I. du Pont believed could be improved by the application of European processes was the leather industry. After learning something of the methods used in the local tanneries he declared them very crude and the labor very dear. In some French technical journals he had read accounts of Armand Seguin's new process of tanning that purported to reduce the time required to tan sole leather from two years to three or four months. The descriptions were rather general, so du Pont wrote his father requesting him to investigate the Seguin process, to find out how widely it was being used in France,

whether it was economical, whether any recent improvements had been made in it, and whether Seguin had made
a fortune out of it.[14] His father complied, sent him several
bulky memoranda, and offered this opinion about American
leather:

> It is certain that your Americans make a great deal of
> leather and make it badly: it absorbs water like a
> sponge and does not wear at all. That is why it pays
> to send shoes, even poor shoes, from Europe as cargo.[15]

His son, however, thought it would pay him to set up a
tannery using the French process, somewhere adjacent to
his powder mills.

It was not until 1815, when the elder du Pont fled France
upon Napoleon's return and came to live with his son at
Eleutherian Mills, that the tannery was begun. It was located between the powder mills and the Duplanty, McCall
Cotton Factory, under the direction of Alexandre Cardon
de Sandrans, a young Frenchman who had been companion
and secretary to the senior du Pont in Paris. M. Chenou,
a tanner experienced in the Seguin process, supervised the
operations for the first year until Cardon became thoroughly
acquainted with the procedure.

The secret of the Seguin process was the addition of carefully measured amounts of sulphuric acid to the tanning
solution. This accelerated the penetration of the tannin into
the hide, thus sharply reducing the number of months the
hides had to lie soaking in the tanning vats. Another aspect
of the process was a different arrangement of the hides and
the bark, and the periodic shifting of the hides in the vats.

In the first flush of successful production Cardon and the
du Ponts felt assured their fortunes would be made by the
tremendous advantage this saving of time and labor costs
gave them. But tanning by the Seguin method sacrificed
quality; though the outer part of the hide was tanned, not
enough time was allowed for the tannic acid thoroughly
to penetrate and tan the inner part. The result was an inferior grade of leather. There were markets for cheaper
leather, but in the 1820's the Cardon tannery fell upon
hard times, induced by general depressed business conditions and by mismanagement within the firm. The enterprise had a spotty career, bolstered from time to time by
fresh capital supplied by E. I. du Pont, but it ceased operations, bankrupt, in 1826.[16]

3. The Powder Mills

The black powder mills that E. I. du Pont erected between 1802 and 1804 were in every respect modeled after those in France. Explosives manufacture in France was a government monopoly under the direction of the Régie des Poudres, with headquarters and laboratory at the Arsenal in Paris and mills in various locations throughout France. While still in his teens, from 1787 to 1790, du Pont had been employed in the Paris Arsenal, then directed by his father's friend, Antoine Lavoisier, after which he had worked for about a year in the mills at Essonne.

From 1791 to 1799 du Pont had no connections with powdermaking; he was assisting his father in the management of a printing and publishing business in Paris. Briefly, in 1794, he was called upon to direct the gathering and refining of saltpeter in one of the districts of Paris, but this was a wartime emergency assignment of brief duration.

When the du Pont family, father and two sons and their families, thirteen in all, first came to the United States in 1799 their plans included neither powdermaking nor publishing as means of livelihood. The father had ambitions to found a colony somewhere on the near-frontier, in western Virginia or Kentucky, based on the philosophy of the Physiocrats, a school of economists of which he was a most eloquent exponent. To this end he raised capital among French sources and organized the Compagnie d'Amérique, with himself at its head, and his sons Victor and Eleuthère as his assistants.

Upon Jefferson's advice the settlement project was shortly given up, and among the dozen or so alternate ventures contemplated for the fruitful investment of their limited funds the "eighth plan" called for the establishment of a powder factory to be directed by the younger son, Eleuthère.[17]

English and Dutch gunpowder nearly monopolized the American market in 1800. The few American mills produced small amounts, and the quality was reputedly not very good. After visiting the mills of Lane and Decatur in Frankford, north of Philadelphia, du Pont noted that the head workman was a Dutchman who was using a process a half century or more out of date, a wasteful method that produced costly but poor-quality powder. Du Pont was certain that using the most recent techniques developed in France, and with French equipment, he could make powder superior to the American and equal in quality to any of the imported.[18]

To prepare himself, du Pont returned to France in 1801 and spent three months "brushing up" on powdermaking technology. He consulted with the heads of the Régie des Poudres, MM. Bottée and Riffault, and the superintendent of the Essonne mills, M. Robin. Methods were discussed; drawings of mill structures and machinery were made available to him for copying; some equipment was obtained from the Arsenal, and orders were left for other items to be made and sent on later. Approximately $4,000 were spent for drawings, castings, eprouvettes, utensils, sieves, presses, laboratory items, and small quantities of French powder. The French authorities also promised they would try to persuade one of their more experienced powdermen to come to America to assist du Pont in getting his mills established.[19]

In this manner du Pont refreshed himself on the improvements and newer processes that the French had developed since he had left Essonne back in 1791. The willingness of the government officials to aid him so generously was due in part to the prestige and associations of his father in high places, but possibly more—as du Pont himself expressed on several occasions—to the realization by the French that a flourishing American powder producer could hurt the English in their gunpowder trade with the United States.[20] The U.S. Navy was also of assistance, for some of the heavier machinery ordered in Paris was brought over later in the year in the sloop-of-war *Maryland,* commanded by Captain John Rodgers.[21]

During the months he was constructing his mills du Pont corresponded frequently with Bottée, Riffault, and Robin, getting a great deal of additional technical information and encouragement. His appreciation was shown by sending them boxes of seeds, shrubs, plants, and small trees for propagation in France.[22] Du Pont had a keen interest in horticulture, and if the choice of vocation had been wholly his own there is good reason to think he would have been a botanist—the occupation he entered on his passport when he had left France in 1799.

Du Pont de Nemours, the father, returned to France in 1802, and by him Eleuthère was kept supplied with technical journals containing articles on explosives—*Bulletins des Sciences, Annales de Chimie,* the *Lycée des Arts,* and the *Journal de L'Ecole Polytechnique.* Treatises, memoranda, and government reports dealing with refining techniques, the making of charcoal, and trials of new machinery were sent to him. In 1811 Bottée and Riffault published a two-volume study entitled *L'Art de Fabriquer la Poudre à Canon;* at his son's urgent request—"its many details of the improvements made in France may be most useful to me" —the elder du Pont sent him two copies of this highly prized work. Without doubt it served as his manual of instructions in his own Brandywine mills. Among du Pont's papers is a lengthy bibliography of articles on gunpowder and gunnery that appeared in the *Transactions* of the Royal Society of London from 1665 to 1800, an indication of his interest in the earlier and the contemporary developments in British powdermaking.[23]

In 1804, after receiving descriptions of an improved French granulating machine, du Pont obtained a patent on a granulating machine that may or may not have been patterned after the French one; we have the description of his, but not of the French machine.[24] One of his powder agents, writing some time later about a patent granted a rival New York powder manufacturer, made this apt remark: "Mr. Rogers' discovery is like many others, a European discovery brought to light here."[25]

Du Pont's success was in large measure due to his diligence in acquiring knowledge of the best methods in use in countries where black powder had been manufactured for over 500 years. Verification of this is found in one of his periodic reports to his father in which he stated, "In the midst of all my work I have neglected nothing that might facilitate or improve the manufacture."[26] The results of his efforts may be discerned in Secretary of the Treasury Galllatin's praise of the Brandywine factory in 1810 as "the most perfect establishment for making powder in America."[27]

4. Papermaking

Most cosmopolitan of the Brandywine manufacturers was Joshua Gilpin, a downstream neighbor of du Pont and member of the Philadelphia Gilpin merchant family. At the suggestion of Benjamin Franklin he and his uncle, Miers Fisher, had converted a Brandywine snuff mill into Delaware's first paper mill in 1787, and here made paper of many types for a variety of purposes for fifty years.[28] Franklin loaned them some French works on papermaking and received from Gilpin some of the first paper made for his use with a request for his opinion of its quality. Brissot de Warville was asked by Fisher during the first year of the mill's operations if he would assist by trying to secure some paper moulds and some experienced workmen from France, since Fisher and Gilpin had not been successful in obtaining men and moulds from England. De Warville praised the Gilpin-made paper as "equal to the finest made in France," both for writing and printing.[29]

With the mills in good running order under the supervision of his younger brother, Thomas Gilpin, who had come into the business in the 1790's, Joshua Gilpin took off on an extended "Grand Tour" of the British Isles and the Continent for a six-year stay beginning in 1795. He made a subsequent, shorter visit from 1811 to 1815. Gilpin was a man of limitless curiosity who wanted to further his education beyond the limits of the grammar school he had attended. His library suggests a man of catholic tastes—a desire to be a "universal" man who knew something about everything—and one who sought the "polish" that came from extensive travel and contacts with interesting people and places. He privately published a volume of poetry and wrote an essay on a more mundane subject of internal improvements, the Chesapeake and Delaware Canal, a project promoted by the Gilpin family.

His practical side as merchant and manufacturer is displayed in some 60 pocket-size memoranda books that he filled with notes and sketches made during his prolonged visits abroad. Possibly representing himself as more merchant than manufacturer, Gilpin seems to have had no difficulty, despite England's "closed door" policy toward foreigners ferreting out its manufacturing secrets, in gaining access to factories, in talking with owners, managers, and mechanics, and in recording what he learned; he made notes on processes, sketched installations and equipment, and commented on their relative merits. Connections with British merchants, his numerous Gilpin relatives in England, and letters of introduction may have gained for him access that would have been denied others less well-connected. The value of Gilpin's voluminous notes lies in the wealth of technical information they contain; in total, they offer a close look at industrial Britain at the end of the eighteenth century.[30]

What elements of British papermaking technology did Gilpin's diligent inquisitiveness allow him to "borrow" for adoption in his own mills? First, he acquired the services of a former paper mill owner, Lawrence Greatrake, who came to Delaware about 1801 to superintend the Gilpin mills. Another Englishman, Thomas Oakes, who had designed and erected mills, was engaged to enlarge and remodel the original Gilpin mill and erect new buildings. In English and Scotch paper mills Gilpin had observed the process of bleaching rags that were pulped for paper stock, and he became convinced that chlorine, rather than muriatic acid and alkalis, was the best chemical for bleaching. In 1804 a chlorine bleachery was installed in his mill, an innovation that marked him as the first American papermaker to use chlorine for taking the color out of rags. According to one authority on the American chemical industry, chlorine was not generally adopted by the American textile industry as a bleaching agent until the 1830's.[31]

The most significant "borrowing," one that revolutionized the American paper industry, was the papermaking machine that Thomas Gilpin constructed in 1816 based upon descriptions and drawings of English machines supplied him by his brother Joshua and by Greatrake. Joshua had familiarized himself with the development of the endless woven belt machine invented in France in 1798 by Louis Robert and subsequently perfected by Bryan Donkin and John Hall, who installed it in the Apsley paper mill of the Fourdrinier brothers in 1804. A second machine employing a wire-covered cylinder was invented by John Dickinson in 1809 and put into operation in Dickinson's Nash Mill. Gilpin and Greatrake visited both mills, and after talking with the owners and seeing the machines operate, realized their possibilities—"Whoever first gets one to work in America . . . will open a sure road to the fortunes of all concerned . . ." enthusiastically wrote Greatrake. Up to a point, Dickinson and the Fourdriniers were free with technical information, but certain essentials were not disclosed. To obtain these the Americans used various means, some proper, others suggesting bribery and "pirating," and got together enough data and drawings from which Thomas Gilpin, the mechanically talented member of the firm, could build his endless papermaking machine—a close resemblance to Dickinson's—which he patented in December 1816 and put into production in February of the following year.[32]

In place of a single sheet of paper of limited size made by the slow hand method, Gilpin's new "wonderworking" machine turned out paper of smooth texture and excellent quality in a roll of any length, at greater speed, and with lower labor costs. It was estimated that the new machine, tended by two men and one boy, could produce as much as the old hand method did with twelve men and six boys, saving $6,000 to $12,000 yearly in wages.[33] It should also be noted that making paper in rolls of great length brought about subsequent changes in the printing trade and the publication of newspapers, periodicals, and books.

The Gilpins enjoyed the monopoly of the endless paper-making machine for about five years, but in 1822 John Ames, a Springfield, Massachusetts, papermaker, inveigled construction details of the machine from a former employee, and about 1827 English machines of the Fourdrinier type were introduced into America.

Conclusion

The migration of industrial technique illustrated by these "borrowings" in the Brandywine mills is but one of numerous streams of influence from Europe that shaped early American society. Our technological indebtedness has not, however, to the best of our knowledge, received the attention that has been given to the cultural transitions of law, religion, education, political philosophies, and the fine arts from the other side of the Atlantic. Possibly our preoccupation with our later inventiveness and eminence as the world's leading industrial nation has made it easy to overlook the first factory era when we were an apprentice nation, learning, imitating, and sometimes improving upon the machine technology that had been developing in Western Europe since the mid-1700's.

If the industrial history of the Brandywine was typical, it follows that similar studies of the half dozen or more river valleys where the first factories appeared would provide a much more comprehensive picture of our technological debt to Europe. I believe we would conclude that Thomas Lea was wrong—the spirit, invention, and genius of Europe were far from decrepit; and Liancourt was right —Americans *were* adopting all the good inventions of Europeans by which they could improve their mills.

REFERENCES

[1] Duc de la Rochefoucault-Liancourt, *Travels Through the United States of North America, 1795, 1796, 1797* (2nd ed.; London, 1800, 4 vols), Vol. 3, pp. 496–497. Hereafter cited as Liancourt, *Travels.*

[2] Liancourt, *Travels,* Vol. 3, pp. 492–493.

[3] Jedidiah Morse, *The American Gazeteer* (Boston, 1804, 1810). Under *Delaware.*

[4] Thomas Rodney, "Propositions for new & Useful Inventions & Improvements by Thomas Rodney Esq. Philosophical Tracts and Journals, July 17 to August 20, 1795." Historical Society of Delaware.

[5] Liancourt, *Travels,* Vol. 3, p. 502.

[6] B. G. du Pont (trans. and ed.), *Life of Eleuthère Irénée du Pont, from Contemporary Correspondence* (Newark, Delaware: 1923–1927; 11 vols. and index), Vol. 7, p. 15. Hereafter cited as *Life of E. I. du Pont.*

[7] Memorandum Book of Deborah Hodgson Jones, privately owned. For details on English restrictions and the industrial immigrant see Herbert Heaton, "The Industrial Immigrant in the United States, 1783–1812," *Proceedings of the American Philosophical Society,* Vol. 95, No. 5 (October 17, 1951), pp. 519–527.

[8] John Vaughan to E. I. du Pont, November 12, 1812, Eleutherian Mills Historical Library; hereafter cited as E.M.H.L. Charles W. Peale to John de Peyster, August 8, 1813, Calendar to C. W. Peale Letter Books, Letter Book XII, 102, American Philosophical Society.

[9] *Life of E. I. du Pont,* Vol. 8, p. 130.

[10] *Ibid.,* pp. 144–145.

[11] *Ibid.,* pp. 160–166.

[12] Du Pont Family Manuscripts, V, E.M.H.L. Anthony and Marie Ravigneaux were engaged in May 1816 to work "at the construction of stocking machinery, to direct the operation of the said machines and to train for the operation of the said machines apprentices," and, "to reel, double, sew, and to hot-press the stockings as well to instruct the apprentices in this work." The combined wage paid to husband and wife was $2.00 a day; they did not remain long in the employ of Duplanty and McCall.

[13] Duplanty, McCall Co., Final Accounts with E. I. du Pont de Nemours & Co., Old Stone Office Records, E.M.H.L.

[14] *Life of E. I. du Pont,* Vol. 6, pp. 211–212.

[15] *Ibid.,* Vol. 8, p. 258.

[16] See Peter C. Welsh, "A. Cardon and Company, Brandywine Tanners, 1815–1826," in *Delaware History,* Vol. VIII, No. 2 (September, 1958), pp. 121–147.

[17] *Life of E. I. du Pont,* Vol. 5, pp. 191–192.

[18] *Ibid.,* pp. 199–200.

[19] *Ibid.,* pp. 212–225, 230, 233–234.

[20] *Ibid.,* Vol. 8, pp. 102, 145. In a letter to his father, October 1, 1808, du Pont said, "The greatest harm that can be done to the English is to destroy their trade; the only way to accomplish that in this country is to establish manufactures that will rival theirs. The French manufacturers will never be able to overcome American prejudice and habit and it is only by American industries that English can be fought. This truth was felt

in France before my journey in 1801 and secured for me all the help that I found there."

[21] Shipping Manifest, dated July 13, 1801, in "Receipted Bills, 1802–1803," E.M.H.L.

[22] *Life of E. I. du Pont,* Vol. 5, pp. 253, 355, 359, 362–365; Vol. 6, pp. 158–159.

[23] *Ibid.,* Vol. 9, pp. 64–65.

[24] *Ibid.,* Vol. 7, pp. 31–35. This was patent No. 590, dated November 23, 1804.

[25] John Vaughan to E. I. du Pont, December 3, 1818, Old Stone Office Records, E. I. du Pont de Nemours and Company, E.M.H.L.

[26] *Life of E. I. du Pont,* Vol. 8, p. 28.

[27] *American Watchman,* June 2, 1810.

[28] See Harold B. Hancock and Norman B. Wilkinson, "Thomas and Joshua Gilpin, Papermakers," in *The Paper Maker,* Vol. 27, No. 2 (1958); also by the same authors, "The Gilpins and their Endless Papermaking Machine," in *The Pennsylvania Magazine of History and Biography,* Vol. 81, No. 4 (October, 1957).

[29] Dard Hunter, *Papermaking in Pioneer America* (Philadelphia, 1952), pp. 83–84; Miers Fisher to Brissot de Warville, November 25, December 11, 1788, Letters to Brissot de Warville, Scioto and Ohio Land Co. Papers, New-York Historical Society; J. P. Brissot de Warville, *New Travels in the United States of America* (London, 1794), Vol. 1, pp. 362–363.

[30] Joshua Gilpin's Journals are in the Pennsylvania State Archives, Harrisburg, Pa. Parts of his journals commenting on a dozen or more industries have been edited and published in the *Transactions* of the Newcomen Society, Vols. 32 and 33.

[31] Sidney M. Edelstein, "Origins of Chlorine Bleaching in America," *American Dyestuff Reporter,* Vol. 49, No. 8 (April 18, 1960), pp. 39–48; Williams Haynes, *The American Chemical Industry* (New York, 1954; 6 volumes), Vol. I, p. 140.

[32] "Richard A. Gilpin's Paper Making Book," Gilpin Collection, Historical Society of Pennsylvania. Scattered through the Gilpin Collection are memoranda with headings, "Dickinson's Machine," "The New Improvements in Paper Making," and "Ideas Relative to a Contemplated Improvement in the Paper Mill at Brandywine Generally," which bear upon the efforts to obtain information about the English machines.

[33] *American Watchman,* March 4, 1818.

Henry Ford's Science and Technology for Rural America

Reynold Millard Wik

Henry Ford profoundly influenced rural life in America during the first half of the twentieth century. As father of the famous Model T car and Fordson tractor, he became a folk hero, a secular saint among people possessing a strong admiration for the self-made man. The editor of the *Rural New Yorker* in 1923 observed a world full of stories about Ford whenever people met and the air smelled of gasoline.[1] "Another Abraham Lincoln," mused Edwin Markham; "The Mussolini of Industry," echoed H. V. Kaltenborn.[2] Chauncey Depew tabbed him "conqueror of the front page," while Theodore Roosevelt complained that he got more publicity than the president of the United States. Will Rogers saw him making more history than his critics ever read. The *Prairie Farmer* claimed Ford received more free advertising than any other person in the United States,[3] and a Chicago editor jibed, "One need not mention Ford . . . he mentions himself."[4]

Rural people regarded the "Flivver King" as one of their kind—one born on a farm who retained an affinity for the land. His simplicity made him more at home with agrarians than with the "Organization Men" of his time. Nor were his agricultural interests merely a rich man's hobby reflect-

Dr. Wik is May Treat Morrison Professor of American History at Mills College and a recipient of the Albert Beveridge Memorial Award of the American Historical Association for his book Steam Power on the American Farm. *He has long been engaged in research on Henry Ford. This article, published in the Summer 1962 issue of* Technology and Culture *(Vol. 3, No. 2): 247–58, is based upon a paper read at the 1961 Mississippi Valley Historical Association meeting.*

ing nostalgia for the bucolic life. His home at Fair Lane on the banks of the River Rouge possessed a rural atmosphere with its woods, pastures, gardens, and hundreds of acres planted to wheat, corn, and alfalfa. His farms in Michigan during the 1920's included 7,000 acres. Perhaps the most accurate statement of his views is carved in stone above the entrance to the old administration building of his company on Shaeffer Road, Dearborn:

> Industrious application of inventive genius to the natural resources of the earth is the groundwork of prosperous civilization.

Unquestionably Ford's greatest contribution to the application of technology to American agriculture lay in his insistence that mechanical power should supersede animal power on the farm. He scorned the pro-horse crowd's bragging about the nobility of old gray mares, these hammerheaded nags which devoured much of the best oats and grass in the country. Machines were superior to horse flesh in efficiency, in ease of handling, and in cost per unit of work produced. Besides, you couldn't fix a dead horse with a monkey wrench.

Although 1,400 auto companies and almost as many tractor firms shared in the demise of the horse, the 15,000,000 Tin Lizzies really demonstrated the feasibility of motor transportation in rural regions. The Model T churned through mud roads, crawled along fence lines, and rattled into farm yards with persistent fortitude. Barnyard mechanics cleaned the spark plugs, ground the valves, tightened up the rods, applied blow-out patches, and changed the transmission bands. Farmers attached pulleys to the crankshaft or rear wheels to utilize the twenty-horsepower motor for grinding feed, sawing wood, churning butter, sharing sheep, pumping water, elevating grain, shelling corn, and even knocking the shells off of walnuts. Meanwhile, cream cans and chicken crates were piled in the back seat or lashed to the running boards and carted off to market. The Model T put America on wheels and Ford among the billionaires. This feat became the greatest sociological phenomenon of the early part of the century. Some wits predicted people would be born in cars, live, marry, and die in them, with their bodies cremated by the heat of the engine, all without having put foot on the ground. Facetiously, Will Rogers quipped, "Brigham Young originated mass pro-

duction, but Ford was the guy who improved on it. . . .
He changed the habits of more people than Caesar, Mussolini, Charlie Chaplin, Clara Bow, Xerxes, Amos n' Andy
and Bernard Shaw."[5]

The story of how Ford's mechanical contraptions changed
rural life in America is well known. Not so well known,
however, are his attempts to introduce scientific technology
into American agriculture.

One of Ford's early scientific efforts to improve agriculture consisted of experiments to produce alcohol as a motor
fuel by distilling it from farm crops. When World War I
threatened to create a gasoline famine, he announced in
1915 that alcohol could be extracted from grain and from
garbage. The new Fordson tractor would be designed to
burn alcohol as well as gasoline; thus the supply of fuel
would be unlimited.[6]

Although far from original in these matters, the Ford
pronouncements received wide publicity. Farmers were ju-
bilant. Now grain, sorghum, pine trees, and potatoes would
be converted into industrial alcohol. Many wrote Ford asking what kind of potatoes were best suited for this purpose,
how many should be planted to the hill, should the eyes of
the potatoes in the ground face up or down, and should
they be planted in the light of the moon?

Again, during the depression of 1920–23, farmers complained of low prices for farm commodities while viewing
with alarm the rise in the cost of gasoline. They disliked
selling livestock to pay fuel bills and accused auto manufacturers and petroleum companies of collusion to keep
efficient carburetors off the market, thereby permitting gasoline prices to remain inordinately high. An irate farmer
near Hawley, Minnesota, in 1923 decried his bondage to
Standard Oil Company while he and his neighbors in the
Red River valley faced bankruptcy because potatoes brought
less than the cost of production. Ford should buy the potatoes, convert them into motor fuel, and rescue farmers
from the clutches of the oil trusts.[7]

Interest in this subject subsided until the Great Depression of 1929 revived the notion of using farm crops to provide cheap motor fuels. In spite of the press stories citing
Ford's intentions to meet the farmer's needs, the practical
progress in research proved negligible. Although the Ford
Motor Company experimented with alcohol distillation in
the Engineering Laboratory in Dearborn from 1915 to
1917, the efforts were feeble and inconsistent. When the

United States entered the war in 1917, a secretary answering a correspondent in New Jersey added, "Ford has kinda lost interest in these recent substitutes."[8] Likewise, the British by 1919 had discovered that alcohol could not be sold at a cheaper price than petrol. During the 1930's the publicity Ford received on the alcohol question outweighed actual achievement. While the press quoted Ford's claims that an acre of potatoes could furnish enough alcohol to plow a hundred acres, his secretary, W. J. Cameron, continued to write letters stating that all research indicated that alcohol blends could not compete with petroleum products.[9] The oil companies pointed to Germany where alcohol fuels were burned only because a high tariff kept out foreign petroleum products. As late as 1938 the Ford experiments were still classified experimental.[10]

Automotive officials became interested in efforts to secure a domestic rubber supply in 1923 when Harvey Firestone insisted the British monopoly of crude rubber had doubled the price of this commodity within the last five years and had added $150 million to the annual cost of rubber used in the United States.[11] Henry Ford and Thomas Edison combined their attempts to extract rubber from plants aside from the *Hevea Brasiliensis* rubber tree and other tropical rubber-producing vegetation. William H. Smith directed the research work in Dearborn, where credence was placed in the work of Harvey M. Hall of the Carnegie Institute of Washington, D.C., during the First World War.[12] Hall concluded rubber could be derived from Rabbit Brush and milkweed plants which yielded from 3 to 6 per cent rubber.[13] The Bureau of Plant Industry of the United States Department of Agriculture encouraged these views.[14] When the press announced that Ford chemists would produce rubber from domestic plants, scores of farmers congratulated the "Motor King" for converting weeds into cash. They said they had been fighting milkweeds, pig weeds, sunflowers, and Russian thistles for years. Now they could be sold on the market. All agreed the raw materials were plentiful.

In 1924, Ford and Edison decided to conduct further experiments in Florida. Over 7,000 acres were purchased near La Belle, Florida, and another 22,000 acre tract in Bryan County near Ft. Meyers, where Ford had maintained a winter home.[15] Thomas Edison set up a laboratory on the Cherry Hill plantation near Way Station, Georgia, in 1929 where 70 plant varieties were tested for rubber-bearing qualities.[16] Even though the goldenrod yielded some rubber in a chemical reduction process, the meager results, the advancing age of Edison, and the advent of the depression all combined to lead Ford to abandon the project.

Meanwhile the Fordlandia experiments on a 2,500,000 acre plot on the Tapajos river in Brazil had begun in 1927. Here the Ford scientists learned they could clear the land, build hospitals, control disease, and install sewage and refrigeration plants, but they could not grow rubber trees successfully on plantations. Jungle trees defied domestication, for when cultivated they died from too much exposure to the weather.[17]

Alert to the value of publicity, Ford exploded one of his journalistic bombshells in the spring of 1921 when he exclaimed, "The cow must go." To replace the milk produced by cows he proposed artificial milk made synthetically. His dislike of dairy cows stemmed from unpleasant experiences on his father's farm where milking had been a most disagreeable and exasperating chore. Now he regarded cows as inefficient and unsanitary animals. At times he refused to drink milk. When interviewed by the editor of *Farm and Fireside* in 1925, Ford expounded, "Why should a farmer spend a lot of time taking care of a bunch of cows? It takes 20 days of actual work to grow and harvest the grain crops on a dairy farm. The rest of the time is spent taking care of animals. It is all wrong."[18] When asked why he owned 150 head of excellent Durham and Ayrshire dairy cows and one of the best dairy barns in the state, he replied that these were kept just to prove they were all wrong.

Mixed reactions followed. Members of the Near East Relief Committee rejoiced, thinking the new synthetic milk would save thousands of babies from starvation. Besides, the tuberculosis carried in cow's milk could be avoided.[19] City people hoped the innovation would reduce the cost of living. However, dairy farmers were alarmed. They could be reconciled to the elimination of the horse, but to liquidate the dairy cows seemed unthinkable. Farm journals ran cartoons picturing Ford's grotesque mechanical cow with captions to the effect that these were produced on an assembly line, sold f.o.b. Detroit, and shipped knocked down in crates, but equipped with interchangeable udders.

Yet, it was the soybean which gave Ford his best opportunity to apply science and technology to farming practice. The "Ford and the Beanstalk" story originated as another panacea to fight the depression following the stock market crash of 1929. Rejecting the Townsend Plan, the Ham and Egg formula, the nostrums of Huey Long, Father Coughlin, and the Technocrats, Ford came up with his own self-help solution which called for closer cooperation between industry and agriculture. If industry could use more agricultural products, this new demand would raise prices of farm crops. Since by-products were going to waste, chemists would discover ways to convert wheat, corn, carrots, cabbages, sunflowers, straw, weeds, and corn cobs into products with commercial value.

After some experimentation in the new Edison Institute of Technology in Dearborn in 1930, Ford chemists chose the soybean as the most promising raw material. The beans could be grown in all parts of the country, they were remarkably drought-resistant, and they added valuable nitrogen to the soil.

Of course, the Ford Motor Company did not discover the soybean. Written records reported their use in China in 2838 B.C. Rich in proteins, fats, and vitamins, they provided a well-balanced diet for people in the Far East. Introduced into the United States in 1804, they were first used as cattle forage. During World War I, the Department of Agriculture, when searching for a cheap source of protein, discovered the bean as a soil fertilizer. Interest languished, however, until Ford officials began to advertise its virtues in the 1930's.

The company spent $1,250,000 on soybean research in 1932 and 1933.[20] Three hundred varieties of the bean were planted on 8,000 acres on the Ford farms.[21] In the laboratory a processing machine weighing 6½ tons was designed to extract the oil from the soybeans, first crushing the beans into thin flakes, then exposing them to a gasoline solvent which absorbed the oil. The solvent was recovered again by distillation and reused. Therefore the machine delivered in a continuous process both the soybean oil and meal. The oil made a superior enamel for painting automobiles and yielded a fluid for shock absorbers. The soybean meal containing almost 50 per cent protein could be molded into horn buttons, gear shift balls, distributor cases, window trim strips, and electrical switch assemblies.[22] The window trim on the Ford cars utilized 144,000 bushels of soybeans annually, equivalent to the crop off 7,200 acres of land. Although Ford officials never claimed any original inventions in these processes, their efforts led to expanded use of farm commodities and demonstrated the advantages of cooperation between agriculture and industry.

In 1937 Ford scientists developed a fiber from soybean protein and were credited as the first to spin a textile filament from protein derived from a vegetable source.[23] This fiber resembled a soft wool with a natural crimp and a high degree of resilience, and it could be used for upholstery in automobiles, for felt hats, and for suits and overcoats. *The Progress Guide* in September 1943 stated, "Go into the field and pick yourself a suit, all wool and acres wide."[24] Henry Ford appeared at a convention with his entire costume made from soybeans except his shoes. He envisioned a new fabric four times cheaper than wool, and since his laboratory produced 5,000 pounds of synthetic wool daily, he thought the importation of wool from foreign countries would become unnecessary. He also tried to induce the federal government to use the fiber to manufacture soldiers' uniforms in World War II; but the schemes to develop soybean fiber on a commercial basis failed because the soybean process proved less economical than conventional means.

Additional publicity accompanied Ford's venture into plastics. News photos showed the Flivver King hitting the rear deck of a plastic car body with an ax without inflicting any damage. Ford suggested that soybean plastics might become his most significant contribution to society. Since plastics were molded rather than cast or stamped, production costs could be reduced. Plastics first were formed from the cellulose in cotton or wood pulp, but soybeans, corn, wheat, hemp, China grass, and ramie could all be used for this purpose. Plastic automobile bodies of the future would weigh only 250 pounds, but they would withstand ten times more shock than steel; hence they could roll over without crushing, making them the safest autos on the road. *Business Week* explained that the Ford plastics could be used for making car bodies, windows, and upholstery.[25] However, after demonstrating some of the uses of plastics, the Ford Motor Company discontinued publicizing their work in this field.

Meanwhile, Henry Ford carried out considerable research on foods for human consumption. He had revealed his food faddist tendencies as early as 1915 when he asked Andrew Fruehoff in the metallurgy laboratory to investigate the values of various foods. The construction of the Ford Hospital in Dearborn gave him an opportunity to explain his views on health. He insisted Americans ate three times too much, that gluttony produced a sluggish brain, and that people should eat only when hungry. "Cut down on the rations and you won't need doctors," he admonished. "Eat your dessert first and you won't eat so much," he advised.[26]

At age 52 he demonstrated his own agility by kicking his foot above his head like a chorus girl and running a mile on his birthday.

Following this curiosity, Ford had hired in 1926 his boyhood companion, Dr. Edsel A. Ruddiman, to experiment with foods. Installed in a laboratory with his white rats, Ruddiman was instructed by Ford to prepare a biscuit possessing all the vitamins required for good health. Like a good soldier, the chemist created a biscuit out of soybeans and various chemicals. Ford ate these, even professed to like them, and offered them to friends who were in no position to refuse them. Harold M. Cordell, one of Ford's secretaries, later recalled the soybean all-purpose vitamin biscuit as the "most vile thing ever put into human mouths."[27] On occasion, Ford invited guests to dinner where he served nothing but a soybean menu: soybean soup, soybean bread, soybean beans, soybean croquettes, soybean pie, soybean coffee, and soybean ice cream. The Ford Motor Company published booklets with dozens of recipes explaining how to prepare these delicacies.

The good word spread. Dietitians and home economics teachers testified that soybeans were high in vitamins A and D, that they were almost 100 per cent digestible, that one pound of soybeans contained as much protein as two pounds of beef steak, and that soybean milk proved easier for babies to digest than cow's milk and with a diminished degree of burping.[28] Ruddiman later admitted responsibility for the soybean milk. He simply rubbed the beans with water to produce a white liquid rich in protein, then added constituents to make it equivalent to cow's milk. There was talk of putting it on the market, but this never occurred. However, these projects in the early 1930's gave Ford another chance to reiterate his earlier cry that "The Cow Must Go." Again the days of the cow as the foster mother of the world seemed numbered. The *Southern Dairy Products Journal,* in an editorial entitled "Can the Soybean Say Moo?", wondered if Ford had estimated all his costs when he announced that soybean milk could be produced for three cents a quart.[29] Paul V. Chapman, dean of the College of Agriculture at the University of Georgia, defended the cow and his own vested interest by saying Americans five years hence would not accept a vegetable drink as a substitute for a glass of cold cow's milk; neither would a plate of sprouting soybeans replace a nice thick juicy steak.[30] The *Detroit Free Press* reported the whole world had once laughed about Henry's mechanical cow, but now companies were making candy out of soybeans and sweet potatoes. The paper pointed out that George Washington Carver had extracted 300 products from the peanut, and that soybeans were in everything from high-class foods to doorknobs.[31]

Much of Ford's interest in agriculture during the 1930's centered in chemurgy—a science putting chemistry to work in industry for the direct benefit of farmers and indirectly for society in general. To fight the depression, 300 leaders of agriculture, industry, and technology met at Dearborn, Michigan, in May 1935, with Henry Ford and Francis P. Garvin, president of the Chemical Foundation, as joint hosts. Here was born the National Farm Chemurgic Council, an organization to cope with the farm-crop surplus. These leaders decided that some destination other than the human stomach must be found for the surplus products of the soil. Industrialists, farmers, and scientists would seek to advance the industrial use of farm crops through applied science and technological innovations. Henry Ford opened the conference saying men's sustenance issued from the soil and not from merchants' shelves. The conference adopted a constitution calling for less reliance on man's vain intellect and placing more on nature's laws. When industrial centralization caused harmful congestion in the cities and hard times, men must turn to the soil. Science as a new frontier could conquer the agricultural depression.

With the assistance of twenty-five state chemurgic councils, almost a million dollars were raised to encourage research with farm products. Efforts were made to convert pine trees in the Southern states into cellulose and paper pulp; to extract starch from sweet potatoes; and to secure industrial alcohol from corn, barley, rye, wheat, potatoes, and grain sorghums. Argol, an alcohol blend with gasoline, might utilize the crops grown on 50 million acres and bring farmers half a billion dollars annually. New jobs for half a million workers would be created. The Argol plant at Atchinson, Kansas, went into production, but athough observers claimed the alcohol additive made motors easier to start, produced less knock, and gave smoother performance than gasoline, the product could not compete with petroleum.[32]

Naturally, these efforts stirred up interest. The United States Department of Agriculture established a soybean industrial research laboratory at Urbana, Illinois. Katherine Dos Passos in the *Women's Home Companion* stated that state chemurgic councils were cooperating with federal government laboratories, state universities, and privately owned laboratories.[33] Since the Nazis had manufactured antifatigue pills from soybean flour which could sustain soldiers on thirty-mile hikes, it was urged that the United States should develop more products from the same source.[34] Soybeans could be used as filler in sausage, as a base for water paints, as body in beer and for making enamels, glycerine, linoleum, oleomargarine, soaps, varnish, and for the hardening of steel. Experts at Iowa State College at Ames recommended soybeans for fattening beef cattle.[35] Various business firms used the beans for manufacturing glue, printing ink, candy, breakfast foods, and insecticides. The Ford Motor Company sold soybean meal to farmers for feeding livestock and poultry, while Ford processing plants at Saline and Milan, Michigan, exchanged the farmer's beans from the field for 44 per cent protein soybean meal. These developments help explain why the soybean acreage in the United States jumped from 1 million acres in 1934 to 12 million acres in 1944.[36]

Ford's scientific interest in agriculture is also reflected in his notion that farmers should derive more benefits from the available water power in rural areas. If water mills could grind wheat into flour and drive sawmills, why couldn't they power rural factories? By bringing some industries to the farm, rural people would gain additional economic security. They could work their farm lands in the summer and the village factories in the winter time. Ford stated, "We cannot eat or wear our machines. If the world were one vast machine shop it would die. When it comes to sustaining life we go to the fields. With one foot in agriculture and the other in industry, America is safe."[37]

With these objectives in mind in 1921, Ford began talking about the decentralization of industry. He argued that cities were cluttered with hideous factories which belched out smoke and that grimy workers were crowded into hovels. Therefore, much of industry should be scattered about the countryside where factories could be placed in leafy bowers surrounded with flowering shrubs and shade trees. In the early twenties, the Ford Motor Company established seven village industries on small water sites within twenty miles of Dearborn. This number increased to twenty plants by 1934, employing 2,400 part-time farmers. During the depression of the thirties, the "Back to the Land" movement appealed to the unemployed lined up in soup kitchen lines.

However, these plans fell short of Ford's initial expectations. Stanley Ruddiman in recent years explained that the program did not work out as Ford expected because factories had to operate throughout the year. Laborers had neither the time nor the energy to work the land after their stint in the factory. Yet the Ford village industries stood as an experiment for combining rural and city life. In 1951 the Ford Motor Company still owned the village industries at Northville, Dundee, Manchester, Brooklyn, and Ypsilanti.

It is difficult to measure Ford's scientific and technological contributions to American rural life because techniques for measuring them are often unreliable. The voices of a few need not represent the views of the multitudes. Although Ford's reputation impressed many, not everyone worshiped "the Great White Father of Fair Lane." Some critics claimed Ford's words outran his deeds, that his press clippings overshadowed his performance, that he dabbled in science instead of initiating well-planned research programs. Those who hated men of wealth frequently accused the Detroit Motor King of pursuing hobbies because he had more money than he could invest in his business. Still others accused him of being ignorant of the cultural arts and devoid of sympathy for philanthropy; hence he turned to things he knew best—technology and the practical aspects of science.

On the other hand, a careful examination of the materials in the Ford Motor Company archives will demonstrate Ford's interest in improving agriculture. His personal papers in 1907, for example, show receipts for the purchase of seed grain, farm implements, poultry, and bull rings. He asked to be put on the mailing list of all agricultural experiment stations in the country. He collected the bulletins published by the Government Printing Office, read widely in farm journals, and instructed his secretaries to correspond with private firms manufacturing farm products. Thirty years later, the same records reveal Ford's intensive search to find out if a short-stemmed, high-producing wheat could be located in the farm lands of Honan province in China. Over the span of years he never apologized for his agrarian bent. This inclination encouraged him to attempt to buy Muscle Shoals in 1921 to break the British-Chilean nitrate monopoly and produce cheaper fertilizers and obtain lower rural electrical rates. Again, he established a cooperative farm near Chelmsford, England, in 1934 to prove that modern farming practices could bring a profit to those engaged in farming. Furthermore, he spent $23,000,000 on his Dearborn museum, filling large sections of it with steam engines and other farm implements to show the evolution of rural technology. Meanwhile, he gave money to encourage vocational training programs at the Berry School near Rome, Georgia. Scores of trucks, cars, and tractors were given to agricultural colleges and industrial schools. During the 1940's he established a Farm Youth Organization to encourage farm boys to make more intelligent use of power machinery.

Henry Ford believed that science and technology led to Progress. Convinced of this fact, he became a propagandist for the cause. He talked science constantly. Technological endeavors would lead to the promised land. To the skeptical, he preached faith in the machine; to the doubter, he suggested innovation; and to the conservative, he advocated change. Thus the automobile, truck, and tractor must replace the horse. The cow, too, must eventually succumb to the machine. Farmers fully mechanized would do all their farm work in twenty days of the year. Chemurgy would develop industrial uses for farm commodities. Domestic crops should be converted into manufactured goods. Waste must be eliminated in farming and in industry. Solutions to the farm problem lay in individual enterprise, not with paternal governmental agencies. By today's standards, Ford's attitudes in later life toward national politics, labor organizations, and social responsibility can be condemned as naïve. Yet, his views on technology and the mechanization of rural America were generally enlightened, progressive, and often far ahead of his times.

REFERENCES

[1] *Rural New Yorker,* June 9, 1923, p. 832.

[2] *Brooklyn Eagle,* March 14, 1932, p. 2.

[3] *Prairie Farmer,* June 1, 1918, p. 514.

[4] *Journal of Commerce* (Chicago), February 10, 1921, p. 1.

[5] Will Rogers, *Wit and Philosophy from Radio Talks of Will Rogers* (New York, 1930), p. 33.

[6] Walter H. Pay, "Industrial Alcohol," *The Natal Agricultural Journal* (London), Vol. II (March, 1909), pp. 577–583. As early as 1908, the British attempted to secure motor fuel for transport in Africa by distilling alcohol from sugar cane. In the same year, the Hart-Parr Company of Charles City, Iowa, equipped some of their tractors in Idaho, Colorado, and Cuba with alcohol-burning carburetors.

[7] *Fair Lane Papers* (Ford Motor Company Archives, Dearborn, Michigan), Ford-Ferguson File, No. 07825, Knud Weford, Hawley, Minnesota, to Henry Ford, Detroit, March 17, 1923.

[8] *Fair Lane Papers,* Ernest G. Liebold, Dearborn, Michigan, to Rudolph Schroeder, Hoboken, New Jersey, May 19, 1917.

[9] *Fair Lane Papers,* Ford-Ferguson File, Accession 380, Box 25.

[10] *Fair Lane Papers,* Ford-Ferguson File, Floyd Radford, Detroit, Michigan, to G. W. Nash, Del Norte, Colorado, July 13, 1938.

[11] Allan Nevins and Frank Hill, *Ford: Expansion and Challenge* (New York, 1957), p. 231.

[12] *Fair Lane Papers,* C. A. Bullwinkle, San Francisco, to Ford Motor Company, Dearborn, Michigan, June 10, 1928. See also *Carnegie Institution Bulletin,* no. 313.

[13] *Botany* (University of California, Berkeley) Vol. 7, pp. 159–278. Harvey M. Hall worked with the Carnegie Institution in Washington, D.C., where he did research on substitutes for rubber in 1917 and 1918. After the war he joined the faculty of the University of California at Berkeley. He favored making synthetic rubber from desert Rabbit Brush because the residue could be used to make paper.

[14] *Fair Lane Papers,* Wilson Papenae, Agricultural Explorer for the United States Department of Agriculture, Washington, D.C., to William H. Smith, Ford Motor Company, Dearborn, Michigan, August 15, 1923.

[15] *Fort Meyers Tropical News,* February 6, 1930, p. 1. From Ford Motor Company clipping file.

[16] *Tampa Tribune,* February 22, 1930, p. 1.

[17] Ford Motor Company Archives (Central Office Building, Dearborn, Michigan), J. L. McCloud, Oral History, Vol. II, p. 125.

[18] Andrew S. Wing, "A Farmer Visits Henry Ford," *Farm and Fireside* (September, 1925), p. 46.

[19] *Fair Lane Papers,* Ford-Ferguson File, No. 03139. Evelyn Trastle, Los Angeles, California, to Henry Ford, Detroit, April 4, 1921.

[20] *Fair Lane Papers,* Manuscript for Press Release (no author), March, 1970.

[21] *Fair Lane Papers, Henry Ford Imprints Reflecting His Views,* C.O.B. Pamphlet (1934), pp. 1–9.

[22] *Ibid.,* p. 4.

[23] R. A. Boyer, "Soybean Protein Fibers," *Modern Plastics* (February, 1942), p. 42.

[24] *Progress Guide* (Chicago), Vol. 5, No. 2 (September, 1943), p. 343.

[25] *Business Week* (March 2, 1940), p. 20.

[26] Ford Motor Company Archives (Central Office Building, Dearborn, Michigan), J. L. McCloud, Oral History, p. 197. See also *Detroit Journal,* July 7, 1915, p. 1.

[27] *Ibid.,* Harold M. Cordell, Oral History, p. 3.

[28] "The Soybean as a Food," *The Madison Survey* (Nashville

Agricultural Normal Institute, Madison, Tennessee), Vol. 13, No. 28 (December 9, 1931), pp. 99–100.

[29] *Southern Dairy Products Journal,* April, 1944, p. 2.

[30] Paul W. Chapman, "Cowless Milk," *Atlantic Journal,* April 16, 1944, p. 3.

[31] *Detroit Free Press,* January 17, 1938, p. 1.

[32] Carl B. Fritsche, "Chemurgy—A Scientific Solution to Agricultural Problems." An address before the Twelfth Annual Canadian Chemical Convention, Vancouver, B.C., June 18, 1937 (manuscript), pp. 1–52. *Fair Lane Papers,* Ford-Ferguson File, Accession 380, Box 20.

[33] Katherine Dos Passos, "Science and the Beanstalk," *Women's Home Companion* (September, 1942), pp. 14–15.

[34] *Atlanta Journal,* April 16, 1944, p. 10.

[35] *Fair Lane Papers,* Accession 13, Box 26, August, 1943.

[36] United States Bureau of the Census, *Statistical Abstract of the United States: 1955* (Washington, D.C., 1955), p. 664.

[37] *Henry Ford and Farm and Factory* (Printed Bulletin, Ford Motor Company, May, 1932), p. 2.